APPLICATIONS OF
PROBABILITY AND
RANDOM VARIABLES

McGRAW-HILL SERIES IN
PROBABILITY AND STATISTICS

David Blackwell and Herbert Solomon, Consulting Editors

**McGRAW-HILL
BOOK COMPANY**

New York
St. Louis
San Francisco
Düsseldorf
Johannesburg
Kuala Lumpur
London
Mexico
Montreal
New Delhi
Panama
Rio de Janeiro
Singapore
Sydney
Toronto

GEORGE P. WADSWORTH

Professor of Mathematics
Massachusetts Institute of Technology

JOSEPH G. BRYAN

Senior Research Scientist
Casualty-Property Department
The Travelers Insurance Companies

Adjunct Professor of Mathematics
Rensselaer Polytechnic Institute of Connecticut

Applications of Probability and Random Variables

SECOND EDITION

This book was set in Times New Roman.
The editors were Robert H. Summersgill and Michael Gardner;
the production supervisor was Bill Greenwood.
New illustrations were done by Textart Service, Inc.
The printer and binder was Kingsport Press, Inc.

5 19.2
W 124a

Library of Congress Cataloging in Publication Data

Wadsworth, George Proctor, date
    Applications of probability and random variables.

    (McGraw-Hill series in probability and statistics)
    First ed. published in 1960 under title: Introduction
    to probability and random variables.
    1. Probabilities.   I. Bryan, Joseph G., joint author.   II. Title.
QA273. W2   1974   519.2   73–6732
ISBN 0–07–067651–8

APPLICATIONS OF
PROBABILITY AND
RANDOM VARIABLES

1 2 3 4 5 6 7 8 9 0 K P K P 7 9 8 7 6 5 4

CONTENTS

32516

PREFACE TO THE SECOND EDITION

Since the publication of the first edition, the importance of probability as a tool for the solution of problems in applied mathematics has become accepted fact for the scientist and engineer, and the use of stochastic models and techniques has therefore become very common. In order to update the necessary background for the student in the field of stochastic processes, we have made basic changes in the text, mainly in the direction of added material which is now considered indispensible for the practitioner of applied mathematics, as well as changes of emphasis in other topics included in the previous edition.

Three new chapters have been added to the text. The applicability of difference equations to certain classes of stochastic processes particular to random walks is considered in Chap. 3. The use of generating functions, both as a theoretical tool in distribution theory and as a means of analysis of stochastic situations, is developed in some detail (Chap. 8) so as to indicate the importance of this technique as a basic method of approach. The third new chapter (Chap. 9) discusses Markov processes, both discrete and continuous. Many examples of such processes are offered, along with an explanation of their potential value when used in cases where rewards for various transitions are given. An optimum strategy when alternative choices are possible is also suggested.

Besides these new chapters, several additions and alterations have been made in the original text. The treatment of hypothesis testing has been expanded, and a completely new section on the Bayesian technique has been added to update the discussion of statistical analysis. A new introduction to the theory of search and best distribution of effort has been provided to give the reader a better concept of recently developed stochastic techniques, and to suggest viable answers to important problems.

Many new examples involving applications in this field have been interspersed in the text, and numerous illustrative problems, both new and revised, have been included in the problem section at the end of each chapter. It is hoped that these additions and modifications will increase the relevance of the text for students in both the theoretical and applied sciences.

GEORGE P. WADSWORTH
JOSEPH G. BRYAN

PREFACE TO THE FIRST EDITION

Probability, the "laws of chance," has had a colorful history dating back at least to the early Renaissance, when commercial insurance against risks was established in the Italian cities. The mathematical study of probability was stimulated by questions of practical concern: how to hedge against the uncertainty of individual events by taking advantage of the paradoxical rule of regularity en masse—an empirical discovery.

Pioneering work on probability was done in the seventeenth century by Galileo, Pascal, Fermat, Huygens, J. Bernoulli, and Halley—to mention only the most famous. The roster of contributors since then includes the greatest mathematicians and some of the greatest scientists of each generation. Perhaps the most outstanding contribution of any one individual was that of Laplace, whose celebrated work entitled "Théorie analytique des probabilités" was published in 1812 and dedicated to Napoleon.

The astronomer Edmund Halley, distinguished as the first person ever to predict the return of a comet, made an important application of probability to actuarial science in 1693 by showing how premiums for life insurance could be calculated from mortality tables. Laplace, Legendre, and Gauss were cofounders of the theory of errors—an application of probability to the subject of variations among repeated measurements of the same quantity.   The great physicist J. C. Maxwell derived the gas laws in 1860 on the basis of the probability distribution of molecular velocities.   In 1900, Max Planck presented his quantum theory of radiation in terms of probability.   Under the leadership of Karl Pearson and R. A. Fisher, rapid progress has been made in the field of statistics since the beginning of the twentieth century through the application of probability to problems concerning valid inference from limited—but adequate—data.   Problems of this sort arise in quality comparisons of all kinds, sampling inspection of manufactured articles, design of experiments to guide product development when important factors cannot be subjected to laboratory control, market surveys, public-opinion polls, and statistical prediction.   World War II opened new fields of application when probability was applied to the theory of search for enemy submarines, bomber defense against fighters, and questions concerning the best tactics in combat or the best use of new weapons.   The men responsible for these applications designated their work as " operations research "—a rapidly developing field which has since gained wide industrial acceptance.   Thus through operations research, probability, which owes its origin to commerce, is at last being brought to bear vigorously and competently on the uncertainties of business operations.

Despite its long history of engaging the greatest intellects of successive generations and proving its value in many fields, probability has been neglected in the college curriculum until very recently.   Within the last decade, however, it has emerged in its true light as a subject of broad application both to science and industry, and it is now beginning to receive due recognition in the technical training of scientists and engineers.

The authors' primary purpose has been to supply a text for a short undergraduate course, with a prerequisite of elementary calculus.   However, they feel it equally important to present the material in a form which will be suitable for self-study on the part of practicing engineers and research workers whose mathematical background does not include a substantial course in probability. Both for the sake of readers without previous grounding in probability and others who would like a review, much space has been devoted to discussion of concepts and detailed solution of illustrative problems.

The material of this book has been class-tested in mimeographed form for eight years at the Massachusetts Institute of Technology.   There it has served as the text for a one-semester course offered twice a year by the Department of Mathematics and has had the benefit of being used by seven members of the staff.   Considerable revision has taken place during this period, but the contents have remained substantially unaltered for the last two years.   Abbreviations and specialized symbols have been avoided to a large extent because the beginning student finds the use of such shorthand an obstacle to comprehension.

Statistical concepts are developed where they arise as direct applications of the topics in probability under consideration.   In this way the student receives an early but brief introduction to tests of significance, confidence limits, and the major sampling distributions without interruption to the main theme.   The last chapter is the only one exclusively devoted to statistics.   There the student is instructed in the principal uses of four major distributions—the normal, chi-square, t, and F—and is given a clear statement of their relationships and distinct provinces.

It is our hope that this text will serve two purposes: first, instruction both in university classrooms and the technical professions; second, promotion of greater awareness and appreciation of the potential values of probability to science and industry.

GEORGE P. WADSWORTH

JOSEPH G. BRYAN

PROBABILITY CONCEPTS

1-1 INTRODUCTION

A common type of phenomenon observed in many areas of human experience is characterized by the production of a series of events that are rather definitely fixed in their possible distinct forms but are haphazard in the order in which one form follows upon another or, alternatively, by the generation of a quantitative variable having a fairly well-defined range of possible values but no consistent rule of progression. Despite the absence of any discernible law of serial order, there usually exists a striking similarity between the proportionate compositions of any pair of long records of the same phenomenon. The observed regularity of proportion in random series is the empirical basis for the so-called "laws of chance" which are the subject matter of the *theory of probability*.

Controversies over the philosophical foundation of the theory of probability have been going on for many years. Points of view have ranged from the consideration of probability as a mathematical property of an abstract system to its definition as a measure of credibility or as an aspect of human psychology. Notwithstanding this diversity of thought about the philosophical foundation, there has been almost universal agreement on the mathematical superstructure.

The essential mathematical properties which probability must possess (and about which there is no argument) suffice to establish the theory of probability on an axiomatic basis as a rigorous branch of pure mathematics. Like the axioms of euclidean geometry, those of probability represent idealizations of practical experience or extensions of elementary principles of logic. Since the axioms appear arbitrary when stated without reference to their rational background, we shall indicate something of their conceptual origin by pointing out some of the properties of relative frequencies in Sec. 1-3. The axioms, however, are a priori; therefore, this or any other interpretation is actually irrelevant. Nevertheless, an attempt to explain the axioms by any of the various approaches generally used is useful provided the limitations are acknowledged.

We shall use the term *experiment* to denote any clearly defined procedure which is to be carried out (such as the tossing of a coin); the term *trial* will be used to denote a single carrying out of the experiment. An *event* will be a possible result of a single tria!. The following definitions of probability will involve the probability of the occurrence of a specific event. Denoting a specific event by the symbol E, we shall use the notation $\bar{E}$ for the absence of E or the failure of E to occur. (Besides $\bar{E}$, two other symbols for not-E are E' and $\sim E$.) The events E and $\bar{E}$ are said to be complementary, inasmuch as either one or the other is bound to occur; they are also *mutually exclusive*, for they cannot coexist. The symbol $P(E)$ is used to denote the probability of the occurrence of E and $P(\bar{E})$ to denote the probability of the nonoccurrence of E.

1-2 CONCEPT OF PROBABILITY

The reader is well aware from past experience with this subject that probability measure is associated with some number between 0 and 1, where 0 indicates that the event will not occur (in some sense) and 1 that it is certain to occur (in some sense also). Gradations between these two numbers indicate varying degrees of likelihood about the occurrence of the event in question.

One's thinking about probability is clarified by realizing that probability is not defined except in the context of a particular problem, which itself must be uniquely defined. In other words, the *modus operandi* by which the probability is generated in reality defines the probability itself. When different authors arrive at discrepant figures for what is purported to be the probability of the occurrence of a specific event, the disparity usually stems from the fact that each author is solving a different problem by his choice of model, even though the "real-world" problem is the same in all cases. Different mathematical models usually yield different results, and the disagreement among models is the real cause of the discrepancies in the answers.

Let us propose the following conceptualization. An unpredictable event can occur or not occur through the operation of some random process. The inner workings of the random process are not describable, but a mathematical model of its output defines the probability that the stated event occurs. From this mathematical model, it is possible to calculate the probability of any number of occurrences of the event in a long series of independent trials and to show that the percentage of occurrences (relative frequency) out of a very large number of trials approximates the probability of the event. This correspondence between the abstract notion of probability and the experimentally determined relative frequency forms the principal link between theory and application.

Now let us consider some examples.

As the first example, suppose that the object is to find the probability of obtaining either a 1 or a 4 in the toss of a single die. To define the problem, we assume that each side of the die, 1 through 6, is equally likely to occur, implying that the actual tossing of the die does not favor one face over another. We ignore both the practical impossibility of manufacturing a perfectly balanced die and the impossibility of making an unbiased throw; by hypothesis, the mechanics of the problem is uniquely defined. The corresponding mathematical model defines the probability of the event as the proportion of favorable possibilities out of all possibilities: two favorable faces out of six possible faces. Hence the probability, by this model, is $\frac{2}{6}$ or $\frac{1}{3}$.

Again, consider the probability of choosing a point at random on a line segment $\bar{L}$ of length L and falling inside a segment l of length l which is entirely contained in $\bar{L}$. To define the problem, one can visualize a mechanism of the following type. The total segment $\bar{L}$ is divided into a very large number of segments which we will number from 1 to n. A subset of these numbers will be contained in the segment l. This entire group of numbers will now be written as n poker chips and placed in a box and thoroughly mixed. One chip is drawn out at random, and we can visualize the probability of any point so chosen as falling in l to be given by the ratio of the number of segments contained in l to the total number in $\bar{L}$. This experiment is operationally conceivable, and it generates a number which has a limit as $n \to \infty$. The limit may be defined as the probability of choosing a point at random and having it lie in the segment l. In fact, we can visualize the probability of a point falling in the interval dx as dx/L if $l = dx$. In both these cases the defined problem at the same time actually defines the probability. We could equally well define the probability differently by redefining the problem.

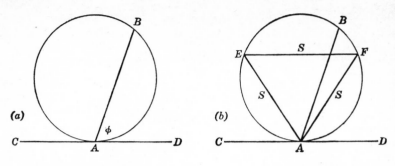

FIGURE 1-1

An example of a problem that is not clearly defined involves a hypothetical experiment in which a large number of people are asked to draw a chord at random inside a circle of given radius R. After the chords are drawn, an equilateral triangle is inscribed in the circle, and the problem is to deduce from probability considerations what percentage of the chords may be expected to be longer than the side of the triangle. We shall consider two analyses of the problem.

Represent the chord as AB (Fig. 1-1a). In our first analysis, we construct the tangent CD at the point A and let ϕ denote the angle which the chord makes with the tangent. Measuring ϕ in the conventional counterclockwise direction, let us make the assumption that all values of ϕ, 0 to 180°, inclusive are equally likely. To determine the conditions under which the length of AB exceeds the length S of the inscribed equilateral triangle, let us construct the triangle so that A is one of the vertices. Then, referring to Fig. 1-1b, we know from plane geometry that side AF makes an angle of 60° with CD and that side AE makes an angle of 120° with CD. Hence the length of AB will exceed S provided that ϕ lies between 60 and 120°. Since this favorable range is one-third the total range, the probability in question is $\frac{1}{3}$. For the reader who wonders what effect the exclusion of the end points 60 and 120° of the favorable range has on the probability, we state here that because of continuity this exclusion is immaterial; the general principle will be expounded in Chap. 4.

In our second analysis, we bisect the chord AB by the radius OH (length R) and let X denote the point of intersection (Fig. 1-2a). Then the distance $\overline{OX}$ can range from zero to R, and we shall now assume that all these values are equally likely. This time, to compare the length of AB with S, we locate the triangle $E'F'G'$ so that one side $F'G'$ is bisected by OH, the point of intersection being Y. From Fig. 1-2b it is clear that the length of AB exceeds S

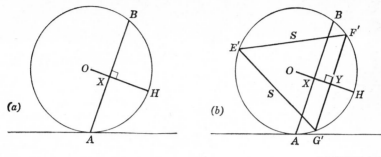

FIGURE 1-2

when $\overline{OX}$ is less than $\overline{OY}$. From plane geometry it is easy to show that OH and $F'G'$ are perpendicular bisectors of each other, whence $\overline{OY} = R/2$. Consequently, the chord length exceeds S for half of the possible range of values of $\overline{OX}$. By this second analysis, therefore, the probability in question is $\frac{1}{2}$.

The fact that these two answers are different implies that the initial assumptions are incompatible. That is, if the values of ϕ are equally likely, the values of $\overline{OX}$ cannot be equally likely, and vice versa. We shall consider this question of functional dependence further in Chap. 6. Obviously an infinite number of answers are possible for this problem. One must be very careful not to fall into the trap of defining a probability without defining the mechanism by which the probability is generated.

1-3 RELATIVE FREQUENCY

Relative frequency is an observational concept involving repeated trials of an experiment. A record is kept of the number of times (say n_1) that a certain event E occurs in n trials of an appropriate experiment, and the ratio n_1/n is called the *relative frequency* of the event E; similarly, the complementary ratio $(n - n_1)/n$ is called the relative frequency of $\bar{E}$. Denoting the two relative frequencies by $R(E)$ and $R(\bar{E})$, respectively, we see that each has a mathematically possible range of 0 to 1 and that $R(E) + R(\bar{E}) \equiv 1$. We have previously made a point of the regularity of proportion in random series. By this we mean the tendency of relative frequencies to stabilize at definite values as the number of trials increases. This tendency has been verified experimentally on numerous occasions and seems to be inherent in the nature of common random phenomena.

As an illustration of the relative-frequency concept, consider a blood-typing survey conducted in Massachusetts by Macready and Manin in 1951. Out of

150,064 persons tested, 125,055 were found to possess the type of blood classified as Rh D-positive. Hence, the relative frequency of Rh D-positive is approximately .833. Insofar as the people tested constituted a representative cross section of all human beings, we may regard the stated relative frequency as an estimate of the true relative frequency of Rh D-positive in the entire human race or, more specifically, as an estimate of the percentage of Rh D-positive among the residents of Massachusetts.

Let us therefore examine some of these properties of relative frequency.

Property 1 For a single event E we have already noted that $R(E)$ is a real number lying somewhere in the range 0 to 1 and such that $R(E) + R(\bar{E}) = 1$.

Now consider two events A and B, which may exist simultaneously. All possible results of a given experiment may be classified under some one of four mutually exclusive categories: $(A \cap B)$, $(A \cap \bar{B})$, $(\bar{A} \cap B)$, $(\bar{A} \cap \bar{B})$, representing the simultaneous occurrence of A and B, the simultaneous occurrence of A and $\bar{B}$, etc. Denoting the respective numbers of occurrences by n_1, n_2, n_3, n_4 and the total by n, we may summarize the results as shown in Table 1-1. The total number of occurrences of A is $n_1 + n_2$, and the total number of occurrences of B is $n_1 + n_3$. Hence the corresponding relative frequencies are

$$R(A) = \frac{n_1 + n_2}{n} \qquad R(B) = \frac{n_1 + n_3}{n}$$

It often happens that a sufficient condition for the occurrence of some contingent result is the single or simultaneous occurrence of two events A, B. (As an analogy, a joint meeting of two organizations is open to all who belong to either one, including, of course, all who belong to both.) The composite event defined by the occurrence of either A or B alone or both together is termed the *union* of A and B and is denoted by the symbol $A \cup B$; an alternative symbol is $A + B$. Since the possible distinct ways in which A and B can occur either together or alone are $A \cap B$, $A \cap \bar{B}$, $\bar{A} \cap B$, and the respective frequencies of these are n_1, n_2, n_3, the relative frequency of $A \cup B$ is

$$R(A \cup B) = \frac{n_1 + n_2 + n_3}{n}$$

Table 1-1

Category	$(A \cap B)$	$(A \cap \bar{B})$	$(\bar{A} \cap B)$	$(\bar{A} \cap \bar{B})$	Total
Number of occurrences	n_1	n_2	n_3	n_4	n

By adding and subtracting n_1/n, the relative frequency of the union of A and B may be rewritten as

$$R(A \cup B) = \frac{n_1 + n_2}{n} + \frac{n_1 + n_3}{n} - \frac{n_1}{n}$$

Hence,

Property 2

$$R(A \cup B) = R(A) + R(B) - R(A \cap B)$$

If the separate quantities n_1, n_2, etc., were known, the most direct method of calculating $R(A \cup B)$ would be simply the one that follows from the definition, namely $(n_1 + n_2 + n_3)/n$. The point of introducing the formula stated in property 2 is that most mathematical problems entail the expression of one variable in terms of others, and the relative frequencies $R(A)$, $R(B)$, $R(A \cap B)$ are usually the ones most readily available from either direct observation or theoretical considerations.

We now introduce the idea of *conditional relative frequencies*, where the word *conditional* signifies restriction to a special class as distinguished from the *unconditional* inclusion of all possibilities. An event E occurring under condition C is denoted by the symbol $E|C$, read "E given C." A conditional relative frequency is calculated in the same manner as an ordinary (unconditional) relative frequency except that the calculation is confined to those events which satisfy the prescribed criterion (condition). The conditional relative frequency of A given B, denoted by $R(A|B)$, is simply the proportionate number of occurrences of A among all occurrences of B, hence the ratio of the number of simultaneous occurrences of A and B to the total number of occurrences of B:

$$R(A|B) = \frac{n_1}{n_1 + n_3} = \frac{n_1/n}{(n_1 + n_3)/n} = \frac{R(A \cap B)}{R(B)}$$

By a similar argument we obtain the conditional relative frequency of B given A as

$$R(B|A) = \frac{n_1}{n_1 + n_2} \equiv \frac{R(A \cap B)}{R(A)}$$

Combining these statements into one chain of relationships, we arrive at

Property 3

$$R(A \cap B) \equiv R(A)R(B|A) \equiv R(B)R(A|B)$$

The object of expressing property 3 in the form of a double identity is that the unknown quantity in a problem varies according to the circumstances; hence it is convenient to bring out all the relationships at once.

1-4   LAWS OF PROBABILITY

A system of axioms for a subject can be formulated in more than one way and yet have the same logical consequences.   Couched in the technical terms of modern mathematical analysis, a complete set of axioms for the theory of probability is intelligible only to readers with considerably more background than we wish to presuppose or supply.   However, the following three properties, related to two events A, B and suggested by the behavior of relative frequencies, suffice for most purposes as an axiomatic basis of probability.   We call them properties rather than axioms because, aside from formality of statement, a system of axioms should be reduced to the fewest possible assertions, whereas we have chosen an extended form of expression of property 3 in the interest of convenience.   Property 1 defines the general character of probability; property 2, the axiomatic relationship between the probability of the union $A \cup B$ of A and B and the probabilities of A, B and $A \cap B$; property 3, the axiomatic relationship connecting the probability of the joint occurrence $A \cap B$ of A and B with the probability of either event and the conditional probability of the other.

Property 1 General character of probability   The probability $P(E)$ of an event E is a real number in the range of 0 to 1.   The probability of an impossible event is 0, that of an event certain to occur is 1, and in general $P(E) + P(\bar{E}) = 1$.

Property 2 Law of total probability

$$P(A \cup B) = P(A) + P(B) - P(A \cap B)$$

Property 3 Law of compound or joint probability   If neither $P(A)$ nor $P(B)$ is zero

$$P(A \cap B) = P(A)P(B|A) = P(B)P(A|B)$$

If either $P(A)$ or $P(B)$ is zero,

$$P(A \cap B) = 0$$

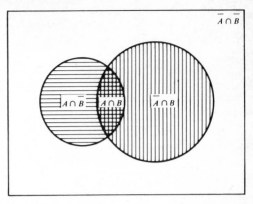

FIGURE 1-3

In a more advanced treatment of this subject, the counterpart of the number of occurrences of an event is a well-defined mathematical quantity known as *measure*. In connection with continuous variables, it is usually possible to interpret measure geometrically as area or volume. One of the paradoxes of an infinite number of possibilities, however, is that a probability of zero does not necessarily imply impossibility, nor does a probability of unity necessarily imply certainty. This may be understood by noting that a relative frequency will approach zero if the numerator remains finite while the denominator approaches infinity, or even if the numerator increases at too slow a rate in comparison with the denominator. Again, from the geometrical point of view, a rectangle may be shrunk to a line, which has zero area, hence zero measure in a two-dimensional context.

Venn diagrams are often used to illustrate probability concepts. In a Venn diagram, the universal set of points (of total probability 1) is depicted by the interior of a rectangle. Arbitrary events are represented by the interior points of closed curves, usually circles, and joint occurrences by the region in common to two or more closed curves. The Venn diagram for two events A, B is shown in Fig. 1-3.

1-5   FUNDAMENTAL THEOREMS

The quantity $P(A \cup B)$ is called the *total probability* of A and B. If two events A, B are *mutually exclusive*, their simultaneous occurrence is impossible, and $P(A \cap B) = 0$. In that case property 2 yields $P(A \cup B) = P(A) + P(B)$.

This result can be generalized to any finite number of mutually exclusive events. Let A, B, C be mutually exclusive and set $S = A \cup B$. Then

$$A \cup B \cup C \equiv S \cup C$$

and $$P(S \cup C) = P(S) + P(C) = P(A) + P(B) + P(C)$$

a result which can be extended by induction to any finite number of mutually exclusive events.   Hence the following theorem:

Theorem 1-1 Law of total probability for mutually exclusive events   If A, B, ..., N are mutually exclusive events, then

$$P(A \cup B \cup \cdots \cup N) = P(A) + P(B) + \cdots + P(N)$$

With regard to an infinite number of events the extension is reasonable but not rigorously demonstrable.   Hence one of the axioms in a technically complete set asserts that the extension does hold for an infinite number of mutually exclusive events.   We shall take this extension for granted.   ////

From property 2 we may also deduce a more general theorem applicable to any finite number of events, whether mutually exclusive or not.   Let A, B, C be any three events that need not be mutually exclusive and, as before, set $S = A \cup B$.   Then

$$P(A \cup B \cup C) \equiv P(S \cup C) = P(S) + P(C) - P(S \cap C)$$

$$= P(A) + P(B) - P(A \cap B) + P(C) - P(S \cap C)$$

Now the symbol $(S \cap C)$ means that C occurs in conjunction with A or B or both and has the same logical import as the expression $(A \cap C) \cup (B \cap C)$ since the simultaneous occurrence of $(A \cap C)$ and $(B \cap C)$ implies nothing more nor less than the simultaneous occurrence of A, B, C.   Accordingly,

$$P(S \cap C) \equiv P[(A \cap C) \cup (B \cap C)] = P(A \cap C) + P(B \cap C) - P(A \cap B \cap C)$$

Therefore

$$P(A \cup B \cup C) = P(A) + P(B) + P(C) - P(A \cap B) - P(A \cap C)$$

$$- P(B \cap C) + P(A \cap B \cap C)$$

When this result is extended by induction, the following theorem is obtained.

Theorem 1-2 General law of total probability   The probability $P(A \cup B \cup \cdots \cup N)$ equals the algebraic sum of the probabilities of the events in all possible distinct combinations: singles, pairs, triples, ...,

N-tuples.   The sign is plus for the odd combinations (singles, triples, etc.) and minus for the even combinations (pairs, quadruples, etc.).   ////

The probability of the simultaneous occurrence of two or more events is called the *compound probability* or, synonymously, the *joint probability* of the events.   The relations stated by property 3 can be generalized to any finite number of events.   Thus, considering the simultaneous occurrence of three events A, B, C, let X denote the simultaneous occurrence of A and B.   Then

$$(A \cap B \cap C) \equiv (X \cap C)$$

and $$P(A \cap B \cap C) = P(X)P(C \mid X) = P(A)P(B \mid A)P(C \mid A \cap B)$$

Since all permutations of the letters A, B, C have the same meaning as regards the simultaneous occurrence of the three events, there are six equivalent expressions of the foregoing type.   For example, we also have

$$P(A \cap B \cap C) \equiv P(B \cap C \cap A) = P(B)P(C \mid B)P(A \mid B \cap C)$$

Continuing in the same vein, we may express the joint probability of any finite number N of events A, B, C, ..., M, N as the product of N factors, the first of which is the unconditional probability of any particular one of the events chosen arbitrarily; the second, the conditional probability of any particular one of the remaining events, given the occurrence of the one first selected; and the general term, the conditional probability of any particular one of the remaining events, given the occurrence of those already chosen.   In all there are $N!$ equivalent expressions for this same joint probability.   Putting this rule in the form of a theorem, we have:

Theorem 1-3 General law of compound probability

$$P(A \cap B \cap C \cap \cdots \cap M \cap N)$$
$$= P(A)P(B \mid A)P(C \mid A \cap B) \cdots P(N \mid A \cap B \cap C \cap \cdots \cap M) \qquad ////$$

A special case of great importance arises when the events are *independent*. In the probability sense, two events, A, B, are said to be independent when and only when neither one affects the probability of the occurrence of the other. The formal definition of independence is that

$$P(B \mid A) = P(B) \qquad P(A \mid B) = P(A)$$

It turns out that each one of these equations implies the other, for if either holds, the joint probability assumes the symmetrical form

$$P(A \cap B) = P(A)P(B)$$

from which the other equation can be deduced. Consequently, a sufficient *definition of independence* of two events is that their joint probability equals the product of their respective unconditional probabilities. An advantage of this form of definition is that it does not break down when one of the events has zero probability. The reader's attention is called to the fact that *mutually exclusive* events are not independent, for the occurrence of one implies that the others are impossible. In general, N events are independent if the probability of each event is unaffected by the occurrence or nonoccurrence of any of the others, either singly or in combination. A formal definition is as follows:

Definition of Independence

The events $A, B, C, \ldots, N$ are mutually independent as a system if the probability of every N-fold combination that can be formed from these events or their complements in any proportion (as $A \cap B \cap C \cap \cdots \cap N; \bar{A} \cap B \cap C \cap \cdots \cap N;$ $A \cap \bar{B} \cap C \cap \cdots \cap N; \ldots; \bar{A} \cap \bar{B} \cap C \cap \cdots \cap N; \ldots; \bar{A} \cap \bar{B} \cap \bar{C} \cap \cdots \cap \bar{N}$) factors into the product of the probabilities of the N components of the combination.

    This definition is, of course, both the necessary and sufficient condition for the events to be independent. If, however, the events *are* independent it follows that the joint probability is equal to the product of the individual probabilities. Hence,

Theorem 1-4 Law of compound probability for independent events

$$P(A \cap B \cap C \cap \cdots \cap N) = P(A)P(B)P(C) \cdots P(N) \qquad ////$$

    Since the 2^N possible N-fold combinations of $A, B, C, \ldots, N$ and $\bar{A}, \bar{B},$ $\bar{C}, \ldots, \bar{N}$ represent a complete system of mutually exclusive categories, their total probability is necessarily equal to unity. Moreover, the factorization of their probabilities in the case of independence guarantees the analogous probability factorization of all possible combinations of any subset of the initial N events and hence the independence of any subset. (This fact may be established by applying the law of total probability to a typical subset.) On the other hand, the independence of subsets does not suffice for the independence of the entire system.

    The definition of independence for a system of events is illustrated by the following example, in which the dependence or independence of three events A, B, C is determined under four different hypothetical situations, each giving rise to its own set of compound probabilities, as shown in Table 1-2; that is, the compound probabilities themselves are hypothesized. Under the first three situations, $P(A) = P(B) = P(C) = \frac{1}{2}$; hence for the independence of A, B, C, the probability of each triple combination $A \cap B \cap C$, $A \cap B \cap \bar{C}$, etc., would have to be $\frac{1}{8}$.

Under situation 1, some of the compound probabilities factor just as they would for independent events — in particular, the triple combination $A \cap B \cap C$, for

$$P(A \cap B \cap C) = \frac{1}{8} = P(A)P(B)P(C)$$

and the same is true of three others: $A \cap \bar{B} \cap C$, $\bar{A} \cap B \cap \bar{C}$, $\bar{A} \cap \bar{B} \cap C$. Furthermore, the two events A, C are independent as a pair, for

$$P(A \cap C) = P(A \cap B \cap C) + P(A \cap \bar{B} \cap C)$$
$$= \frac{1}{8} + \frac{1}{8} = \frac{1}{4} = P(A)P(C)$$

Nevertheless, the three events A, B, C are not mutually independent as a system, because not all the compound probabilities yield the requisite factorization. This failure is evident in the case of the four remaining triple combinations $(A \cap B \cap \bar{C}, A \cap \bar{B} \cap \bar{C}, \bar{A} \cap B \cap C, \bar{A} \cap \bar{B} \cap C)$, and it also occurs in the two pairs of events $A \cap B$ and $B \cap C$ for

$$P(A \cap B) = \frac{1}{8} \neq P(A)P(B)$$
$$P(B \cap C) = \frac{3}{8} \neq P(B)P(C)$$

Under situation 2, the events A, B, C are clearly dependent, for none of the triple combinations yields the factorization required for independence. Nevertheless, all the pairs meet this criterion, for

$$P(A \cap B) = \frac{1}{4} = P(A)P(B)$$
$$P(A \cap C) = \frac{1}{4} = P(A)P(C)$$
$$P(B \cap C) = \frac{1}{4} = P(B)P(C)$$

Table 1-2 COMPOUND PROBABILITIES OF THREE EVENTS UNDER FOUR HYPOTHETICAL SITUATIONS

Joint event	Situation 1	Situation 2	Situation 3	Situation 4
$A \cap B \cap C$	$\frac{1}{8}$	$\frac{1}{4}$	$\frac{1}{8}$	$\alpha\beta\gamma + \varepsilon$
$A \cap B \cap \bar{C}$	0	0	$\frac{1}{8}$	$\alpha\beta(1 - \gamma) - \varepsilon$
$A \cap \bar{B} \cap C$	$\frac{1}{8}$	0	$\frac{1}{8}$	$\alpha(1 - \beta)\gamma - \varepsilon$
$A \cap \bar{B} \cap \bar{C}$	$\frac{1}{4}$	$\frac{1}{4}$	$\frac{1}{8}$	$\alpha(1 - \beta)(1 - \gamma) + \varepsilon$
$\bar{A} \cap B \cap C$	$\frac{1}{4}$	0	$\frac{1}{8}$	$(1 - \alpha)\beta\gamma - \varepsilon$
$\bar{A} \cap B \cap \bar{C}$	$\frac{1}{8}$	$\frac{1}{4}$	$\frac{1}{8}$	$(1 - \alpha)\beta(1 - \gamma) + \varepsilon$
$\bar{A} \cap \bar{B} \cap C$	0	$\frac{1}{4}$	$\frac{1}{8}$	$(1 - \alpha)(1 - \beta)\gamma + \varepsilon$
$\bar{A} \cap \bar{B} \cap \bar{C}$	$\frac{1}{8}$	0	$\frac{1}{8}$	$(1 - \alpha)(1 - \beta)(1 - \gamma) - \varepsilon$

Thus the events are pairwise independent and yet dependent as a whole system.

Under situation 3, the events are independent.

Situation 4 was constructed by setting $P(A) = \alpha$, $P(B) = \beta$, $P(C) = \gamma$ and requiring the events to be pairwise independent, as in situation 2. If, in addition, we require that $P(A \cap B \cap C) = P(A)P(B)P(C) = \alpha\beta\gamma$, we have $\varepsilon = 0$, and the events become independent, as shown by the factorization of the probabilities of all triple combinations. With N events instead of three, we could have independent subsets of all orders less than N and still have the N-fold system dependent.

1-6   GENERAL COMMENTS

Since the outcome of a contemplated action is often in doubt prior to its performance, the major use of probability lies in making judicious guesses, for probability represents the summary and generalization of experience. A considered decision to pursue a stated course of action is usually reached by weighing the odds associated with various possible outcomes, insofar as these eventualities can be foreseen. In several contexts, but conspicuously in this one, certain loosely phrased expressions have become so much a part of the accepted terminology of probability that they have acquired special idiomatic significance of their own. After the fact, the outcome of a particular action is uniquely determined; therefore, with reference to a single situation the probability of any designated event is either one or zero, depending upon whether it does or does not correspond to the actual result. Understood as a ratio, probability has only this trivial interpretation as applied to unique happenings, and we agree not to use the term probability in this sense. Hence, when we say that A has a probability p of succeeding in a proposed enterprise, what we really mean is that, according to available records on comparable cases, the relative frequency of successes is p. How pertinent the records themselves are may be open to question, but improving the estimate of probability by sharpening the classification is a project for research.

Practical applications of the theory of probability call for an intelligent combination of empirical knowledge and mathematical deduction. The basic probabilities are estimated by computing the appropriate relative frequencies from observational data, or in sufficiently clear-cut situations, they are inferred from a priori considerations according to the classical tradition. The probabilities of various composite events are then derived by applying the laws of probability to the component events, single or joint. Separate estimates are needed for joint probabilities unless the assumption of independence is justified. Incidentally, the use of the classical model in practical research is intended merely as a short cut to, and idealization of, results potentially obtainable from relative frequencies. When it is necessary to keep track of the chronological order of events, the natural procedure of setting up chronological classifications or of introducing time as an explicit variable is sometimes analytically inconvenient. By exercising proper care in the definition of terms, it is nearly always possible to regard any particular sequence of events as a simultaneous occurrence in the logical present.

1-7 ILLUSTRATIONS OF GENERAL PRINCIPLES

The ideas thus far developed equip us to solve a wide variety of problems. The following simple examples are presented in order to put the definitions to work and demonstrate the application of general principles.

EXAMPLE 1-1 Alta and Webster City are two weather stations in Iowa. Letting A and W denote the occurrence of rain at Alta and Webster City, respectively, during any 24-hour period in the month of June, it is found that $P(A) = P(W) = .40$ and $P(A \cap W) = .28$. Determine the two conditional probabilities $P(A|W)$ and $P(W|A)$ as well as the total probability $P(A \cup W)$. Are A and W independent? The conditional probabilities are found by using property 3. Here,

$$P(A|W) = \frac{P(A \cap W)}{P(W)} = \frac{.28}{.40} = .70$$

$$P(W|A) = \frac{P(A \cap W)}{P(A)} = \frac{.28}{.40} = .70$$

The total probability is found by using property 2.

$$P(A \cup W) = P(A) + P(W) - P(A \cap W)$$

$$= .40 + .40 - .28 = .52$$

The two events are not independent since $P(A|W) \neq P(A)$ or, equivalently, $P(A \cap B) \neq P(A)P(B)$. ////

EXAMPLE 1-2   Given $P(A) = \alpha$, $P(B) = \beta$, and $P(A \cap B) = \alpha\beta$, prove that $P(A \cap \bar{B})$, $P(\bar{A} \cap B)$, and $P(\bar{A} \cap \bar{B})$ factor as they should in accordance with the general definition of independence.   By property 1 we have $P(\bar{A}) = 1 - P(A) = 1 - \alpha$ and $P(\bar{B}) = 1 - P(B) = 1 - \beta$.   Now there are two exhaustive mutually exclusive ways in which event A can occur, namely, in conjunction with B or in conjunction with $\bar{B}$.   Therefore, by Theorem 1-1, total probability for mutually exclusive events,

$$P(A) = P(A \cap B) + P(A \cap \bar{B})$$

whence, by substitution,

$$\alpha = \alpha\beta + P(A \cap \bar{B})$$

$$P(A \cap \bar{B}) = \alpha - \alpha\beta = \alpha(1 - \beta) = P(A)P(\bar{B})$$

Similarly,

$$P(B) = P(A \cap B) + P(\bar{A} \cap B)$$

$$\beta = \alpha\beta + P(\bar{A} \cap B)$$

$$P(\bar{A} \cap B) = (1 - \alpha)\beta = P(\bar{A})P(B)$$

Finally, since the four pairs of events $A \cap B, A \cap \bar{B}, \bar{A} \cap B, \bar{A} \cap \bar{B}$ are exhaustive, their total probability is unity by property 1, and since they are mutually exclusive, we may apply Theorem 1-1 to obtain

$$P(A \cap B) + P(A \cap \bar{B}) + P(\bar{A} \cap B) + P(\bar{A} \cap \bar{B}) = 1$$

By substitution, therefore,

$$P(\bar{A} \cap \bar{B}) = 1 - \alpha\beta - \alpha(1 - \beta) - (1 - \alpha)\beta$$

$$= 1 - \alpha(\beta + 1 - \beta) - (1 - \alpha)\beta$$

$$= (1 - \alpha)(1 - \beta) = P(\bar{A})P(\bar{B})$$

Alternatively, this last result can be obtained in the same manner as the first two.   Thus

$$P(\bar{A}) = P(\bar{A} \cap B) + P(\bar{A} \cap \bar{B})$$

$$1 - \alpha = (1 - \alpha)\beta + P(\bar{A} \cap \bar{B})$$

$$P(\bar{A} \cap \bar{B}) = (1 - \alpha)(1 - \beta) = P(\bar{A})P(\bar{B})$$

Therefore, the requirement of factorization is satisfied in all cases, and the simple definition of independence as given for two events agrees with the general definition, as of course it should.   ////

EXAMPLE 1-3 An electrical mechanism containing four switches will fail to operate unless all of them are closed. In the probability sense, the switches are independent with regard to proper closing or failure to close, and for each switch the probability of failure is .1. Find the probability of failure of the whole mechanism, neglecting all sources of failure except switches. Letting F denote failure and $\bar{F}$ the complementary event that the mechanism operates, we apply property 1 and write

$$P(F) = 1 - P(\bar{F})$$

Letting S_1 denote the event that switch 1 is closed and $\bar{S}_1$ the complementary event that it is open, we are given that $P(\bar{S}_1) = .1$, whence

$$P(S_1) = 1 - P(\bar{S}_1) = .9$$

and similarly for the other switches. The mechanism operates only when all switches are closed, and this corresponds to the compound event $S_1 \cap S_2 \cap S_3 \cap S_4$ whence

$$P(\bar{F}) = P(S_1 \cap S_2 \cap S_3 \cap S_4)$$

Applying Theorem 1-4, the law of compound probability for independent events, we have

$$P(S_1 \cap S_2 \cap S_3 \cap S_4) = P(S_1)P(S_2)P(S_3)P(S_4)$$
$$= (.9)(.9)(.9)(.9) = (.9^4) = .6561$$

Hence $P(\bar{F}) = .6561$ and

$$P(F) = 1 - .6561 = .3439$$

This is the simplest way of solving the problem, but it is instructive to solve it in another way, using Theorem 1-2. Since the mechanism will fail whenever one or more of the switches fail to close, the event F is equivalent to the composite event $\bar{S}_1 \cup \bar{S}_2 \cup \bar{S}_3 \cup \bar{S}_4$. We have to use the general law of total probability, because the events are independent and, therefore, certainly not mutually exclusive. By Theorem 1-2, then,

$$P(F) = P(\bar{S}_1 \cup \bar{S}_2 \cup \bar{S}_3 \cup \bar{S}_4) = P(\bar{S}_1) + P(\bar{S}_2) + P(\bar{S}_3) + P(\bar{S}_4)$$
$$- P(\bar{S}_1 \cap \bar{S}_2) - P(\bar{S}_1 \cap \bar{S}_3) - P(\bar{S}_1 \cap \bar{S}_4) - P(\bar{S}_2 \cap \bar{S}_3)$$
$$- P(\bar{S}_2 \cap \bar{S}_4) - P(\bar{S}_3 \cap \bar{S}_4) + P(\bar{S}_1 \cap \bar{S}_2 \cap \bar{S}_3)$$
$$+ P(\bar{S}_1 \cap \bar{S}_2 \cap \bar{S}_4) + P(\bar{S}_1 \cap \bar{S}_3 \cap \bar{S}_4)$$
$$+ P(\bar{S}_2 \cap \bar{S}_3 \cap \bar{S}_4) - P(\bar{S}_1 \cap \bar{S}_2 \cap \bar{S}_3 \cap \bar{S}_4)$$

By inspection we note that there are four single events, six pairs, four triples, and one quadruple. Since the events are independent, the compound probabilities equal the products of the corresponding single probabilities, and since these are uniform we may collect terms of like degree as follows:

$$P(F) = 4(.1) - 6(.1^2) + 4(.1^3) - (.1^4)$$
$$= .4 - .06 + .004 - .0001$$
$$= .3439$$

Accordingly, although this alternative method of solution is far more complicated than the first and not to be recommended in a situation to which that one applies, it yields the correct answer and illustrates the general fact that all methods employing appropriate mathematical principles in valid ways will lead to the same results.   ////

EXAMPLE 1-4   Sixty orders are to be filled in a warehouse, and it is known that five of them are for a certain item A.   If these 60 orders are filled at random, what is the probability that the first and fourth orders are for item A and that simultaneously the second and third are not?   What is the probability that at least two orders of item A are in the first four orders to be filled?

Let A denote the event that a particular order being filled is for item A and let $\bar{A}$ denote the complementary event that it is not for item A.   Since the probability that a particular order falls in a specified class (whether A or $\bar{A}$) is affected by the number of orders of the same class already filled, this problem illustrates the general law of compound probability (Theorem 1-3).   A good way to attack the problem is to imagine a deck of 60 cards, all alike except that 5 of them are marked A and 55 marked $\bar{A}$.   The random filling of orders may be thought of as dealing the cards from a thoroughly shuffled deck, so that all cards available on a given trial have equal chances of being picked.   The event that the first and fourth orders are for item A and the second and third are not corresponds to dealing the card sequence $A, \bar{A}, \bar{A}, A$.   Since there are 5 cards marked A, the probability that the first card will be an A is $\frac{5}{60}$.   On the second trial, there are 59 cards in the deck, and 55 are marked $\bar{A}$.   Hence the conditional probability that the second card will be an $\bar{A}$ is $\frac{55}{59}$.   On the third trial, there are 58 cards, 54 of which are marked $\bar{A}$; consequently, the conditional probability that the third card is an $\bar{A}$ is $\frac{54}{58}$.   Finally, on the fourth trial, there remain 57 cards, 4 of which are marked A, and so the conditional probability that the fourth card is an A is $\frac{4}{57}$.   Therefore, multiplying these together in accordance with Theorem 1-3, we get

$$P(A, \bar{A}, \bar{A}, A) = (\tfrac{5}{60})(\tfrac{55}{59})(\tfrac{54}{58})(\tfrac{4}{57}) = \frac{59,400}{11,703,240} = .0051$$

The probability of the event E that at least two orders of item A are among the first four to be filled is equal to $1 - P(\bar{E})$, where $\bar{E}$ denotes the event that the first four orders contain less than two A's, hence zero or one. Now the probability of no A is given by

$$P(0) = P(\bar{A}, \bar{A}, \bar{A}, \bar{A}) = ({}^{55}\!/_{60})({}^{54}\!/_{59})({}^{53}\!/_{58})({}^{52}\!/_{57}) = \frac{8,185,320}{11,703,240} = .6994$$

Since one A can occur in four mutually exclusive ways, its total probability is

$$P(1) = P(A, \bar{A}, \bar{A}, \bar{A}) + P(\bar{A}, A, \bar{A}, \bar{A}) + P(\bar{A}, \bar{A}, A, \bar{A}) + P(\bar{A}, \bar{A}, \bar{A}, A)$$

$$= ({}^{5}\!/_{60})({}^{55}\!/_{59})({}^{54}\!/_{58})({}^{53}\!/_{57}) + ({}^{55}\!/_{60})({}^{5}\!/_{59})({}^{54}\!/_{58})({}^{53}\!/_{57})$$

$$+ ({}^{55}\!/_{60})({}^{54}\!/_{59})({}^{5}\!/_{58})({}^{53}\!/_{57}) + ({}^{55}\!/_{60})({}^{54}\!/_{59})({}^{53}\!/_{58})({}^{5}\!/_{57})$$

Noticing that the denominators are the same in all cases and that the numerators differ only in the order in which the numbers appear, we conclude that all of these probabilities are equal and that

$$P(1) = \frac{4(787,050)}{11,703,240} = \frac{3,148,200}{11,703,240} = .2690$$

Therefore, $$P(\bar{E}) = P(0) + P(1) = .9684$$

and the required probability is

$$P(E) = 1 - P(\bar{E}) = .0316 \qquad ////$$

EXAMPLE 1-5   A batch of N items contains k defective ones, although the majority $N - k$ are up to specification. If n items are picked out at random, what is the probability that the first c of them $(c < k)$ are defective and the remaining $n - c$ items are not? What is the total probability that exactly c out of n items picked at random will be defective? The principles involved in this problem are much the same as those in Example 1-4. From Theorem 1-3, the probability $P(E)$ of obtaining a sequence of c defective items followed by $n - c$ standard items is given by the product of fractions, such that each numerator equals the available number of items in the stated class at the commencement of the trial represented, and each denominator equals the corresponding total number of all items. Hence

$$P(E) = \frac{k}{N} \frac{k-1}{N-1} \cdots \frac{k+1-c}{N+1-c} \frac{N-k}{N-c} \frac{N-k-1}{N-c-1} \cdots \frac{N-k+1-n+c}{N+1-n}$$

We may express this result more compactly if we multiply and divide it by the quantity

$$\frac{(k - c)!(N - k - n + c)!}{(N - n)!}$$

which we then incorporate with the original expression to obtain

$$P(E) = \frac{k!(N - k)!(N - n)!}{N!(k - c)!(N - k - n + c)!} \tag{1-1}$$

The probability of any other sequence with c defectives in n trials would be computed in the same fashion. As we observed in Example 1-4, the numerator factors, though occurring in a different order, would involve the same set of numbers and thus yield the same product as before, and the denominator factors, being unaltered, would also yield the same product as before. Hence all sequences with c defectives in n trials have the same probability, given by Eq. (1-1). Students are often surprised by this fact, because they erroneously attach special significance to the pattern itself.

The number S of distinct sequences containing exactly c defective items in a total of n is equal to the number of permutations $Pm(n; c, n - c)$ of n things taken n at a time with c being of one type and the rest of another type; that is,

$$S = Pm(n; c, n - c) = \frac{n!}{c!(n - c)!} = C(n, c) \tag{1-2}$$

(See Appendix on Permutations and Combinations at the end of this chapter.) The total probability, say $P(c)$, of exactly c defectives in n trials is equal to the number of sequences S times their common probability $P(E)$:

$$P(c) = SP(E) = \frac{n!k!(N - k)!(N - n)!}{c!(n - c)!N!(k - c)!(N - k - n + c)!}$$

For algebraic elegance, we regroup these terms, obtaining the formula

$$P(c) = \frac{C(k, c)C(N - k, n - c)}{C(N, n)} \tag{1-3}$$

This is known as the *hypergeometric law of probability*.

In the solution of this problem, we have visualized the sampling process as the drawing of items in sequence, one by one. That this serial concept can sometimes be avoided through an appropriate model is demonstrated by the following alternative method of solving the second part. Fixing our attention on a whole sample of n items, we note that out of a batch of N items there are $C(N, n)$ possible samples which are distinct, one from another, in the sense of differing by at least one individual item. From k defects, we can choose c individuals in $C(k, c)$ different ways, and for each way we must make up the deficit $n - c$ by drawing from the store of $N - k$ standard items. This we can do in $C(N - k, n - c)$ different ways, and so (by the mn rule) there exist $C(k, c)$ times $C(N - k, n - c)$ different samples of n items containing exactly c defectives. Therefore, assuming that all possible distinct samples are equally likely, the required probability is

$$P(c) = \frac{C(k, c)C(N - k, n - c)}{C(N, n)}$$

which agrees with Eq. (1-3).   ////

EXAMPLE 1-6 An experiment can yield k possible mutually exclusive results $R_1, R_2, \ldots, R_k$, the respective probabilities of which are $p_1, p_2, \ldots, p_k$, and their total probability is unity, that is, $p_1 + p_2 + \cdots + p_k = 1$. In N independent trials of the experiment, what is the total probability of obtaining exactly n_1 results of the first type, n_2 of the second, $\ldots$, and n_k of the kth, where $n_1 + n_2 + \cdots + n_k = N$? Since the trials are independent, the probability of obtaining any particular result R_i on a given trial is wholly unaffected by the results of other trials. Therefore the probability P_s of any stated sequence of results is equal to the product of their separate unconditional probabilities, and so

$$P_s = p_1{}^{n_1}p_2{}^{n_2} \cdots p_k{}^{n_k}$$

The number S of distinct sequences yielding the stated number of results of each kind is equal to $Pm(N; n_1, n_2, \ldots, n_k)$, and the total probability, say $P(n_1, n_2, \ldots, n_k)$, is given by the product SP_s; whence

$$P(n_1, n_2, \ldots, n_k) = \frac{N!}{n_1!n_2!\cdots n_k!} p_1{}^{n_1}p_2{}^{n_2} \cdots p_k{}^{n_k} \qquad (1\text{-}4)$$

This is known as the *multinomial law of probability*. The name is derived from the fact that the general term of the expansion of the multinomial $(x_1 + x_2 + \cdots + x_k)^N$ is given by an expression of the same form as Eq. (1-4) except that the powers of $p_1, p_2, \ldots, p_k$ in the latter are replaced by corresponding powers of $x_1, x_2, \ldots, x_k$.

The special case with $k = 2$, corresponding to two alternatives $(E, \bar{E})$, is known as the *binomial law of probability*. Noting that $n_2 = N - n_1$ and $p_2 = 1 - p_1$, we have

$$P(n_1, n_2) = \frac{N!}{n_1!(N - n_1)!} p_1{}^{n_1}(1 - p_1)^{N - n_1} \qquad (1\text{-}5)$$

Since the permutation formula $Pm(N; n_1, n_2)$ for two alternatives reduces to the combination formula $C(N, n_1)$, we may write the binomial law of probability as

$$P(n_1, n_2) = C(N, n_1)p_1{}^{n_1}(1 - p_1)^{N - n_1} \qquad (1\text{-}6)$$

////

EXAMPLE 1-7   An engineer's report on the causes of failure of hot water heaters for home use revealed that 90 percent of the failures could be ascribed to one of three defects: leaking seams, leaking connections, or pinpoint corrosion —the respective probabilities being .4, .3, .2.   Neglecting the remote possibility of simultaneous defects and assuming independent trials, what is the probability that a random sample of five failures would contain two instances of leaking seams, two of leaking connections, one of pinpoint corrosion, and none due to miscellaneous causes?   Under the hypothesis that simultaneous defects are negligible, we may regard the different types of failure as mutually exclusive events, and by including the miscellaneous category, we obtain an exhaustive system.   Hence the multinomial law of probability directly applies, and the answer is given by

$$P(2, 2, 1, 0) = \frac{5!}{2!\,2!\,1!\,0!} (.4^2)(.3^2)(.2^1)(.1^0)$$

$$= (30)(.00288) = .0864 \qquad ////$$

EXAMPLE 1-8   The probability of a salesman's making a sale on a single call is $\frac{1}{6}$.   What is the probability that he will make at least one sale in the next five calls?   What is the probability that he will make four or more sales in these same five calls?   Assuming that one call does not affect another (which, of course, may not be strictly true), we apply the binomial law of probability. Letting $P(S, R)$ denote the probability of exactly S sales and R refusals in $S + R$ calls, the probability $P(E)$ of the event that the salesman makes at least one sale is equal to $1 - P(\bar{E})$, where $\bar{E}$ denotes the complementary event that no sales are made.   Here $S + R = 5$ and $\bar{E}$ is equivalent to $(0, 5)$ whence

$$P(\bar{E}) = P(0, 5) = \frac{5!}{0!\,5!} (\tfrac{1}{6})^0(\tfrac{5}{6})^5 = .402$$

and the probability of making at least one sale is

$$P(E) = 1 - P(\bar{E}) = .598$$

The event E' of making four or more sales can occur in two mutually exclusive ways, namely, by making exactly four sales or by making exactly five sales.

Hence

$$P(E') = P(4, 1) + P(5, 0)$$
$$= C(5, 4)(\tfrac{1}{6})^4(\tfrac{5}{6})^1 + C(5, 5)(\tfrac{1}{6})^5(\tfrac{5}{6})^0$$
$$= .0032 + .0001 = .0033 \qquad\qquad ////$$

EXAMPLE 1-9   In matching pennies three people A, B, C flip one coin each at a given signal and then compare results.   If one coin shows a different side from the other two, the "odd man" wins; if all three are alike the toss is repeated until one is different.   A "match" is defined as any situation with one coin different from the other two.   Assuming fair play with unbiased coins, show that all players have equal chances of winning.   Find the probability that, in a series of six such matches, C would lose at least five times.   Indicating the side of a coin by H or T and listing results in the order of players A, B, C we may exhibit the possible match situations and designate the winners as follows:

$$(H, H, T; C) \qquad (H, T, H; B) \qquad (T, H, H; A)$$
$$(T, T, H; C) \qquad (T, H, T; B) \qquad (H, T, T; A)$$

Since there are six equally likely match situations and each player wins in two of them, all players have equal chances of winning, namely $\tfrac{1}{3}$.   Letting (W, L) signify that player C wins exactly W times and loses exactly L times in $W + L$ trials, the event E of his losing at least five out of six matches can occur in two mutually exclusive ways, $(1, 5)$ and $(0, 6)$, the respective probabilities of which are given by the binomial law.   Therefore,

$$P(E) = P(1, 5) + P(0, 6)$$
$$= C(6, 1)(\tfrac{1}{3})^1(\tfrac{2}{3})^5 + C(6, 0)(\tfrac{1}{3})^0(\tfrac{2}{3})^6$$
$$= 6(.0439) + 1(.0878) = .351 \qquad\qquad ////$$

(*Suggested Problems 1-1 to 1-23*)

1-8   BAYES' THEOREM

Given a set of mutually exclusive events B_1, B_2, ..., B_n, let us assume that the occurrence of one or another of them is a necessary condition for the occurrence of an event A.   Depending on the circumstances, the B's may precede A in time or may occur simultaneously with A.   From the viewpoint of the logical present, both types of association may be regarded as constituting joint events, and we shall denote the corresponding probabilities by the same symbol $P(B_i \cap A)$, where B_i is a specific one of the B's.

From property 3 we have

$$P(B_i \cap A) = P(B_i)P(A|B_i) = P(A)P(B_i|A)$$

Therefore, the conditional probability of B_i, given A, is

$$P(B_i|A) = \frac{P(B_i)P(A|B_i)}{P(A)}$$

This conditional probability is understood to mean (in elementary terms) the proportionate number of times that the antecedent (or concomitant) of A is B_i. In connection with the use of probability in deciding upon a course of action, we have previously noted an idiomatic contraction of the precise formulation of the probability statement, and a similar idiom is used in this context. Thus, the following type of question is often propounded: Having observed A on a particular trial, what is the probability that the antecedent was B_i? Taken literally, this question refers to a unique situation and has only the trivial answer of one or zero, as the case may be; but it is really meant to refer (as it were) to the relative frequency of B_i among all situations characterized by the occurrence of A.

Although the probability $P(A)$ could be estimated directly as a relative frequency if records were available, it is often more feasible to depend upon mathematical synthesis. For instance, one might have access to a large amount of data on the relative frequencies of the B's but comparatively little data on A. This would be the case if A were a new development in the economic world or perhaps a newly discovered symptom in medical research. Under such circumstances it might be possible to deduce the conditional probabilities $P(A|B_i)$ ($i = 1, 2, \ldots, n$) from theoretical considerations (typically true of kinematic problems) or to design small-scale experiments by which the conditional probabilities could be estimated. When this is possible, $P(A)$ can be computed from the other information. Since the B's are mutually exclusive, the event A is logically equivalent to the following union of mutually exclusive events:

$$(B_1 \cap A) \cup (B_2 \cap A) \cup \cdots \cup (B_n \cap A)$$

Hence, by Theorem 1-1,

$$P(A) = P(B_1 \cap A) + P(B_2 \cap A) + \cdots + P(B_n \cap A)$$

which, by property 3, may be expressed as

$$P(A) = P(B_1)P(A|B_1) + P(B_2)P(A|B_2) + \cdots + P(B_n)P(A|B_n)$$

Therefore, substituting this result in the denominator of the expression for $P(B_i | A)$, we obtain what is known as *Bayes' theorem:*

$$P(B_i | A) = \frac{P(B_i)P(A | B_i)}{P(B_1)P(A | B_1) + P(B_2)P(A | B_2) + \cdots + P(B_n)P(A | B_n)}$$

Although Bayes' theorem can be developed rigorously from the basic laws of probability, it is one of the most controversial topics in all probability theory. One point of attack is the uniqueness of a particular event. However, this point is not peculiar to Bayes' theorem and is wholly avoidable by exercising care in the precise phrasing of statements. Another criticism, which applies only to inappropriate uses of Bayes' theorem, is that under some conditions a multiplicity of possible antecedents does not exist; there is one and only one B, but its nature is unknown. Again, the theorem has been put in an unfavorable light because many writers on the classical theory have been prone to make gratuitous assumptions about the probabilities of the respective B's—generally that they were equally likely. This, of course, is no fault of the theorem itself. These objections notwithstanding, if there does exist a variety of possible antecedents for the event A, and if the component probabilities are properly estimated from data or inferred from legitimate hypotheses, then the use of Bayes' theorem is perfectly sound.

EXAMPLE 1-10 Three urns, U_1, U_2, U_3, contain white balls, black balls, and red balls, in different proportions. U_1 contains one white, two black, and three red balls; U_2 contains two white, one black, and one red ball; and U_3 contains four white, five black, and three red balls. We reach into one urn and draw out two balls, without knowing which urn was sampled. If it turns out that one ball is white and the other is red, find the respective probabilities that the urn sampled was U_1, that it was U_2, that it was U_3. Since it is reasonable, in this case, to assume that all three urns are equally likely to have been chosen, we have $P(U_i) = \frac{1}{3}$ ($i = 1, 2, 3$). The conditional probabilities for the event A are now easily computed on the basis of the knowledge of what is actually present in each urn, so that $P(A | U_1) = \frac{1}{5}$, $P(A | U_2) = \frac{1}{3}$, and $P(A | U_3) = \frac{2}{11}$. Substituting these values in Bayes' formula we obtain the respective probabilities, which, of course, add up to unity.

$$P(U_1 | A) = \frac{(\frac{1}{3})(\frac{1}{5})}{(\frac{1}{3})(\frac{1}{5}) + (\frac{1}{3})(\frac{1}{3}) + (\frac{1}{3})(\frac{2}{11})} = \frac{33}{118}$$

$$P(U_2 | A) = \frac{(\frac{1}{3})(\frac{1}{3})}{(\frac{1}{3})(\frac{1}{5}) + (\frac{1}{3})(\frac{1}{3}) + (\frac{1}{3})(\frac{2}{11})} = \frac{55}{118}$$

$$P(U_3 | A) = \frac{(\frac{1}{3})(\frac{2}{11})}{(\frac{1}{3})(\frac{1}{5}) + (\frac{1}{3})(\frac{1}{3}) + (\frac{1}{3})(\frac{2}{11})} = \frac{30}{118} \qquad ////$$

EXAMPLE 1-11   In the experience of a certain insurance company, customers who have sufficient funds in their checking accounts postdate a check by mistake once in 1,000 times, whereas customers who write checks on insufficient funds invariably postdate them.   The latter group constitutes 1 percent of the total. A cashier receives a postdated check from a customer.   What is the probability that such a customer has insufficient funds?   There are two possible antecedents to this postdated check; namely, the client came from the population of customers with sufficient funds (B_1), where $P(B_1) = .99$, or alternatively from the group of customers with insufficient funds (B_2), where $P(B_2) = .01$.   The two conditional probabilities for the event A are now given by the statement of the problem as $P(A|B_1) = .001$ and $P(A|B_2) = 1.00$.   Therefore, using Bayes' theorem

$$P(B_1|A) = \frac{(.99)(.001)}{(.99)(.001) + (.01)(1.00)} = .09$$

and
$$P(B_2|A) = \frac{(.01)(1.00)}{(.99)(.001) + (.01)(1.00)} = .91$$

Thus the probability that the customer has insufficient funds in his account is .91.   ////

EXAMPLE 1-12   An unbiased coin is tossed, and if it comes up heads, a black ball is placed in an urn, but if tails, a white ball.   This is done four times. Another person now samples the urn by drawing out two balls simultaneously, which turn out to be black.   What is the probability that there were two black and two white balls in the urn?   Because of the method used in filling the urn, there exist five possibilities for the final color distribution of the four balls, and the probability of the occurrence of each color combination can be computed.   They are as follows:

Four white (B_1): $P(B_1) = \frac{1}{16}$
Three white and one black (B_2): $P(B_2) = \frac{4}{16}$
Two white and two black (B_3): $P(B_3) = \frac{6}{16}$
One white and three black (B_4): $P(B_4) = \frac{4}{16}$
Four black (B_5): $P(B_5) = \frac{1}{16}$

Since in this case the event (A) cannot occur with antecedents B_1 and B_2, the conditional probabilities $P(A|B_i)$ $(i = 1, 2)$ must equal zero.   The other three conditional probabilities are computed as

$$P(A|B_3) = \frac{1}{C(4, 2)} = \frac{1}{6} \quad P(A|B_4) = \frac{C(3, 2)}{C(4, 2)} = \frac{1}{2} \quad P(A|B_5) = 1$$

and therefore the required probability is given by Bayes' formula:

$$P(B_3 \mid A) = \frac{(\%_{16})(\%_6)}{(\%_{16})(\%_6) + (\%_{16})(\frac{1}{2}) + (\frac{1}{16})(1)} = \frac{6}{6 + 12 + 6} = \frac{1}{4} \qquad ////$$

(*Suggested Problems 1-24 to 1-26*)

1-9 GEOMETRICAL PROBABILITY

Many problems of a probability nature present facets which can be solved geometrically. For this reason, we shall now discuss a few simple examples of geometrical probability. This subject introduces in a natural way a broader concept of probability which goes beyond the elementary notion of the number of occurrences of discrete events.

In the nature of geometrical probability, all points under consideration lie within prescribed boundaries, and the probability measure is so defined that the total probability of the admissible region is unity, whereas that of all exterior space is zero. Since any finite portion of space, no matter how small, contains an infinite number of points, probability cannot be defined in terms of the number of points included. Instead, it is defined in terms of the geometrical measure appropriate to the dimensionality of the admissible region: in one dimension, length; in two dimensions, area; in three dimensions, volume. Although it is perfectly possible to define probability as a variable function of position, in geometrical problems it is usually assumed that the probability measure of any subdivision of the admissible region is directly proportional to the size of the subdivision. Thus, in one dimension, if the admissible region is a line segment of length R, the probability measure of any included segment of length r is r/R, that of an infinitesimal segment of length dr is dr/R, and that of the whole segment is $R/R = 1$. The statement "a point is chosen at random" is a conventional expression for the fact that the chosen point can be any point within the region designated, and the probability of its falling within any stated portion of that region is equal to the probability measure of the portion itself.

EXAMPLE 1-13 On a straight line segment AB, a point C somewhere between A and B is so located that the distance $a = AC$ is greater than the distance $b = CB$. Let us choose a point X at random in the segment AC and a point Y at random in segment CB. What is the probability that the lengths AX, XY,

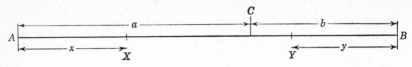

FIGURE 1-4

and YB could form a triangle?   Referring to Fig. 1-4, let x be the distance from A to X, and y the distance from Y to B.   Without regard to whether or not the specified segments could form a triangle, all possible values of x must lie in the range 0 to a, and all possible values of y must lie in the range 0 to b. We shall call the possible ranges of these two variables the *admissible region*, and by choosing two rectangular coordinates, as shown in Fig. 1-5, we may represent this admissible region by a rectangle of base a and height b.   The problem is to determine that portion of the admissible region which includes all points such that the segments may form a triangle.   If the area of this portion is r, the required probability is r/ab, since the total area of the admissible region is ab.

In order that the three lengths may correspond to the sides of a triangle, it is necessary and sufficient that each length will be less than the sum of the other two lengths.   This fact leads to three conditions:

Condition 1:

$$x < (a + b - x - y) + y \qquad \text{that is, } x < \frac{a + b}{2}$$

Condition 2:

$$(a + b - x - y) < x + y \qquad \text{that is, } x + y > \frac{a + b}{2}$$

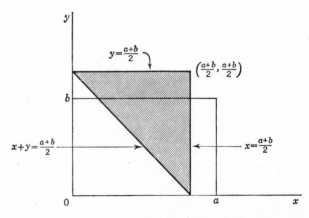

FIGURE 1-5

Condition 3:

$$y < (a + b - x - y) + x \qquad \text{that is, } y < \frac{a+b}{2}$$

Since these three conditions have to be satisfied simultaneously, it is obvious that x must be taken to the left of the straight line $x = (a + b)/2$, that y must be chosen below the straight line $y = (a + b)/2$, and at the same time x and y simultaneously must lie above the straight line $x + y = (a + b)/2$. All of this can be seen geometrically by reference to Fig. 1-5, where the shaded area represents the points which satisfy the three inequalities. Only that portion of the shaded area which overlaps the rectangle will include all points which satisfy all conditions of the problem. Since the area of overlap is $b^2/2$, the probability of the event is

$$P(E) = \frac{b^2/2}{ab} = \frac{b}{2a} \qquad ////$$

EXAMPLE 1-14 Consider a set of ruled lines infinite in extent and infinite in number. They are parallel, and the distance from one to the next is d. A needle of length a (where a is less than d) is tossed randomly on the set of ruled lines. What is the probability that the needle will intersect one of the ruled lines? This is known as the *Buffon needle problem*. Referring to Fig. 1-6, let the variable x be the perpendicular distance from the midpoint of the needle to the nearest line, and let the angle θ be the angle that the needle makes with the perpendicular. Thus x takes on a value at random somewhere between 0 and $d/2$ and the angle θ assumes a value at random somewhere between $-\pi/2$ and $+\pi/2$. From Fig. 1-6, it is apparent that if x is less than $(a/2) \cos \theta$ the needle will intersect one of the straight lines, and if x is greater than the amount, it will fail to intersect. Therefore, the curve $x = (a/2) \cos \theta$ is the boundary between the regions of intersection and nonintersection. Referring to Fig. 1-7, we now have a rectangle of height $d/2$ and base π whose area represents the total field of possible values of x and θ, and the area under the cosine curve is

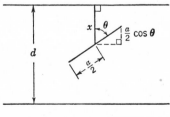

FIGURE 1-6

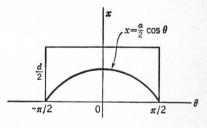

FIGURE 1-7

that portion of the admissible region for which the needle intersects the straight lines. Therefore the probability is

$$P(E) = \frac{\int_{-\pi/2}^{\pi/2} (a/2) \cos \theta \, d\theta}{(d/2)\pi} = \frac{2a}{\pi d} \qquad ////$$

EXAMPLE 1-15 The base x and altitude y of a triangle are obtained by picking points X and Y at random on two line segments of length a and b, respectively (see Fig. 1-8). What is the probability that the area of the triangle with base x and altitude y is less than $ab/4$? Since the area of the triangle is given by $xy/2$, this quantity is required to be less than $ab/4$. Now the hyperbola $xy = ab/2$ divides the admissible area, which is a rectangle with base a and altitude b, into parts I and II (see Fig. 1-9). Pairs of values for x and y which fall in area I

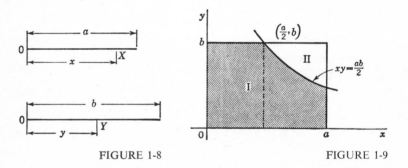

FIGURE 1-8   FIGURE 1-9

will determine triangles with area less than $ab/4$ and, of course, those points in II will determine triangles with areas greater than $ab/4$. Therefore, the desired probability is

$$P = \frac{ab/2 + \int_{a/2}^{a} (ab/2x) \, dx}{ab} = \frac{1}{2}(1 + \ln 2) = .85 \qquad ////$$

EXAMPLE 1-16 A triangle is cut from a piece of thin sheet metal. This triangle is to be tossed at random upon a ruled board as in Buffon's needle problem. The distance between rulings d is greater than any side of the triangle. What is the probability that the triangular figure will cover any portion of any line? Let the sides of the triangle be denoted by A, B, and C, and their actual lengths by a, b, and c, respectively. If a ruled line cuts through the boundary of the triangle it must cut two and only two of the sides. Therefore if it intersects at all, it cuts either A and B, or A and C, or B and C. Let $P(A, B)$, $P(A, C)$, and $P(B, C)$ be the respective probabilities of these happenings. Then the total probability of intersection will be

$$P = P(A, B) + P(A, C) + P(B, C)$$

But the respective probabilities $P(A)$, $P(B)$, $P(C)$ that it intersects any particular one of the sides A, B, C can also be expressed as follows from the law of total probability:

$$P(A) = P(A, B) + P(A, C)$$
$$P(B) = P(A, B) + P(B, C)$$
$$P(C) = P(A, C) + P(B, C)$$

If these three equations are added together, we obtain

$$P(A) + P(B) + P(C) = 2P(A, B) + 2P(A, C) + 2P(B, C) = 2P$$

and therefore

$$P = \frac{P(A) + P(B) + P(C)}{2}$$

The result of the Buffon needle problem can now be applied to the evaluation of $P(A)$, $P(B)$, and $P(C)$ since these probabilities depend on the lengths of the sides. Consequently,

$$P = \frac{2a/\pi d + 2b/\pi d + 2c/\pi d}{2} = \frac{a + b + c}{\pi d} = \frac{\text{perimeter}}{\pi d} \qquad ////$$

EXAMPLE 1-17   A variation of the Buffon needle problem is the Laplace problem. Instead of ruled lines running in one direction, we have two sets of perpendicular ruled lines which divide the plane into rectangles of length a and width b. A needle of length l is tossed on to the grid. What is the probability that the needle will intersect the boundaries of a rectangle? Assume that l is less than a or b, whichever is smaller. The center of the needle must fall within some rectangle, and we shall measure the position of the center of the needle from the lower left-hand corner of the rectangle (see Fig. 1-10).

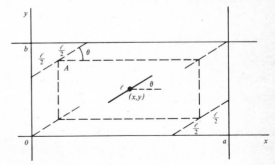

FIGURE 1-10

As a model, we shall assume also that x is a randomly chosen point on the line segment O to a, that y is a randomly chosen point on the segment O to b, and θ a randomly chosen angle between 0 and π. If θ is fixed and if the center of the needle is located anywhere inside the rectangle A, the needle will not intersect the grid lines, but if the center is outside A an intersection will take place. Thus for a fixed θ the conditional probability of no intersection will be the area of A divided by ab or

$$P = \frac{(a - l|\cos \theta|)(b - l \sin \theta)}{ab} \qquad 0 < \theta < \pi$$

The probability that θ takes on a specific value is $d\theta/\pi$ and so the joint occurrence is $P \, d\theta/\pi$; therefore the total probability of no intersection will be

$$\frac{1}{ab} \int_0^\pi (a - l|\cos \theta|)(b - l \sin \theta) \frac{d\theta}{\pi}$$

and thus the probability of an intersection is

$$1 - \frac{2}{\pi ab} \int_0^{\pi/2} (ab - bl \cos \theta - al \sin \theta + l^2 \sin \theta \cos \theta) \, d\theta$$

$$= \frac{2bl + 2al - l^2}{\pi ab} = \frac{l}{\pi ab} [2(a + b) - l]$$

Notice that, as $b \to \infty$, the probability is $2l/\pi a$ as before, and if $l = b \leq a$, we have $(2a + b)/\pi a$ or $3/\pi$ if $a = b$. ////

(*Suggested Problems 1-27 to 1-31*)

APPENDIX TO CHAPTER 1

Permutations and Combinations

1-10 PERMUTATIONS AND COMBINATIONS

Permutations and combinations relate to the possible subgroups and arrangements which can be formed from a stated collection of objects. A *permutation* is a particular sequence of a given set of objects, whereas a *combination* is the *set itself* without reference to order. Thus *ABC, ACB, BAC, BCA, CAB, CBA* are all different permutations but the same combination.

The relevant question in most applications is simply how many different permutations or combinations exist under the conditions assumed. An obvious but fundamental principle upon which the answer to this question rests is the so-called *mn* rule.

mn Rule If an event A can take place in m different ways and afterward an event B can take place in n different ways for *each* instance in which A occurs, the sequence AB can take place in mn different ways. The following three formulas are in constant use. In all cases the phrase "r at a time" means that the permutation or combination must contain exactly r objects.

1 Let $Pm(n, r)$ denote the number of different permutations which can be formed from n distinct objects taken r at a time. Then

$$Pm(n, r) = n(n - 1)(n - 2)\cdots(n - r + 1) = \frac{n!}{(n - r)!} \tag{1-7}$$

There are n objects with which the first position can be filled; when any one of these is definitely chosen, there are $n - 1$ objects left from which the second position can be filled; thus the first two positions, by the *mn* rule, can be filled in $n(n - 1)$ ways. Having fixed the first two positions, we can fill the third with any one of the $n - 2$ objects that remain; thus the first three positions can be filled in $n(n - 1)(n - 2)$ ways. In general, after k positions have been filled, the next position can be filled in $n - k$ ways; hence by successive use of the *mn* rule, the total number of permutations is the continued product $n(n - 1)(n - 2)\cdots(n - r + 1)$. In case $r = n$, this product becomes equal to $n!$. If $r < n$, the factors missing from $n!$ are $n - r, n - r - 1, \ldots, 3, 2, 1$, and their product exactly equals $(n - r)!$. Therefore, we may obtain the correct result in all cases if we divide $n!$ by $(n - r)!$, remembering that $0! = 1$ by definition.

2 Let $C(n, r)$ denote the number of different combinations which can be formed from n distinct objects taken r at a time. Then

$$C(n, r) = \frac{Pm(n, r)}{r!} = \frac{n!}{r!(n - r)!} \tag{1-8}$$

Since the formula $Pm(r, r)$ reduces to $r!$, it follows that every distinct combination of r objects yields $r!$ distinct permutations. Hence there are $r!$ times as many permutations as combinations, and so $C(n, r) = Pm(n, r)/r!$. The quantity $C(n, r)$ is often called a *binomial coefficient*, inasmuch as the binomial expansion of $(a + b)^n$ is given by

$$(a + b)^n = \sum_{r=0}^{n} C(n, r)a^{n-r}b^r \tag{1-9}$$

Sometimes groups of objects having a common characteristic can be regarded as conceptually identical for the purposes at hand, and there arises a problem of how many effectively different arrangements exist. When the n objects are not all distinct, the foregoing formulas do not hold.   Suppose there are n_1 objects of type 1, n_2 objects of type 2, ..., n_k objects of type k, where $n_1 + n_2 + \cdots + n_k = n$.   In this case, we cannot give simple formulas for the number of permutations or combinations for an arbitrary number of objects r; but it so happens that the only question ordinarily considered concerns the number of different permutations of all n objects at once, and for this a simple formula can be given. 3   Let $Pm(n; n_1, n_2, \ldots, n_k)$ denote the number of different permutations of n objects taken n at a time, given that n_i are of type i, where $i = 1, 2, \ldots,$ k and $\Sigma n_i = n$.   Then

$$Pm(n; n_1, n_2, \ldots, n_k) = \frac{n!}{n_1!n_2!\cdots n_k!} \qquad (1\text{-}10)$$

If all objects were distinguishable, there would be $n!$ different permutations. For the sake of argument, imagine the separate objects of each type tagged so as to be distinguishable as individual entities but still having the relevant characteristic in common, and consider the total number of permutations which are identical as to the order of all individuals of types $2, 3, \ldots, k$ but different as to the particular objects of type 1 which occupy the remaining positions.   Since there are $n_1!$ possible orders of all n_1 objects of type 1, there must be $n_1!$ permutations which are identical with respect to types 2, 3, ..., k.   Therefore, when the objects of type 1 are rendered indistinguishable, this group of permutations is reduced to a single pattern, and the total number of different permutations is then divided by $n_1!$.   The same argument applies to all other types, and so the total divisor is the product $n_1!n_2!\cdots n_k!$.

We conclude with a few typical applications of the foregoing formulas on permutations and combinations.

EXAMPLE 1-18   How many different sequences of three notes can be formed from an eight-note scale if no note is repeated?   Since the order is relevant, the answer is given by the number of permutations of eight things taken three at a time:

$$Pm(8, 3) = \frac{8!}{5!} = 8 \times 7 \times 6 = 336 \qquad ////$$

EXAMPLE 1-19   How many different 5-card hands can be made up from a deck of 52 cards?   Here the order is immaterial, and so the answer is given by the number of combinations of 52 things taken 5 at a time:

$$C(52, 5) = \frac{52!}{5!47!} = \frac{52 \times 51 \times 50 \times 49 \times 48}{5 \times 4 \times 3 \times 2 \times 1} = 2,598,960 \qquad ////$$

EXAMPLE 1-20   A light panel has six bulbs in a horizontal row.   How many different signals, each showing six lights simultaneously, can be formed from three red lights, two green, and one yellow?   In this case the answer is given by the number of permutations of six things taken six at a time, with three of one type, two of another, and one of still another type:

$$Pm(6; 3, 2, 1) = \frac{6!}{3!2!1!} = \frac{6 \times 5 \times 4 \times 3 \times 2 \times 1}{3 \times 2 \times 1 \times 2 \times 1 \times 1} = 60$$

Using three colors, how should six lights be divided among them in order to obtain the maximum number of signals of the type considered?   Here the object is to choose three integers x, y, z, with $x + y + z = 6$, and such that $Pm(6; x, y, z)$ is greatest.   Clearly the maximum is obtained by choosing $x = y = z = 2$, for if any number is taken less than 2, another must be made larger, and since the total number of factors multiplied together in the denominator of the fraction is constant (6, in this case), an increase in the size of any number will replace a smaller factor by a larger one.   With two lights of each color, the number of possible signals is

$$Pm(6; 2, 2, 2) = \frac{6!}{2!2!2!} = \frac{6 \times 5 \times 4 \times 3 \times 2 \times 1}{2 \times 1 \times 2 \times 1 \times 2 \times 1} = 90 \qquad ////$$

PROBLEMS

1-1   The individual cards in a special deck of 20 are marked as follows:

$$A \cap \bar{B}, A \cap B, \bar{A} \cap \bar{B}, \bar{A} \cap B, A \cap \bar{B}, \bar{A} \cap \bar{B}, A \cap B, \bar{A} \cap B, \bar{A} \cap \bar{B}, A \cap \bar{B}$$
$$A \cap B, \bar{A} \cap B, A \cap \bar{B}, \bar{A} \cap \bar{B}, \bar{A} \cap B, A \cap B, \bar{A} \cap \bar{B}, A \cap B, \bar{A} \cap B, A \cap B$$

Assuming all cards equally likely, find $p(A)$, $p(B)$, $p(A \cap B)$, $p(A \cup B)$, $p(A|B)$, $p(B|A)$, by count of favorable cases.   By substitution, verify the laws of total and compound probability, and state whether or not A and B are independent.

1-2   There are eight possible outcomes to the toss of three distinguishable coins. What is the probability that only one head comes up?   What is the conditional probability of one head, given that the three coins do not come up the same?

1-3   What is the probability that the last digit of a phone number is greater than 3? Divisible by 4? *Ans.* .6, .2.

1-4 A has a batting average of .333; B's average is .250. What is the probability that they will each get a hit in their next at bat? That at least one of them will get a hit? That neither will? *Ans.* .0833, .500, .500.

1-5   Private Jones will be happy if dinner includes either baked potatoes or ice cream. If there is a .2 chance that either will be served, what is the probability that he will be satisfied? *Ans.* .36.

1-6   In a batch of 40 light bulbs, three are bad. If five are picked at random, what is the probability that all will be good? That two will be bad? *Ans.* .662, .035.

1-7   Consider all possible outcomes of the throw of two dice. What is the probability that at least one 3 will appear? That doubles will be thrown? That the sum of the dice will be 7? *Ans.* 11/36, 1/6, 1/6.

1-8   What is the probability that a player will throw doubles at least once in four throws of two dice? *Ans.* .518.

1-9   What is the probability that any two people chosen at random were born on the same day of the week? That two or more of a random group of four were? *Ans.* 1/7, .647.

1-10 N people are in a room; F of them are female. Two people leave, one at a time. Show that the total probability that the second person to leave is female is F/N.

1-11   What is the probability that the first head will occur on the nth toss of a coin? Before the sixth toss? On an even-numbered toss? *Ans.* $1/2^n$, 31/32, 1/3.

1-12   What is more likely, one 4 in four throws of a die or two 4s in eight throws? What is the probability of no 4s in n throws? Of one or less 4s in $2n$ throws?

1-13   An inspection rule reads: Choose four pears at random. If all four are ripe, all the pears will be used for dessert in the mess hall. What is the probability that a shipment of pears only 80 percent of which are ripe will be used for dinner? *Ans.* 256/625.

1-14   Ten streets in a city need repaving; two are in the business district, four in each of two residential areas. If the city can afford to repave only four streets chosen at random, what is the probability that they will be two in each of the residential areas? That one will be in the business district, two in residential area A, and one in residential area B? *Ans.* 6/35, 24/105.

1-15   The 2^N relations implied by the general definition of independence of N events are not all independent. Show that three events A, B, C are mutually independent as a system if the following four relations are satisfied:

$$p(A \cap B) = p(A)p(B) \qquad p(A \cap C) = p(A)p(C)$$
$$p(B \cap C) = p(B)p(C) \qquad p(A \cap B \cap C) = p(A)p(B)p(C)$$

1-16   Three people A, B, C who live in the same neighborhood work in different parts of the city but utilize the same bus line. They aim to leave their offices regularly at 5 P.M. and each takes the first bus that comes along. In the rush hour, the buses leave at 5-minute intervals. Random variations of travel time to the bus terminal work out in such a way that each person has a probability of $\frac{1}{4}$ of making the 5:10 bus, a probability of $\frac{1}{2}$ of making the 5:15, and a probability of $\frac{1}{4}$ of having to wait for the 5:20. Assuming independence, what is the probability that they will all take the same bus home?

1-17 A boy plays football every Saturday unless it is raining. If the probability of rain is .3, what is the probability that he will play four times in 6 weeks?

1-18 A bag contains five white, three red, and four black balls. Three are drawn at random without replacement. Find the probability that no ball is red; exactly one is red; at least one is red; all are of the same color; no two are of the same color. *Ans.* 21/55, 27/55, 34/55, 3/44, 3/55.

1-19 Two players A, B toss coins and compare results; B pays A a fixed amount if both coins show the same side, but A pays B the same amount if the sides are different. Suppose that A is allowed to choose whichever side he wishes but that B is required to determine his side at random by fair flipping of the coin (assumed to be unbiased). Show that no system which A might employ can give him an advantage over B.

1-20 To discover the probability of getting exactly one pair in a poker hand we note that we may select the denomination of the pair in $C(13, 1)$ ways. For any denomination we may choose two cards $C(4, 2)$ ways. This done, we may select three other denominations in $C(12, 3)$ ways and choose a card from each of these three in $C(4, 1)$ ways. Since the total number of distinct five-card hands is equal to $C(52, 5)$, the required probability is

$$P = \frac{C(13, 1)C(4, 2)C(12, 3)(C(4, 1))^3}{C(52, 5)} = .4226$$

In a like manner find the probability of getting (*a*) two pair, (*b*) three of a kind, (*c*) a flush (five cards of the same suit), (*d*) a full house (three of one kind, two of another kind).

1-21 On any one day, the probability of rain without hail in a given area is .2, whereas that of rain with hail is .1. What is the probability that rain with hail will occur before a storm of rain alone? (Assume independence; sum infinite series.)

1-22 The probability of winning on a single toss of the dice is p. A starts, and if he fails, he passes the dice to B. They continue taking turns until one of them wins. What are their respective probabilities of winning? (Let $q = 1 - p$.)

1-23 The probability of winning on a single toss is p. Each of N contestants takes his turn throwing the dice once, passing to the next man if he fails. This procedure continues until someone wins. If all contestants fail on the first round, the first man starts again. What is the probability of the kth contestant winning? What is the limiting value of this probability as $p \to 0$?

1-24 Two men and one woman are seated in waiting room W_1 of a doctor's office. One man and five women are in waiting room W_2. One person leaves W_1

to wait in W_2 before the doctor calls in one person at random from W_2. This person happens to be a man. What is the probability that the person who changed rooms was a man? *Ans.* 4/5.

1-25 Stores A, B, and C under single directorship have 50, 75, 100 employees, and respectively 50, 60, and 70 percent of these are women. Resignations are equally likely among all employees, regardless of sex. One employee resigns, and this is a woman. What is the probability that she works in store C? *Ans.* 1/2.

1-26 Model X accounts for one-third, two-thirds, and one-half of the cars in used car lots A, B, and C respectively. All cars are equally likely to be sold. One man chooses a lot at random and chooses one car at random to test drive. If this car happens to be model X, what is the probability that he is at lot A or lot B? *Ans.* 2/3.

1-27 A line AB with middle point O and total length a is given. A point X is chosen at random on the line. What is the probability that AX, BX, and AO can form a triangle? *Ans.* 1/2.

1-28 Three points X_1, X_2, and X_3 are taken at random on AB. What is the probability that X_3 lies between X_1 and X_2? *Ans.* 1/3.

1-29 Three points are chosen at random on the circumference of a circle. Assuming all arcs are equally likely, what is the probability that they lie on the same semicircle? *Ans.* 3/4.

1-30 Given $AX + B = 0$. The coefficient A is chosen at random between 1 and 2; the coefficient B at random between -1 and $+1$. All possible values are equally likely. What is the probability that the solution of the equation will be greater than .25. *Ans.* 5/16.

1-31 In the triangle shown below, the length of side AB is fixed and is equal to two units. Side AC has variable length x determined by picking a point at random on a line segment of four units length so that all values of x from 0 to 4 inclusive are equally likely. The angle 0 is also determined by a random process, and all values from 0 to π inclusive are equally likely. From geometrical considerations, what is the probability that the area of triangle ABC is less than two square units? *Ans.* .753.

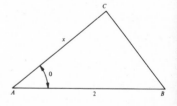

PROB. 1-31

2
DISCRETE RANDOM VARIABLES

2-1 INTRODUCTION

Basic concepts of probability were developed in Chap. 1 in terms of events. Although these were defined as the possible outcomes of an experiment and were interpreted as numerical quantities in some instances (geometrical probability, in particular), the main emphasis was upon events of a qualitative nature. In this chapter and most of those to follow, the emphasis will be placed upon quantitative material because a great many applications of probability theory concern the numerical values of appropriate variables rather than qualitative events, such as "heads or tails," "standard or defective," "fair, cloudy, or rainy," etc. Very often the variable under consideration will be capable of taking on any one of an infinite number of possible values. Nevertheless, provided that probability can be defined at all for the results of a given type of experiment, the total probability of all possible results will necessarily be unity because one result or another will be certain to occur. Our study from now on will deal with whole systems of numerical quantities, such as the set of possible numbers of times certain events may occur in a fixed number of trials or the entire range of values which a physical measurement or observation can yield.

In other words, we shall consider that class of situations in which all the possible outcomes of any particular experiment can be placed in a meaningful one-to-one correspondence with points on a line, a plane, 3-space, etc., depending upon the number of variables necessary to characterize such outcomes.

On an abstract level, ordinary mathematics is concerned with independent variables, the values of which may be chosen arbitrarily, and dependent variables which are determined by the values assigned to the former.   In the concrete domain, science aims at the discovery of laws whereby natural phenomena are interrelated, and the value of a particular variable can be determined when pertinent conditions are prescribed.   Nevertheless, there exist enormous areas of objective reality characterized by changes which do not seem to follow any definite pattern or have any connection with recognizable antecedents.   We do not mean to suggest an absence of causality.   However, from the viewpoint of an observer who cannot look behind the scenes, a variable produced by the interplay of a complex system of causes exhibits irregular (though not necessarily discontinuous) variations which are, to all intents and purposes, random.   Broadly speaking, a variable that eludes predictability in assuming its different possible values is called a *random variable* or, synonymously, a *variate*.   More precisely, a random variable must have a specific range or set of possible values and a definable probability associated with each value.   In general, we shall refer to the random variable itself as X and a particular value in the domain of the random variable X as x.

A few examples of random variables are:

1   The number of points scored by a basketball team from game to game
2   The daily number of absentees from an office
3   The weekly grocery bill of a family
4   The daily or weekly volume of gasoline used by an individual driver
5   The amount of time spent daily (or weekly, or monthly, etc.) by one person (or any group of people) in commuting

A random variable is said to be *discrete* if it is capable of assuming only a finite (or denumerably infinite) number of distinct values, whereas it is said to be continuous in a given range if it can assume any value lying in that range. In the examples cited, the first three variates are discrete, and the last two are continuous.

2-2   DISTRIBUTIONS

The distribution of a random variable X is the definition of its mathematical behavior, that is, the specification of its possible values (domain) with the respective probabilities associated with every point in the domain.   This is in conformity with the usual definition of a function as consisting of a set of ordered pairs (x, y) with only one value of y (the membership property) for every x.

In probability theory, when one speaks of the distribution of X, he may be referring to the distribution function of X, the probability function of X, or the probability density function of X. The domain and character of the function will quickly identify which one is referred to in each case. Distribution functions apply equally well to a discrete or a continuous variable as well as to a variable which is a combination of both classes. Probability functions usually apply to discrete variables, and probability density functions to continuous variables, although carelessly one sometimes refers to probability density functions as just probability functions.

2-3 DISTRIBUTION FUNCTIONS

The basic starting point of distribution theory in the field of probability is the distribution function F which always has as its domain the entire axis of reals $-\infty < x < \infty$. Its value $F(x)$ is the probability that the random variable X will take on a value less than or equal to a particular value of X, say x. Thus in the usual notation

$$F(x) = P(X \leq x) \qquad -\infty < x < \infty$$

It is clear that $F(x)$ must be a monotonically increasing function (i.e., never decreasing as x increases) because of the nature of its definition. The range of $F(x)$ must be from 0 to 1, and its definition also implies that $F(-\infty)$ must equal 0 and $F(+\infty)$ must equal 1 in all cases. By starting with the distribution function as the fundamental concept of the behavior of a random variable, we are able to unify the approach to random variables that are discrete, continuous, and combinations of both in one mathematical approach. If we consider a specific value of x, the inequality $X \leq x$ defines an event in the sense used in Chap. 1, and any particular observation belongs to this category or not, depending upon its value. If x were considered a set of discrete values along the x axis as $x_1 < x_2 < x_3 \cdots < x_n$, the probability of each of these situations $F(x_1) = P(X \leq x_1)$, $F(x_2) = P(X \leq x_2)$, etc., would offer no difficulty, from our previous consideration of concepts of probability and the occurrence of an event. Obviously, a continuous set of values of x is more subtle from this point of view and will be considered further in Chap. 4. Let us assume for the present, however, that continuous variables can be brought within the scope of probability definition also.

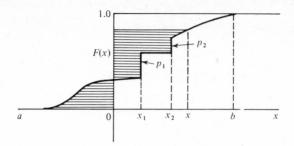

FIGURE 2-1

A possible distribution function is shown in Fig. 2-1. Here $F(x) = 0$ $(-\infty < x < a)$. It is continuous from $x = a$ out to $x = x_1$, where there is a jump in the function of magnitude p_1. $F(x)$ assumes the same value in the region between x_1 and x_2 where it jumps again in amount p_2 and then proceeds continuously to b where it reaches 1. $F(x) = 1$ for $b < x < \infty$. Obviously this function satisfies all the requirements for F. Since the definition of any function requires it to be single-valued, it is necessary to define F at points such as x_1 and x_2. In probability theory we define the value of the function $F(x)$ to always be continuous on the right so that $F(x + 0) = F(x)$. It is now clear why the probability $F(x)$ was defined to be the probability of obtaining a value less than or equal to a specific x. If we choose a specific value of x, there corresponds one value of $F(x)$. The range between 0 and this value of $F(x)$ may be divided into a large number of subintervals, equal or unequal, and each of width $\Delta F(x_i)$. This is, of course, equivalent to dividing the y axis into a large number of intervals between these two points. We shall also define, from Fig. 2-1,

$$F(x) = \sum_{\substack{0 \\ \text{limit} \\ n \to \infty}}^{n} \Delta F(x_i) = \int_{-\infty}^{x} dF(x)$$

and each $\Delta F(x_i) \to 0$.

Again, this is possible since F is monotonic. It can be shown that the limit is unique and represents the simplest definition of a Stieltjes integral. We observe that the limit of the $\Delta F(x_i)$ for the point x_1 is just p_1 or the value of the jump at the discontinuity; this is the amount by which the integral is augmented at the point. When x increases from x_1 to x_2, the integral is not increased but when x_2 is reached the integral again increases in value by the amount p_2.

If a variable is discrete so that the domain consists of a finite or denumerably infinite set of points $x_1, x_2, x_3, \ldots, x_i, \ldots$, then

$$F(x_i) = P(X \le x_i)$$

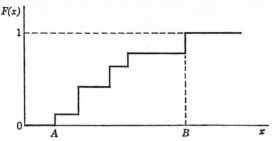

Typical distribution function of a discrete variate. FIGURE 2-2

must be entirely made up of these jumps or what is termed a *step function*. A typical step function or distribution function of a discrete random variable X is shown in Fig. 2-2.

2-4 PROBABILITY FUNCTIONS

In the case of discrete random variates, it is necessary to know only the domain of the variable and the value of the probability at these points. For convenience we shall adopt a shortened notation $f(x_i)$ for the probability $p(x = x_i)$ that a variate assumes one of its possible values x_i.

With reference to a discrete variate, therefore, the function $f(x)$, which we shall call the *probability function* of x, yields the probability that the variate in question takes on any given one of its admissible values. The *distribution* of x is then defined simply by stating the equation for $f(x)$ and indicating the allowable values that x may assume.

EXAMPLE 2-1 As a simple example of a discrete variate with only a small number of admissible values, consider the toss of an unbiased die and let the occurrence of an ace or a six be considered a success. If the die is tossed three times (or if three different dice are tossed once), the number of successes can assume any one of four values: 0, 1, 2, 3. From the classical theory, the probability of success on a single trial is $\frac{2}{6} = \frac{1}{3}$; that of failure is $\frac{2}{3}$. The numbers of distinct sequences of successes and failures which yield 0, 1, 2, 3 successes, respectively, in three tosses are 1, 3, 3, 1. Therefore, if x denotes the number of successes, the total probabilities (assuming independence) are

$$f(0) = P(x = 0) = (\tfrac{2}{3})^3 = \tfrac{8}{27} \qquad f(1) = P(x = 1) = (3)(\tfrac{2}{3})^2(\tfrac{1}{3}) = \tfrac{12}{27}$$
$$f(2) = P(x = 2) = (3)(\tfrac{2}{3})(\tfrac{1}{3})^2 = \tfrac{6}{27} \qquad f(3) = P(x = 3) = (\tfrac{1}{3})^3 = \tfrac{1}{27}$$

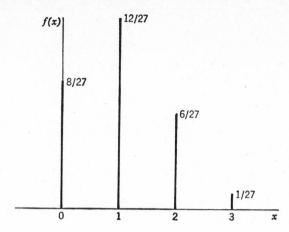

FIGURE 2-3
The probability function $f(x)$ of the discrete variate x considered in Example 2-1.

and
$$f(0) + f(1) + f(2) + f(3) = {}^{27}\!/_{27} = 1$$

showing (as must always be the case) that the sum of the probabilities of all admissible values is unity. Referring to Example 1-6, Eq. (1-6), we note that all these probabilities may be obtained from a single formula by using the binomial law of probability. Accordingly, we may express the distribution of x as follows:

$$f(x) = C(3, x)(\tfrac{1}{3})^{x}(\tfrac{2}{3})^{3-x} \qquad x = 0, 1, 2, 3$$

The graph of this function is shown in Fig. 2-3. We shall call this kind of graph a *line diagram*.   ////

For the variate considered in Example 2-1, corresponding values of the probability function $f(x)$ and the distribution function $F(x)$ are given in Table 2-1. The graph of the distribution function is shown in Fig. 2-4.

Table 2-1   CORRESPONDING VALUES OF $f(x)$ AND $F(x)$ IN EXAMPLE 2-1

x	0	1	2	3
$f(x)$	$\tfrac{8}{27}$	$\tfrac{12}{27}$	$\tfrac{6}{27}$	$\tfrac{1}{27}$
$F(x)$	$\tfrac{8}{27}$	$\tfrac{20}{27}$	$\tfrac{26}{27}$	$\tfrac{27}{27}$

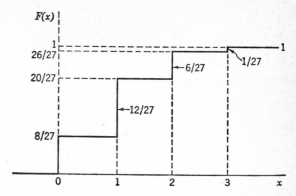

FIGURE 2-4
The distribution function $F(x)$ of the discrete variate x considered in Example 2-1.

(*Suggested Problems 2-1 to 2-4*)

2-5 RANDOM SAMPLING

Already in the course of this book we have touched upon the subject of random sampling without making explicit reference to the distribution of a random variable; we are now prepared for a more detailed statement. We shall limit this discussion to the distribution of discrete random variables since it is more appropriate at this stage. However, the argument is applicable to any random variable with a distribution function.

The concept of random sampling is the operational link between the abstract theory of probability and its practical application to observable phenomena. From the sampling viewpoint, an observation yields a particular value of a random variable, which has a distribution of potential values. Sampling is said to be *independent* when one observation does not affect another in a probability sense. Typically, observations are made on individual physical objects belonging to a definite set or collection of objects technically termed a *population* or *universe*, and it often happens that the act of taking the observation involves the withdrawal of the object observed. If the withdrawal of an object disturbs the probability distribution in the remainder of the population from which it is taken (as in a deck of cards), then in order to keep the probabilities constant, the observed item must be put back before another drawing is made. With an infinite population sampling depletions may be disregarded but in sampling from a finite population they must be taken into account. When replacement in the initial finite population is ruled out by the conditions of the problem (as in dealing card hands), we may imagine a synthetic population of all possible distinct samples and then regard the process as one of drawing at random from that synthetic population. Thus the effect of depletion can be overcome through an appropriate model.

One basic postulate of random sampling is that the empirical-distribution function of a sequence of independent observations will tend toward a fixed form as the number of observations increases without limit. We *postulate* the *existence* of an *observational process* having the properties of independence and convergence toward a fixed form of distribution, and we *name* the process *random sampling*. Conceivably, some observational process could yield a fixed limiting form in a systematic way. For instance, we could deal one card at a time from the top of a deck and return it to the bottom, thus picking 52 different cards in a fixed order, and the cycle could be repeated indefinitely. A random process has no preassigned order of individual values, nor is the order definable in any way except by complete enumeration of the series, term by term. Random sampling is a process of obtaining observations in such a way that one observation has the same chance as another of taking on (or in the case of a continuous variate, falling in the neighborhood of) any value which the variate sampled is capable of assuming, and the limiting relative frequency of any class of values, empirically realized in an infinite sequence, is identically equal to the probability of the same class in the idealized distribution of potential values. The logical role of the distribution of potential values is to provide a foundation for an axiomatic treatment of probability; this is necessary for mathematical rigor, inasmuch as mathematics is essentially deductive. Since all observations are governed independently by the same distribution, all can be identified with (in the sense of being particular values of) independent, identically distributed random variables.

2-6 SPECIFIC DISCRETE DISTRIBUTIONS

The remainder of this chapter is devoted to the introduction of certain discrete distributions. Although the number of such possible probability functions is very large, there exist a certain few so basic in the analysis of a large group of problems that they have become essential tools which should be readily available to the reader.

2-7 THE BINOMIAL DISTRIBUTION

The first probability function to be considered is the *binomial distribution*.[1]

Assuming that the occurrence of an event E is subject to chance, the number x of occurrences in n independent trials is a discrete random variable

---

[1] This is also called the Bernoulli distribution after J. Bernoulli (1654–1705), who was one of the pioneers in the theory of probability.

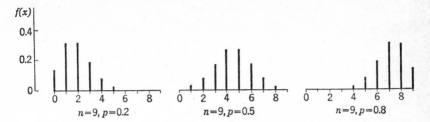

FIGURE 2-5
Binomial distributions with fixed n.

which takes on the possible values 0, 1, 2, ..., n. If the probability p of the occurrence E on an individual trial is constant, x is called a binomial variate and its distribution is known as the binomial distribution. Since the failure of E to occur on a given trial automatically implies the occurrence of the complementary event $\bar{E}$, the probability of which is $q = 1 - p$, any possible sequence of n trials is necessarily made up of x instances of E ("successes") and $n - x$ instances of $\bar{E}$ ("failures"), where x may take on any one of its admissible values 0, 1, 2, ..., n. Applying the binomial law of probability [Example 1-6, Eq. (1-6)] we may write the binomial distribution at once,

Binomial distribution:

$$f(x) = C(n, x)p^x q^{n-x} \qquad x = 0, 1, 2, \ldots, n; q = 1 - p \qquad (2\text{-}1)$$

The binomial distribution with $n = 3$, $p = \frac{1}{3}$ has already been presented in Example 2-1. Different selections of n and p give rise to various geometric shapes. Adjustable constants, such as n and p in the binomial distribution, are called *parameters*, and empirical estimates of parameters are called *statistics*. In Fig. 2-5 the parameter n is held constant ($n = 9$) and the parameter p is allowed to vary ($p = .2, .5, .8$), whereas in Fig. 2-6 p is held constant and n is allowed to vary ($n = 9, 16, 25$).

Figure 2-5 illustrates the fact that the binomial distribution is symmetrical if $p = .5$, but nonsymmetrical otherwise. A nonsymmetrical distribution is said to be *skewed* in the direction toward which it trails off; thus the distribution with $p = .2$ is skewed toward the right, and that with $p = .8$ is skewed toward the left. Incidentally, the interchange of p and q in any binomial distribution yields its mirror image, as demonstrated by the first and third graphs of this figure. Figure 2-6 bears out the general fact that, given p, the skewness of the binomial distribution becomes less pronounced as n increases.

The probability functions and distribution functions of these binomial

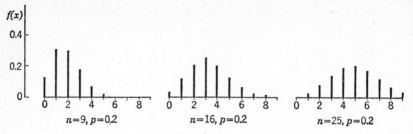

FIGURE 2-6
Binomial distributions with fixed p.

variates are exhibited numerically in Tables 2-2 and 2-3 to four decimals. We shall refer to these later on. The successive values of the F's are given by the corresponding progressive sums of the f's except where rounding errors in the latter figures violate the logical requirements that $F(x) < 1$ when $x < n$, and $F(n) = 1$.

Any probability distribution must yield unit total probability for the entire set of admissible values of the variate; this fact serves as a check on the validity of alleged distributions. We may ascertain that $\sum_x f(x) = 1$ for the binomial distribution by applying the binomial theorem. Thus

$$\sum_{x=0}^{n} f(x) = q^n + nq^{n-1}p + \cdots + nqp^{n-1} + p^n = (q + p)^n = 1^n = 1$$

Table 2-2 PROBABILITY AND DISTRIBUTION FUNCTIONS OF CERTAIN BINOMIAL VARIATES WITH FIXED n ($n = 9$)

x	$p = .2$		$p = .5$		$p = .8$	
	$f(x)$	$F(x)$	$f(x)$	$F(x)$	$f(x)$	$F(x)$
0	.1342	.1342	.0020	.0020	.0000	.0000
1	.3020	.4362	.0176	.0196	.0000	.0000
2	.3020	.7382	.0703	.0899	.0003	.0003
3	.1762	.9144	.1641	.2540	.0028	.0031
4	.0661	.9805	.2461	.5001	.0165	.0196
5	.0165	.9970	.2461	.7462	.0661	.0857
6	.0028	.9998	.1641	.9103	.1762	.2619
7	.0003	1.0000	.0703	.9806	.3020	.5639
8	.0000	1.0000	.0176	.9982	.3020	.8659
9	.0000	1.0000	.0020	1.0000	.1342	1.0000

Table 2-3 PROBABILITY AND DISTRIBUTION FUNCTIONS
OF CERTAIN BINOMIAL VARIATES WITH FIXED
p ($p = .2$)

x	$n = 9$		$n = 16$		$n = 25$	
	$f(x)$	$F(x)$	$f(x)$	$F(x)$	$f(x)$	$F(x)$
0	.1342	.1342	.0281	.0281	.0038	.0038
1	.3020	.4362	.1126	.1407	.0236	.0274
2	.3020	.7382	.2111	.3518	.0708	.0982
3	.1762	.9144	.2463	.5981	.1358	.2340
4	.0661	.9805	.2001	.7982	.1867	.4207
5	.0165	.9970	.1201	.9183	.1960	.6167
6	.0028	.9998	.0550	.9733	.1633	.7800
7	.0003	1.0000	.0197	.9930	.1108	.8908
8	.0000	......	.0055	.9985	.0623	.9531
9	.0000	......	.0012	.9997	.0294	.9825
10	.....	......	.0002	.9999	.0118	.9943
11	.....	......	.0000	.9999	.0040	.9983
12	.....	......	.0000	.9999	.0012	.9995
13	.....	......	.....	.9999	.0003	.9998
14	.....	......	.....	.9999	.0001	.9999
15	.....	......	.....	.9999	.0000	.9999
16	.....	......	.....	1.0000	.0000	.9999
..	.....	......	.....	......	.....	.9999
..	.....	......	.....	......	.....	.9999
..	.....	......	.....	......	.....	.9999
25	.0000	1.0000	.0000	1.0000	.0000	1.0000

EXAMPLE 2-2 The green light at an intersection is on for 15 seconds at a time, the yellow for 5, and the red for 55. Assuming that traffic conditions induce random variations in arrival times, so that "making the green light" is a chance event, let us find the distribution of successes in, respectively, 5 and 25 independent trials. Here we have a problem in geometric probability in the time domain. Letting G, Y, R stand for green, yellow, red, the complete cycle of lights is G, Y, R; and if T denotes the total length of a cycle, we have $T = 15 + 5 + 55 = 75$ seconds. Hence the probability p that a random point of time will fall in the interval G is

$$p = \frac{G}{T} = \frac{15}{75} = .2$$

and $q = 1 - p = .8$. For any number n of independent trials, the number x of successes has the binomial distribution with $p = .2$. The distribution of successes for 5 independent trials is given to four places of decimals in Table 2-4. The distribution for 25 independent trials was previously given in Table 2-3 and was displayed graphically in Fig. 2-6.

Table 2-4 DISTRIBUTION OF SUCCESSES IN FIVE INDEPENDENT TRIALS WITH $p = .2$

x	0	1	2	3	4	5
$f(x)$	.3277	.4096	.2048	.0512	.0064	.0003
$F(x)$	.3277	.7373	.9421	.9933	.9997	1.0000

In both cases ($n = 5$, $n = 25$) let us find the probability that the relative frequency of successes in n trials will be greater than twice the probability p of success on a single trial. Since the relative frequency is x/n and $2p = .4$, the required condition is that $x/n > .4$, whence $x > (.4)n$. When $n = 5$ this means that $x > 2$, and the corresponding probability is

$$P(x > 2) = 1 - F(2) = 1 - .942 = .058$$

When $n = 25$, we have $x > 10$ and

$$p(x > 10) = 1 - F(10) = 1 - .994 = .006$$

Therefore, with the larger number of trials, the chances are much smaller that the relative frequency obtained experimentally will deviate greatly from the true probability of success.   ////

EXAMPLE 2-3   Supposing that in the manufacture of stationery a defective sheet of paper occurs at random with probability $p = \frac{1}{500}$, let us consider the distribution of the number x of defects in a ream of 500 sheets. This distribution, which is binomial with $n = 500$ and $p = \frac{1}{500}$, is presented numerically (to three places of decimals) in Table 2-5 and graphically in Fig. 2-7. From these exhibits, we see that reams with no defects or only one defect are fairly common, that on the average fewer than one in five reams would have two defects, and that not more than five reams in one thousand would have more

Table 2-5 DISTRIBUTION OF THE NUMBER x OF DEFECTIVE SHEETS IN A REAM OF 500, WHERE $p = \frac{1}{500}$

x	0	1	2	3	4	5	6	7
$f(x)$	.367	.368	.184	.061	.015	.003	.001	.000
$F(x)$	.367	.735	.919	.980	.995	.998	.999	.999

† $P(x > 7) < .00006$.

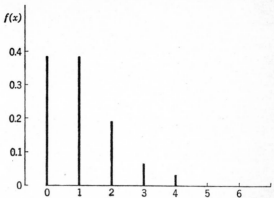

FIGURE 2-7
Line diagram of distribution in Example 2-3. Binomial with $n = 500$, $p = 1/500$.

than four defects, if we computed the relative frequency over a long series of trials. ////

The computation of $f(x)$ and $F(x)$ for the binomial distribution is laborious if n is at all large. Excellent tables giving p at intervals of .01 and n up to 49 were published in 1950 by the National Bureau of Standards under the title "Tables of the Binomial Probability Distribution." Many of the binomial exhibits of this chapter were prepared with their aid. These tables, in turn, were adapted from the well-known "Tables of the Incomplete Beta-function," which were published in 1934 by the Biometric Laboratory, then under the direction of Karl Pearson. Although the evaluation of binomial probabilities is only one of many uses of the incomplete beta function $I_p(u, v)$, it is indeed a remarkable property of the binomial distribution that partial sums of $f(x)$ (which, of course, are sums of discrete terms) are exactly given by integrals of a continuous variable. Using integration by parts, the reader may verify the following equation:

$$1 - F(k - 1) = \sum_{x=k}^{n} C(n, x)p^x q^{n-x} = \frac{\int_0^p y^{k-1}(1 - y)^{n-k}\, dy}{\int_0^1 y^{k-1}(1 - y)^{n-k}\, dy} \tag{2-2}$$

In Pearson's notation, the latter ratio is denoted by $I_p(k, n - k + 1)$.

Outside the range of available tables, it is usually sufficient in practice to approximate the binomial distribution by either (depending on the circumstances) the Poisson distribution or the normal distribution, both of which will be discussed later on. If, despite tables and approximations, there should arise an occasion for the computation of binomial probabilities, the arithmetic can be shortened by the following recurrence formulas, which the reader is invited to verify by substitution.

$$f(x + 1) = \frac{(n - x)p}{(x + 1)q} f(x) \qquad f(x - 1) = \frac{xq}{(n - x + 1)p} f(x) \qquad (2\text{-}3)$$

For large n, it is advisable to choose a value of x in the neighborhood of np, and after evaluating $f(x)$ at this point by logarithms (or logarithms of factorials) to apply Eq. (2-3) on either side of the chosen value of x. This technique was used in Example 2-3 and also in the following example.

EXAMPLE 2-4   When we discuss the approximation of the binomial distribution by the Poisson distribution or the normal, we shall want the exact values

Table 2-6   PORTION OF BINOMIAL DISTRIBUTION
WITH $n = 2,500,\ p = .02$

x	$f(x)$	$F(x)$	x	$f(x)$	$F(x)$
(0–25)	.0000	.0001†	52	.0536	.6469
26	.0001	.0002	53	.0505	.6974
27	.0001	.0003	54	.0467	.7441
28	.0002	.0005	55	.0424	.7865
29	.0004	.0009	56	.0378	.8243
30	.0006	.0015	57	.0331	.8574
31	.0010	.0025	58	.0284	.8858
32	.0016	.0041	59	.0240	.9098
33	.0025	.0066	60	.0199	.9297
34	.0036	.0102	61	.0163	.9460
35	.0052	.0154	62	.0131	.9591
36	.0073	.0227	63	.0103	.9694
37	.0099	.0326	64	.0080	.9774
38	.0131	.0457	65	.0061	.9835
39	.0169	.0626	66	.0046	.9881
40	.0212	.0838	67	.0034	.9915
41	.0260	.1098	68	.0025	.9940
42	.0311	.1409	69	.0018	.9958
43	.0363	.1772	70	.0013	.9971
44	.0413	.2185	71	.0009	.9980
45	.0460	.2645	72	.0006	.9986
46	.0501	.3146	73	.0004	.9990
47	.0534	.3680	74	.0003	.9993
48	.0557	.4237	75	.0002	.9995
49	.0569	.4806	76	.0001	.9996
50	.0569	.5375	77	.0001	.9997
51	.0558	.5933	78	.0000	.9997

† Although the values of $f(x)$ in this range are zero to four decimals, their sum rounds to .0001.

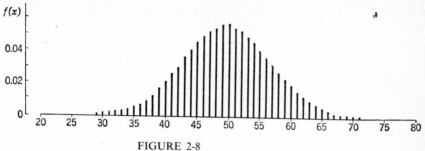

FIGURE 2-8
Line diagram of binomial distribution with $n = 2{,}500$, $p = .02$.

for purposes of comparison.   For future reference, we present in Table 2-6 a portion of the binomial distribution with $n = 2{,}500$, $p = .02$.   From Fig. 2-8, we see that the distribution is nearly symmetrical, on account of the large value of n, even though p is small.   ////

(Suggested Problems 2-5 to 2-10)

2-8   THE MULTINOMIAL DISTRIBUTION

Whereas the binomial distribution pertains to two alternative results E, $\bar{E}$ of an experiment, the *multinomial distribution* applies to any exhaustive set of mutually exclusive results R_1, R_2, ..., R_k.   If n independent trials are made, and the probability of R_i on a single trial is p_i (where $p_1 + p_2 \cdots + p_k = 1$), the multinomial distribution is the compound or joint distribution of the numbers of results of each type, say x_1 instances of R_1, x_2 of R_2, ..., x_k of R_k, where $x_1 + x_2 + \cdots + x_k = n$.   This distribution is obtained immediately from the multinomial law of probability [Example 1-6, Eq. (1-4)].   Hence,

Multinomial distribution:

$$f(x_1, x_2, \ldots, x_k) = \frac{n!}{x_1! x_2! \cdots x_k!} p_1{}^{x_1} p_2{}^{x_2} \cdots p_k{}^{x_k} \qquad (2\text{-}4)$$

where

$$p_1 + p_2 + \cdots + p_k = 1 \qquad x_1 + x_2 + \cdots + x_k = n \qquad x_i = 0, 1, 2, \ldots, n$$

Inasmuch as the x's have a fixed total, any one of them can be eliminated and replaced by the expression n minus the sum of the others; similarly any one of the p's can be expressed as unity minus the sum of the others. Hence, we may regard the multinomial distribution as a joint distribution of $k - 1$ discrete variates and containing k independent parameters. However, the symmetrical form used in Eq. (2-4) is the simplest algebraic statement. The fact that the sum of $f(x_1, x_2, \ldots, x_k)$ taken over all admissible values of the x's is unity is established by noticing that $f(x_1, x_2, \ldots, x_k)$ is simply the general term of the multinomial expansion of $(p_1 + p_2 + \cdots + p_k)^n$ and the latter expression reduces to 1^n or 1.

EXAMPLE 2-5 Competitors A, B, C control respectively 50, 30, and 20 percent of the market for a certain commodity. In a random sample of four independent buyers, what is the distribution of trade division among the three companies? This is an example of the multinomial distribution with $n = 4$, $p_1 = .5$, $p_2 = .3$, $p_3 = .2$. Letting x_1, x_2, x_3 denote the respective numbers of customers of A, B, C, the distribution is

$$f(x_1, x_2, x_3) = \frac{4!}{x_1! x_2! x_3!} (.5)^{x_1}.(.3)^{x_2}(.2)^{x_3}$$

$$x_1 + x_2 + x_3 = 4; \quad x_i = 0, 1, \ldots, 4$$

The numerical values of this joint-probability function are given in Table 2-7.

////

EXAMPLE 2-6 Show that the respective total probabilities of the various values of one of the x's in the multinomial distribution follow the binomial distribution. This fact is intuitively obvious, since the total probability of a stated value of a particular x is obtained by adding the probabilities of all

Table 2-7 MULTINOMIAL DISTRIBUTION WITH $n = 4$, $p_1 = .5$, $p_2 = .3$, $p_3 = .2$; JOINT-PROBABILITY FUNCTION OF EXAMPLE 2-5

(x_1, x_2, x_3)	$f(x_1, x_2, x_3)$	(x_1, x_2, x_3)	$f(x_1, x_2, x_3)$	(x_1, x_2, x_3)	$f(x_1, x_2, x_3)$
4, 0, 0	.0625	2, 0, 2	.0600	0, 4, 0	.0081
3, 1, 0	.1500	1, 3, 0	.0540	0, 3, 1	.0216
3, 0, 1	.1000	1, 2, 1	.1080	0, 2, 2	.0216
2, 2, 0	.1350	1, 1, 2	.0720	0, 1, 3	.0096
2, 1, 1	.1800	1, 0, 3	.0160	0, 0, 4	.0016

Overall sum: 1.0000

distinct ways in which that value can occur; this amounts to consolidating these ways into one composite alternative. However, it is instructive to work out an analytic proof. Holding the value of x_1 fixed, let us put $m = n - x_1$ and write $p_2' = p_2/(1 - p_1)$, $p_3' = p_3/(1 - p_1)$, ..., $p_k' = p_k/(1 - p_1)$. Then $p_2' + p_3' + \cdots + p_k' = 1$ because $p_2 + p_3 + \cdots + p_k = 1 - p_1$; also $x_2 + x_3 + \cdots + x_k = m$. If we divide and multiply by $(n - x_1)! = m!$ we may revamp the joint-probability function as follows:

$$f(x_1, x_2, \ldots, x_k) = f_1(x_1)g(x_2, x_3, \ldots, x_k) \tag{2-5}$$

where

$$f_1(x_1) = \frac{n!}{x_1!(n - x_1)!} p_1^{x_1}(1 - p_1)^{n - x_1} \tag{2-6}$$

$$g(x_2, x_3, \ldots, x_k) = \frac{m!}{x_2! x_3! \cdots x_k!} p_2'^{x_2} p_3'^{x_3} \cdots p_k'^{x_k} \tag{2-7}$$

Now the function $g(x_2, x_3, \ldots, x_k)$ satisfies all the conditions for the multi-nomial distribution with parameters $m, p_2', \ldots, p_k'$. Hence, the sum of g over all admissible values of $x_2, x_3, \ldots, x_k$ is unity, and so the corresponding sum of $f(x_1, x_2, \ldots, x_k)$ reduces to $f_1(x_1)$, as was to be proved. As we shall explain in Chap. 5, the resultant distribution $f_1(x_1)$ is known as the *marginal distribution* of x_1, and we have proved that this marginal distribution of x_1 is binomial with parameters n, p_1. By the same token, the function $g(x_2, x_3, \ldots, x_k)$ constitutes the *conditional* joint-probability function of $x_2, x_3, \ldots, x_k$ for a given value of x_1, and we have found that the conditional distribution of these variates is multinomial with parameters $m = n - x_1, p_2' = p_2/(1 - p_1), \ldots, p_k' = p_k/(1 - p_1)$. By mere change of notation, corresponding propositions can be established for each of the other variates $x_2, x_3, \ldots, x_k$ in turn.   ////

(*Suggested Problems 2-11 to 2-14*)

2-9 THE HYPERGEOMETRIC DISTRIBUTION

The independence of successive trials, assumed in deriving the binomial and multinomial distributions, is realizable in an infinite population, but in the case of a finite population, it necessitates the continual replacement of each unit withdrawn. The *hypergeometric distribution* applies to sampling from a finite population without replacement. Consider a population of N objects and let a denote the number that are of type A. Suppose that the sampling process is random to the extent that all objects still available after any number of removals are equally likely to be chosen on the next drawing; we shall describe this sampling situation as *conditionally random*.   Now, if such a conditionally random sample of n objects is withdrawn, and x denotes the number that are of type A, the distribution of x is called the *hypergeometric distribution*. If $n \leq a$, the admissible values of x are obviously 0, 1, 2, ..., n whereas if $n > a$, the admissible values stop short at a.   By applying the hypergeometric law of probability [Example 1-5, Eq. (1-3)], we may write the distribution as follows,

Hypergeometric distribution:

$$f(x) = \frac{C(a, x)C(N - a, n - x)}{C(N, n)} \qquad x = 0, 1, 2, \ldots, [n, a] \qquad (2\text{-}8)$$

where $[n, a]$ means the smaller of the two numbers n, a if they are different, or, of course, their common value if they are the same.   In this distribution, there are three parameters N, a, n.   Later on in this section, we shall show that $f(x)$ sums to unity.

EXAMPLE 2-7   As the couples enter the ballroom for a college dance, each person is given a ticket bearing a number, no two of which are alike.   A duplicate set of tickets is thoroughly mixed during the intermission, and three are drawn at random, prizes being awarded to the holders of the matching numbers. If 50 people attend the dance, what is the probability that at least one person out of a certain group of three couples wins a prize?   What is the probability that this group receives two prizes?   Here the 50 tickets constitute a population with $N = 50$, the six tickets which duplicate those held by the three couples make up objects of type A, with $a = 6$, and the three tickets drawn provide the sample, with $n = 3$.   Accordingly, the questions are answered by calculating terms of the hypergeometric distribution with $N = 50$, $a = 6$, $n = 3$, the general term of which is

$$f(x) = \frac{C(6, x)C(44, 3 - x)}{C(50, 3)}$$

The calculation of these terms is exhibited in Table 2-8.   The probability that at least one person in the group wins a prize is equal to the complement of the probability that no person wins, hence

$$P(x \geq 1) = 1 - f(0) = .3243$$

The probability that this group receives two prizes is simply $f(2) = .0337$.

Analyzing this problem from another viewpoint, we may regard the three lucky numbers as constituting the objects of type A and the numbers held by the three couples as constituting the sample.   True, the numbers might not be assigned in a random order as the couples enter the ballroom, but the fact that the duplicate set of tickets has been randomized produces essentially the same effect.   Thus we conclude that the probabilities in question could be obtained equally well from an alternative hypergeometric distribution with $N = 50$, $a = 3$, $n = 6$, that is,

$$f(x) = \frac{C(3, x)C(47, 6 - x)}{C(50, 6)}$$

This conclusion is perfectly correct, for [as the reader may easily verify by expressing $f(x)$ in terms of factorials] the interchange of a and n in the hypergeometric distribution has no effect on the numerical value of $f(x)$.   For comparison with the previous distribution, the calculation of $f(x)$ with $a = 3$, $n = 6$ is exhibited in Table 2-9.   The respective terms of the two distributions are identical.   ////

Table 2-8   CALCULATION OF TERMS OF HYPERGEOMETRIC DISTRIBUTION WITH $N = 50$, $a = 6$, $n = 3$

x	$3 - x$	$C(6, x)$	$C(44, 3 - x)$	Product†	$f(x)$	$F(x)$
0	3	1	13,244	13,244	.6757	.6757
1	2	6	946	5,676	.2896	.9653
2	1	15	44	660	.0337	.9990
3	0	20	1	20	.0010	1.0000

† $C(50, 3) = 19,600$; sum of products $= 19,600$.

Table 2-9   CALCULATION OF TERMS OF HYPERGEOMETRIC DISTRIBUTION WITH $N = 50$, $a = 3$, $n = 6$

x	$6 - x$	$C(3, x)$	$C(47, 6 - x)$	Product†	$f(x)$	$F(x)$
0	6	1	10,737,573	10,737,573	.6757	.6757
1	5	3	1,533,939	4,601,817	.2896	.9653
2	4	3	178,365	535,095	.0337	.9990
3	3	1	16,215	16,215	.0010	1.0000

† $C(50, 6) = 15,890,700$; sum of products $= 15,890,700$.

EXAMPLE 2-8   A certain number a out of a pack of N cards, which are otherwise alike, are marked with the letter A.   The pack is then thoroughly shuffled, and b of the cards are marked with the letter B, without reference to whether or not a particular card was previously marked A.   Letting x denote the number of cards bearing the double mark A, B, what is the distribution of x? From N cards we can choose b cards to be marked with the letter B in $C(N, b)$ distinct ways.   Any such assortment containing x cards already marked A will now have x double-marked cards, and the rest, $b - x$, will bear the single mark B.   Now there are $C(a, x)$ distinct ways of selecting x of the A cards to be used in this assortment, and for any given selection of A cards, there are $C(N - a, b - x)$ ways of choosing the remaining $b - x$ cards from the $N - a$ unmarked cards.   Therefore the number of distinct ways of obtaining x double-marked cards is given by the product $C(a, x)C(N - a, b - x)$, and the distribution of x is

$$f(x) = \frac{C(a, x)C(N - a, b - x)}{C(N, b)} \qquad x = 0, 1, 2, \ldots, [b, a]$$

which is hypergeometric with parameters N, a, b.   The logical equivalence of a and b, which is evident in this context, points once again to the fact that the distribution would not be altered if these two parameters were interchanged.

$////$

For large values of N, the computation of the terms of the hypergeometric distribution becomes laborious, and because there are three parameters, convenient tables have not been compiled.   However, the work is shortened considerably by means of the following recursion formulas, which are derived by simple division.

$$f(x + 1) = \frac{(n - x)(a - x)}{(x + 1)(N - a - n + 1 + x)} f(x)$$

$$f(x - 1) = \frac{x(N - a - n + x)}{(n + 1 - x)(a + 1 - x)} f(x)$$

(2-9)

In addition we also have

$$f(0) = \frac{C(N - a, n)}{C(N, n)} \qquad (2\text{-}10)$$

which can be expressed either as the quotient of two products of n factors or as the quotient of two products of a factors, whichever is easier.   This idea was used in computing Table 2-9.

One may readily surmise that the hypergeometric distribution approaches the binomial as N and a become sufficiently large, for sampling depletion can be considered negligible if n is small compared with a (and a fortiori small compared with N).   In that case, we may divide both numerator and denominator of the quotient $(a - x)/(N - a - n + 1 + x)$ by N, obtaining, in the limit, the quotient p/q, where $p = a/N$, $q = (N - a)/N$. Therefore,

$$f(x + 1) \to \frac{(n - x)}{(x + 1)} \frac{p}{q} f(x)$$

in agreement with the binomial recursion formula.   Moreover,

$$f(0) = \frac{(N - a)(N - a - 1) \cdots (N - a + 1 - n)}{(N)(N - 1) \cdots (N + 1 - n)}$$

and under the conditions stated, $f(0) \to q^n$.   Consequently, it follows by induction that the successive terms of the hypergeometric distribution approach the corresponding terms of the binomial for very large values of the parameters N, a, and comparatively small values of n.

We have yet to prove that the terms $f(x)$ of the hypergeometric probability function sum to unity.   This fact may be established by employing the identity

$$(1 + u)^N \equiv (1 + u)^a (1 + u)^{N - a} \qquad (2\text{-}11)$$

Expanding both sides of (2-11) by the binomial theorem we have

$$\sum_{n=0}^{N} C(N, n)u^n \equiv \left[\sum_{r=0}^{a} C(a, r)u^r \right] \left[\sum_{s=0}^{N-a} C(N - a, s)u^s \right] \qquad (2\text{-}12)$$

Multiplying the two sums on the right side of Eq. (2-12) together and collecting terms of degree n in u, we find that the result is

$$u^n[C(a, 0)C(N - a, n) + C(a, 1)C(N - a, n - 1) + \cdots + C(a, n)C(N - a, 0)]$$

Therefore, equating coefficients of u^n on both sides of Eq. (2-12) we get

$$C(N, n) = \sum_{x=0}^{n} C(a, x)C(N - a, n - x) \qquad (2\text{-}13)$$

whence, from Eq. (2-8),

$$\sum_{x} f(x) = 1$$

In this proof we have taken it for granted that $n \leq a$, but this can be done without loss of generality, because these two parameters are interchangeable.

(*Suggested Problems 2-15 to 2-17*)

2-10   THE GEOMETRIC AND PASCAL DISTRIBUTIONS

If a chance event E has a constant probability p of occurring on any given trial, the number of trials x needed for the first occurrence of E is a discrete random variable which can take on any one of an infinite number of positive integral values 1, 2, 3, ... The distribution of x is called the *geometric distribution* because of its mathematical form. The only way in which trial x can yield the first occurrence of E is to have an unbroken run of $x - 1$ failures in the first $x - 1$ trials, followed by a success on trial x. The compound probability of $x - 1$ failures is q^{x-1}, and so that of $x - 1$ failures followed by a success is $q^{x-1}p$. Therefore, the geometric distribution is

Geometric distribution:

$$f(x) = q^{x-1}p = pq^{x-1} \qquad x = 1, 2, 3, \dots \qquad (2\text{-}14)$$

Thus the geometric distribution is characterized by the single parameter p, since $q = 1 - p$. From the familiar formula for the sum of an infinite geometric series, we have

$$\sum_{x=1}^{n} f(x) = \frac{p}{1-q} = \frac{p}{p} = 1 \qquad (2\text{-}15)$$

By the same token, the distribution function $F(x)$ is given by

$$F(x) = \sum_{t=1}^{x} f(t) = 1 - q^x \qquad (2\text{-}16)$$

EXAMPLE 2-9   What are the chances that it will take more than six tosses to throw a 7 with a pair of dice? The probability of throwing a 7 on any toss is $\frac{1}{6}$, and so the probability that more than six trials will be required for the first success is

$$P(x > 6) = 1 - F(6) = 1 - [1 - (\tfrac{5}{6})^6] = (\tfrac{5}{6})^6 = .3349$$

Hence the probability is greater than $\frac{1}{3}$ that it will take more than six tosses to throw a 7 with a pair of dice, despite the fact that the probability of obtaining a 7 on a single toss is $\frac{1}{6}$. The first 10 terms of this geometric distribution are shown in Table 2-10. Here the distribution function $F(x)$ has been computed independently by Eq. (2-16).   ////

Table 2-10   GEOMETRIC DISTRIBUTION WITH $p = \frac{1}{6}$

x	1	2	3	4	5	6	7	8	9	10
$f(x)$	.1667	.1389	.1157	.0965	.0804	.0670	.0558	.0465	.0388	.0323
$F(x)$	.1667	.3056	.4213	.5177	.5981	.6651	.7209	.7674	.8062	.8385

A generalization of the geometric distribution is the Pascal distribution. Here we have a chance event E which has a constant probability of occurring on any trial as p.   If we now let the random variable X be the number of trials necessary to obtain the rth success, then

$$f(x) = C(x - 1, r - 1)p^r q^{x-r} \qquad x = r, r + 1, r + 2, \ldots \qquad (2\text{-}17)$$

Since the last observation must be a success in order to complete the event of exactly r successes, it is only the first $x - 1$ that can be permuted and thus Eq. (2-17).

(Suggested Problems 2-18 to 2-20)

2-11 THE POISSON DISTRIBUTION

Many random phenomena of interest in science and industry yield a discrete variate x having an infinite number of possible integral values 0, 1, 2, 3, ... and satisfying conditions which lead to the *Poisson distribution,*

Poisson distribution:

$$f(x) = e^{-\mu} \frac{\mu^x}{x!} \qquad \mu > 0; x = 0, 1, 2, 3, \ldots \qquad (2\text{-}18)$$

Typical examples of Poisson variates include (1) the number of inert particles of stated type suspended per cubic centimeter in a well-mixed liquid vehicle, (2) the number of misprints per page in a large volume of comparable printed material, (3) the number of responses per day to a large number of question-naires, (4) the number of calls per minute at a telephone switchboard, (5) the number of accidents of a given kind per year in a large community under stable conditions, (6) the number of α particles emitted per unit time by a radioactive element, (7) the number of rocket-bomb hits within a specified small portion of a comparatively large area under prolonged bombardment.   Let us now derive this distribution on the basis of a few simple hypotheses concerning the behavior of the following physical situation.

Events appear to occur at random in time and can be thus represented as points along the time axis (see Fig. 2-9).   A time interval of length T is chosen with the starting point picked at random anywhere on the line.   On the basis of the following four assumptions, let us find the probability of exactly n of these events occurring in the random time interval T.

1 Statistical equilibrium.   This hypothesis demands that the proba-bility of n events occurring in this period of time T is exactly the same,

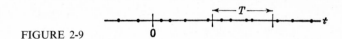

FIGURE 2-9

regardless of where the interval T commences. It is also equivalent to stating that the average number of events per unit time is the same, regardless of how the infinite number of unit lengths are chosen on the time axis in order to obtain this average. Thus, if there are λ events per unit length of time, the average number of events in the interval T is λT.

2 The probability that *one* observation will fall in an interval dT is $\lambda\,dT$. Thus, the probability of an observation falling in length dT is proportionate to dT, and doubling dT doubles the probability. This assumption is a basic one for the Poisson distribution, and it is an important concept.

3 The probability of two or more events in dT is of higher order than dT. The consequences of this assumption are the same as those following from the hypothesis that the events are independent.

4 The differentiability with respect to T. This introduces the restriction that the probability of n events occurring in the interval of T increases or decreases continuously as T increases or decreases. Thus, the probability itself has a continuous derivative with respect to T, which exists for all values of T.

Now let $P(n, T)$ be the probability that exactly n events have occurred in the time interval T. Then (since the law of total probability holds when, as in this case, the events are mutually exclusive) n events can occur in the interval $T + dT$ in the following ways, provided assumption 1 is true:

$$P(n, T + dT) = P(n, T)P(0, dT) + P(n - 1, T)P(1, dT)$$
$$+ P(n - 2, T)P(2, dT) + \cdots \qquad (2\text{-}19)$$

where the following relation must exist:

$$P(0, dT) = 1 - P(1, dT) - P(2, dT) - \cdots \qquad (2\text{-}20)$$

Replacing $P(0, dT)$ from Eq. (2-20) in Eq. (2-19) and dividing by dT gives

$$\frac{P(n, T + dT) - P(n, T)}{dT} = \frac{P(1, dT)}{dT}\left[P(n - 1, T) - P(n, T)\right] + \cdots \qquad (2\text{-}21)$$

Now, by assumption 4, the left-hand side of Eq. (2-21) in the limit becomes the derivative of $P(n, T)$ with respect to the interval T. All terms but the first on the right-hand side are zero by assumption 3, and thus by replacing $P(1, dT)$ by

$\lambda\, dT$, according to assumption 2, we have

$$\frac{dP(n, T)}{dT} + \lambda P(n, T) = \lambda P(n - 1, T) \qquad (2\text{-}22)$$

We shall define $P(-1, T) = 0$ so that by setting $n = 0$ in Eq. (2-22) there results

$$\frac{dP(0, T)}{dT} + \lambda P(0, T) = 0$$

which has the solution

$$P(0, T) = ce^{-\lambda T} \qquad \text{where} \quad c = 1 \quad \text{since} \quad P(0, 0) = 1$$

Now by assuming $n = 1$ in Eq. (2-22), we have

$$\frac{dP(1, T)}{dT} + \lambda P(1, T) = \lambda e^{-\lambda T}$$

which has as the solution

$$P(1, T) = \frac{(\lambda T)e^{-\lambda T}}{1!}$$

if we assume the initial condition $P(1, 0) = 0$.

Continuing to operate recursively on Eq. (2-22), we have finally

$$P(n, T) = \frac{e^{-\lambda T}(\lambda T)^n}{n!} \qquad (2\text{-}23)$$

Since λT is the average number of occurrences, the parameter μ in Eq. (2-18) for the Poisson distribution may be interpreted as the average number of occurrences or successes. Thus, if the average number of successes is μ, the probability of x successes is given by

$$f(x) = \frac{e^{-\mu}\mu^x}{x!} \qquad \begin{matrix} x = 0, 1, 2, \ldots \\ \mu > 0 \end{matrix} \qquad (2\text{-}24)$$

In the above derivation the average number of events per unit time was considered a constant. In many situations, however, this is not a good assumption. For example, if the events were the arrival of customers at a restaurant, the rate would certainly be a function of the time of day. Let $\lambda(t)$ be the density of events per unit time at any instant t so that $\lambda(t)$ would change continuously as in Fig. 2-10. Let $P(n, T, t)$ be the probability of exactly n events occurring in a time interval T after a beginning instant t. The previous assumptions will still be made, except of course that the probability that an observation will fall in an interval dT is now $\lambda(t + T)\, dT$, so that

$$P(n, T + dT, t) = P(n, T, t)P(0, dT, t + T) + P(n - 1, T, t)P(1, dT, t + T) + \cdots$$
$$(2\text{-}25)$$

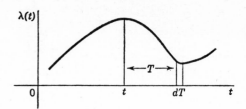

FIGURE 2-10

and

$$P(0, dT, t + T) = 1 - P(1, dT, t + T) - P(2, dT, t + T) \cdots \qquad (2\text{-}26)$$

Since $P(1, dT, t + T) = \lambda(t + T) dT$, we may now substitute $P(0, dT, t + T)$ from Eq. (2-26) into Eq. (2-25) and simplify as before.   Thus

$$\frac{dP(n, T, t)}{dT} + \lambda(t + T)P(n, T, t) = \lambda(t + T)P(n - 1, T, t) \qquad (2\text{-}27)$$

If again we define $P(-1, T, t) = 0$ and also $\mu(t, T) = \int_0^T \lambda(t + T) dT$, the successive integration of Eq. (2-27) gives

$$P(n, T, t) = \frac{\mu^n e^{-\mu}}{n!} \qquad \text{where } \mu = \int_t^{T+t} \lambda(u) \, du \qquad (2\text{-}28)$$

Thus the Poisson distribution is quite broad in its application.

The fact that the Poisson distribution is characterized by the single parameter μ makes it feasible to tabulate it extensively.   Excellent tables have been prepared by E. C. Molina under the title "Poisson's Exponential Binomial Limit."[1]   Molina's tables give $f(x)$ and $1 - F(x - 1)$ to six decimals (seven for very small μ) for values of μ up to 100 by varying decimal intervals of μ from .001 to 15 and unit intervals beyond 15.   Figure 2-11, drawn from Molina's tables, demonstrates the Poisson distribution for $\mu = 1, 2, 3, 5, 10$.   The distribution is highly skewed to the right when $\mu \leq 1$ but becomes more nearly symmetrical as μ increases.

By substituting in Eq. (2-18) it is easy to verify the fact that the Poisson probability function satisfies the following recursion formulas:

---

[1] E. C. Molina, "Poisson's Exponential Binomial Limit," D. Van Nostrand Company, Inc., Princeton, N.J., 1942.

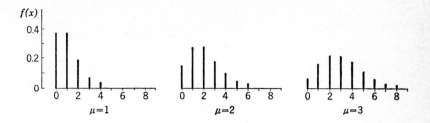

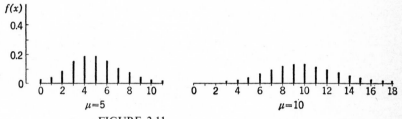

FIGURE 2-11
Line diagram of the Poisson distribution for selected values of μ.

$$f(x + 1) = \frac{\mu}{x + 1} f(x) \qquad f(x - 1) = \frac{x}{\mu} f(x) \qquad (2\text{-}29)$$

In using these formulas to compute successive values of $f(x)$, it is convenient to start at $f(0)$ when μ is small, but when μ is large it is better to choose an initial value of x near μ.

As suggested by some of the examples cited, and by the title of Molina's tables, the Poisson distribution is a limiting form of the binomial under certain conditions.   These conditions are that $p \to 0$ and $n \to \infty$ in such a manner that the product np remains finite and approaches a definite limit μ.   Examining the binomial recurrence formula

$$f(x + 1) = \frac{(n - x)p}{(x + 1)q} f(x) \qquad (2\text{-}30)$$

under these assumptions, we observe that $(n - x)p \to np \to \mu$ while $q \to 1$, whence

$$f(x + 1) \to \frac{\mu}{x + 1} f(x) \qquad (2\text{-}31)$$

in agreement with the Poisson recurrence formula (2-29).   Moreover, we have

$$f(0) = q^n = (1 - p)^n \equiv (1 + h)^{-np/h} \qquad (2\text{-}32)$$

where $h = -p$. Therefore,

$$\lim_{h \to 0} f(0) = \lim(1 + h)^{-np/h} = e^{-\mu} \qquad (2\text{-}33)$$

and this fact together with (2-31) shows inductively that the binomial distribution approaches the Poisson. For any n if p is small, the Poisson distribution with $\mu = np$ may be used to approximate the binomial. The maximum error increases with p but is nearly independent of n.

Illustrations of the Poisson approximation to the binomial are afforded by Examples 2-3 and 2-4. In Example 2-3, $n = 500$ and $p = \frac{1}{500}$, so that $\mu = np = 1$. In Example 2-4, $n = 2,500$, $p = .02$, and $\mu = np = 50$. Corresponding values of $f(x)$ for the binomial and Poisson distributions are presented for Example 2-3 in Table 2-11 and for Example 2-4 in Table 2-12. In the first comparison, the greatest difference amounts to four units in the fourth decimal place, and in the second, the greatest difference is six units in the fourth decimal place. The important question in many applications of probability theory is whether or not the probability of a certain event is less than some prescribed critical value, for instance, .05 or .01. Therefore, with regard to an approximation of this probability, it is ordinarily the absolute value of the error rather than the relative error which affects our decision.

The fact that the Poisson distribution is so widely used as an approximation to the binomial sometimes gives the false impression that it has no independent status of its own. It should be clearly understood that the Poisson distribution is a true probability distribution, mathematically derivable from definite premises, and having properties and applications unique to itself.

Table 2-11   COMPARISON OF PROBABILITY FUNCTIONS $f(x)$ OF BINOMIAL DISTRIBUTION WITH $n = 500$, $p = \frac{1}{500}$ AND POISSON DISTRIBUTION WITH $\mu = np = 1$

x	Binomial	Poisson
0	.3675	.3679
1	.3682	.3679
2	.1841	.1839
3	.0612	.0613
4	.0153	.0153
5	.00303	.00307
6	.00050	.00051
7	.000071	.000073

Table 2-12 COMPARISON OF PROBABILITY FUNCTIONS $f(x)$ OF BINOMIAL DISTRIBUTION WITH $n = 2,500$, $p = .02$ AND POISSON DISTRIBUTION WITH $\mu = np = 50$

x	Binomial	Poisson	x	Binomial	Poisson
(0–25)	.0000	.0000	52	.0536	.0531
26	.0001	.0001	53	.0505	.0501
27	.0001	.0001	54	.0467	.0464
28	.0002	.0002	55	.0424	.0422
29	.0004	.0004	56	.0378	.0376
30	.0006	.0007	57	.0331	.0330
31	.0010	.0011	58	.0284	.0285
32	.0016	.0017	59	.0240	.0241
33	.0025	.0026	60	.0199	.0201
34	.0036	.0038	61	.0163	.0165
35	.0052	.0054	62	.0131	.0133
36	.0073	.0075	63	.0103	.0105
37	.0099	.0102	64	.0080	.0082
38	.0131	.0134	65	.0061	.0063
39	.0169	.0172	66	.0046	.0048
40	.0212	.0215	67	.0034	.0036
41	.0260	.0262	68	.0025	.0026
42	.0311	.0312	69	.0018	.0019
43	.0363	.0363	70	.0013	.0014
44	.0413	.0412	71	.0009	.0010
45	.0460	.0458	72	.0006	.0007
46	.0501	.0498	73	.0004	.0005
47	.0534	.0530	74	.0003	.0003
48	.0557	.0552	75	.0002	.0002
49	.0569	.0563	76	.0001	.0001
50	.0569	.0563	77	.0001	.0001
51	.0558	.0552	78	.0000	.0001

EXAMPLE 2-10 A plane is flying over an area A looking for an object that is continuously moving around on a random course inside A. (See Fig. 2-12.) The total area covered is, of course, dependent upon the speed of the plane and the width of the swath within which a contact is possible by visual means or by the use of equipment. What is the probability of finding the object as a function of the area searched? The probability that the plane has not sighted the object after having covered an area of α is denoted by $P(0, \alpha)$. The change in this probability, $dP(0, \alpha)$, is negative and is equal to the probability of no sighting during a completed search of area α times the probability of finding the object in $d\alpha$ since

$$dP(0, \alpha) = P(0, \alpha + d\alpha) - P(0, \alpha) = P(0, \alpha)\left(1 - \frac{d\alpha}{A}\right) - P(0, \alpha)$$

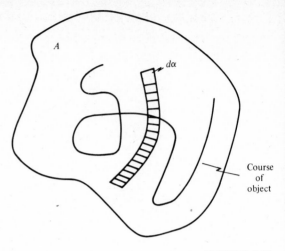

FIGURE 2-12

or

$$dP(0, \alpha) = -P(0, \alpha)\frac{d\alpha}{A}$$

Thus

$$P(0, \alpha) = C_1 e^{-\alpha/A} = e^{-\alpha/A} \qquad C_1 = 1 \quad \text{since} \quad P(0, 0) = 1$$

It is interesting to note that $P(0, A) = 1/e$ and that $1 - e^{-\alpha/A}$ is the probability of at least one sighting of the object after an area of size α has been examined. (See Fig. 2-13.) The value $e^{-\alpha/A}$ is the first term ($n = 0$) of a Poisson distribution with $\lambda = \alpha/A$. ////

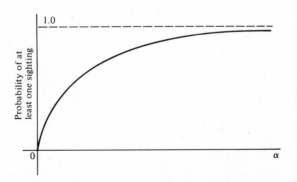

FIGURE 2-13

EXAMPLE 2-11 Our plane flies back and forth across a line of length L looking for an object which is located somewhere on the line segment. If the object is within plus or minus a (usually called the sweep width) of the point where the plane crosses the line, the plane will sight the object. If the density of the search varies along the segment L, approximately what is the probability of sighting the object as a function of this density? The probability of observing the object on one pass is thus $2a/L$ and of missing it is $1 - 2a/L$. In n random passes (or, equivalently, the object is moving at random) the probability of sighting the object at least once is $1 - (1 - 2a/L)^n$.

Comparing the series expansion of $(1 - 2a/L)^n$ and $e^{-2an/L}$ as

$$\left(1 - \frac{2a}{L}\right)^n = 1 - \frac{(n)(2a/L)}{1!} + \frac{(n)(n-1)(2a/L)^2}{2!} - \cdots$$

$$e^{-2an/L} = 1 - \frac{(n)(2a/L)}{1!} + \frac{n^2(2a/L)^2}{2!} - \cdots$$

it is evident that these two series are reasonably close numerically for large n and thus asymptotically $e^{-2an/L} \approx (1 - 2a/L)^n$ so that the probability of observing the object at least once in n trials (n large) is approximately $1 - e^{-2an/L}$. The quantity n/L is the number of passes of the plane per unit length of line L and is thus a rate or density. If n and L are large and the number of passes varies with position x along the line segment (x measured from some arbitrary point), then $2an/L$ can be thought of as a density function of x, $\phi(x)$ (i.e., the density of search at x) so that the probability of at least one sighting in the vicinity of x is $1 - e^{-\phi(x)}$. This corresponds somewhat analogously to $1 - e^{-a/A}$ which we found previously.   ////

EXAMPLE 2-12 In the development of the Poisson distribution, we assumed a constant probability for the occurrence of an event in time dt as $\lambda\,dt$, where λ is the occurence rate per unit time. If a finite population of size N is now considered and if we now assume also that the λ is proportional to the size of the population at any instant of time, what is the probability of exactly n occurrences of an event in a time τ following a time t after the process starts?

Let $P(n, \tau, t)$ be the probability of n occurrences of the event in an interval of time τ following a time t, and $P(n, \tau, t|l)$ be the probability of n occurences in the time τ given that l have already occurred up to time t. We shall also assume, as in the Poisson case, that the probability of more than one event in time $d\tau$ is of higher order than $d\tau$ and, of course, that statistical equilibrium exists throughout the process.

Since the probability of the event happening in dt is $Kn\, dt$, where n is the number in the population at t, we have as in Eq. (2-19)

$$P(n, \tau + d\tau, 0) = P(n, \tau, 0)[1 - (N - n)\, Kd\tau]$$
$$+ P(n - 1, \tau, 0)[(N - n + 1)]\, Kd\tau + \cdots$$

which immediately gives the differential equation

$$\frac{dP(n, \tau, 0)}{d\tau} + (N - n)KP(n, \tau, 0) = (N - n + 1)KP(n - 1, \tau, 0)$$

Since $P(-1, \tau, 0) = 0$, it follows that

$$\frac{dP(0, \tau, 0)}{d\tau} = -NKP(0, \tau, 0) \qquad \text{or} \qquad P(0, \tau, 0) = e^{-NK\tau}$$

$$\text{since} \quad P(0, 0, 0) = 1$$

For $n \geq 1$, we recursively have, since $P(n, 0, 0) = 0$,

$$e^{(N-n)K\tau}P(n, \tau, 0) = \int_0^\tau e^{(N-n)K\tau}(N - n + 1)P(n - 1, \tau, 0)\, Kd\tau$$

where N is the total size of the original population. Integrating successively yields

$$P(n, \tau, 0) = \frac{N!}{(N - n)!\, n!}\, e^{-NK\tau}(e^{K\tau} - 1)^n$$

It is now possible to obtain $P(n, \tau, t|l)$ by replacing N by $N - l$ and by making t equivalent to zero in $P(n, \tau, 0)$ as

$$P(n, \tau, t|l) = \frac{(N - l)!}{n!\, (N - l - n)!}\, e^{-(N-l)K\tau}(e^{K\tau} - 1)^n$$

Since $P(n, \tau, t)$ is made up of $n + 1$ mutually exclusive situations,

$$P(n, \tau, t) = \sum_{l=0}^{N-n} P(n, \tau, t|l)P(l, t, 0)$$

$$= \sum_{l=0}^{N-n} \frac{(N - l)!}{n!\, (N - l - n)!}\, e^{-(N-l)K\tau}(e^{K\tau} - 1)^n\, \frac{N!}{l!\, (N - l)!}\, e^{-NKt}(e^{Kt} - 1)^l$$

$$= \frac{N!}{n!\, (N - n)!}\, e^{-NK(t+\tau)}(e^{K\tau} - 1)^n \sum_{l=0}^{N-n} \frac{(N - n)!}{l!\, (N - l - n)!}\, (e^{K(t+\tau)} - e^{K\tau})^l$$

and so recognizing the binomial expansion,

$$P(n, \tau, t) = \frac{N!}{n!\, (N - n)!}\, (e^{K\tau} - 1)^n e^{-NK(t+\tau)}(1 - e^{K\tau} + e^{K(t+\tau)})^{N-n} \qquad (2\text{-}34)$$

If $N \to \infty$ in such a way that $NK = \lambda$, this expression reduces to $e^{-\lambda\tau}(\lambda\tau)^n/n!$ and is independent of t.   ////

(*Suggested Problems 2-20 to 2-27*)

PROBLEMS

2-1   Graph the probability and distribution functions for the random variable defined by the sum of the digits in a throw of two dice.

2-2   For an unbiased coin, the probability of heads is $\frac{1}{2}$; let x denote the number of heads in three independent tosses. Write an equation for the probability function $f(x)$ and show that the separate probabilities add up to unity.

2-3   Each of two control units, operating independently of one another, has a probability of p_1 of operating when called upon. Connected to each of these units are two subcontrol units (a total of four subunits), each of which acts independently of the other control and subcontrol units. Each subcontrol unit has a probability of p_2 of operating when called upon but can be called on only if the control unit to which it is attached is working. What is the probability function of the number x of subcontrol units which can operate at any arbitrary time?

2-4   A coin is tossed until the same result appears twice in succession. Find the probability that (*a*) the experiment ends before the fifth toss and (*b*) an odd number of tosses is required.   *Ans.* 7/8, 1/3.

2-5   Assuming that the sexes are equally likely, what is the distribution of the number of girls in a family of four children? What is the probability that a family of four children will have at least two girls?   *Ans.* $C(4, x)(1/2)^4$, 11/16.

2-6   A group has ten puptents to use on camping trips but needs only eight for next weekend. If two of the tents are leaky, what is the distribution of good tents in this conditionally random sample of eight? What is the probability of using no defective tents?   *Ans.* $f(8) = 2/90, f(7) = 32/90, f(6) = 56/90$, 1/45.

2-7   If A, B, C match pennies and odd man wins, what is the distribution of the number of times that A wins in four matches? Assume that there must be an odd man to constitute a match. Plot the probability function and distribution function. To three decimals, compute the probability (*a*) that A wins exactly twice, (*b*) that A loses at least twice, (*c*) that A either wins all matches or loses all matches (total probability).   *Ans.* .296, .889, .210.

2-8   The probability of a particular person obtaining a bull's-eye is $\frac{1}{8}$. Ten shots are fired. (*a*) What is the probability of obtaining at least three bull's-eyes? (*b*) What is the conditional probability of obtaining at least three bull's-eyes, knowing that at least one bull's-eye has been scored?   *Ans.* .701, .714.

2-9   How many times must a die be cast in order that the probability of 3 occurring will be at least .8?   *Ans.* 9.

2-10   In research on drugs to counteract the intoxicating effects of alcohol, 10 subjects were used to test the relative merits of Benzedrine and caffeine. Benzedrine brought about more rapid recovery in nine cases. Basing your analysis on a binomial distribution with $p = .5$, determine the probability of obtaining results by chance favoring either drug to this extent or more. Assume that there was no a priori reason to suppose one drug more effective than the other.

2-11   Human blood types have been classified into four exhaustive, mutually exclusive categories O, A, B, AB. In a certain very large population, the respective probabilities of these four types are .46, .40, .11, .03. Given that the population is so large that sampling depletions may be ignored, find the probability that a random sample of five individuals will contain (a) two cases of type O and one case of each of the others; (b) three cases of type O and two type A; (c) no case of type AB.   Ans. .0168, .1557, .8587·

2-12   Using the information of Prob. 2-11, compute the entire distribution for random samples of two. (a) How many distinct allocations are there? (b) What is the total probability of the two most likely allocations? (c) What is the total probability of the two least likely allocations?

2-13   Using the information of Prob. 2-11, find the sample size needed so that the probability of yielding at least one case of blood type AB will be at least .95. Solve this problem from the viewpoint of the binomial distribution.

2-14   A finite population contains N objects, of which the number a belong to type A, b to type B, and c to type C, where $a + b + c = N$. If a conditionally random sample of n objects is taken (without replacements), find the compound distribution of the number of objects (say x_1, x_2, x_3) of each type.

2-15   A box contains ten light bulbs, two of which are burned out. (a) What is the distribution of the number of good bulbs in a conditionally random sample of eight? (b) What is the probability of drawing no defective bulb? (c) Two defective bulbs?   Ans. (b) 1/45, (c) 28/45.

2-16   A five-card hand contains three aces. If another player is allowed to withdraw any two cards sight unseen, what is the distribution (a) of the number of aces withdrawn? (b) The number of aces left? Compute each distribution in full.

2-17   Under the usual assumption of randomness, what is the distribution of the number of diamonds in a 13-card hand?

2-18   The probability of "making the green light" at a certain intersection is .2. (a) In a series of independent trials, what is the distribution of number of trials required for the first success? (b) What is the probability of obtaining a success in less than four trials?   Ans. (a) $.2(.8)^{x-1}$ $x = 1, 2, \ldots$, (b) .488.

2-19   If the probability of success on a single trial is .20, what is the probability that the third success will be obtained before the tenth trial?   Ans. About .26.

2-20   A weighted coin with probability p of heads and q of tails is tossed until both a head and a tail have appeared. What is the probability distribution of this event?   Ans. $pq^{x-1} + qp^{x-1}$, $x = 2, 3, \ldots$.

2-21   The probability that the secretary will have to answer the phone in any short interval of time *dt* hours is 4 *dt*. What is the probability that she will receive exactly two calls between 8:00 and 10:00? *Ans.* About .01.

2-22 Given that the distribution of *x* is Poisson and (to six decimals) $f(0) = .049787$, find $P(x \geq 4)$. *Ans.* .35.

2-23 A magazine has 100 pages and 100 misprints distributed at random. Assuming a Poisson distribution with μ equal to the average number of misprints per page, estimate the probability that a page contains at least two misprints.

2-24 Two coeds get many phone calls. Sue notes that in a sample of 10 calls she has two requests for dates. Jenny notes that in a sample of 15 calls she has five requests for dates. Noting the difference between these rates, Sue begins to wonder about her popularity as compared with Jenny's. If we assume that they are equally popular, that is, both average seven date requests out of 25 calls, what is the probability that Sue would get two or less requests in 10 calls?

2-25 In four independent rolls of a die, find the probability of obtaining (*a*) at least one 6, (*b*) exactly one 6, (*c*) exactly two 6's. Compare these results with the Poisson approximations.

2-26 Assume that the number of cars passing an intersection obeys a Poisson distribution. If the probability of no cars in 1 minute is .20, what is the probability of more than one car in 2 minutes? *Ans.* .831.

2-27 If *x* has the Poisson distribution, and μ is an integer, show that there is some value of *x* for which $f(x) = f(x + 1)$.

3

PROBABILITY AND DIFFERENCE EQUATIONS

3-1 DIFFERENCE EQUATIONS

When functional relationships exist between continuous variables, these relationships can often be expressed by means of a differential equation or a system of differential equations. However, the variables considered in probability theory and stochastic processes are often discrete rather than continuous, and their relationships commonly lend themselves to expression by difference equations.

In this section, we shall consider a function f which has the value $f(x)$ at a set of integers $x = 0, 1, 2, 3, \ldots, n, \ldots$.

The formulation of the difference equation in a specific instance is the determination of the appropriate relationship between the value of the function at a given point $x = k$ with value $f(k)$ and the values of the function at several other points usually in the vicinity of $x = k$. When this concept is applied in probability theory, the value of the function at a point $x = k$ will often be associated with the probability of a certain event occurring on or before the kth trial.

The difference $f(x + 1) - f(x)$ $(x = 0, 1, 2, 3, \ldots)$ between two consecutive values of the function $f(x)$ is called the first difference of the function and is denoted by $\Delta f(x)$. The difference between consecutive values of $\Delta f(x)$ is called the second difference and is denoted by $\Delta^2 f(x)$, and a similar definition is used for higher-order differences. Thus

$$\Delta f(x) = f(x + 1) - f(x)$$
$$\Delta^2 f(x) = \Delta[\Delta f(x)] = [f(x + 2) - f(x + 1)] - [f(x + 1) - f(x)]$$
$$= f(x + 2) - 2f(x + 1) + f(x)$$
$$\Delta^3 f(x) = \Delta[\Delta^2 f(x)] = [f(x + 3) - 2f(x + 2) + f(x + 1)]$$
$$- [f(x + 2) - 2f(x + 1) + f(x)]$$
$$= f(x + 3) - 3f(x + 2) + 3f(x + 1) - f(x)$$

By examining the construction of the $\Delta^k f(x)$, it is apparent that the coefficients are "binomial" so that

$$\Delta^n f(x) = f(x + n) - C(n, 1)f(x + n - 1) + C(n, 2)f(x + n - 2)$$
$$- \cdots + (-1)^n f(x)$$

and therefore the value of $\Delta^n f(x)$ is a linear combination of values of the function at unit intervals starting with $f(x + n)$ and ending with $f(x)$.

An nth-order difference equation (corresponding to an nth-order differential equation) is some functional relationship between the nth-order difference and those differences of lower order as

$$\Delta^n f(x) = F(x, \Delta f(x), \Delta^2 f(x), \ldots, \Delta^{n-1} f(x)) \qquad (3\text{-}1)$$

In this text, we shall consider only linear difference equations with constant coefficients so that Eq. (3-1) becomes

$$b_n \Delta^n f(x) + b_{n-1} \Delta^{n-1} f(x) + \cdots + b_1 \Delta f(x) + b_0 f(x) = g(x) \qquad (3\text{-}2)$$

where the b_i are constants and where $g(x)$ is any function of x.

If the differences of all orders in Eq. (3-2) are replaced in terms of their functional values, the equation becomes

$$a_n f(x + n) + a_{n-1} f(x + n - 1) + \cdots + a_0 f(x) = g(x) \qquad (3\text{-}3)$$

where the a_i are new constants and are all computable from the b_i's. When $g(x) = 0$, Eq. (3-3) is a linear, homogeneous difference equation of the nth order. The order of the equation is thus determined by the maximum difference in the arguments of $f(x)$.

Let us first consider a first-order, homogeneous, linear difference equation with constant coefficients as

$$f(x + 1) - af(x) = 0$$

where the coefficient of $f(x + 1)$ in Eq. (3-3) is different from zero and so may be divided out.

This equation implies

$$f(1) = af(0) \qquad f(2) = af(1) = a^2f(0) \qquad f(3) = af(2) = a^3f(0)$$

etc., which indicates the solution of the equation to be

$$f(x) = a^x f(0) = Aa^x$$

where A is an arbitrary constant.

This form of solution compares with that of an equivalent differential equation $dy/dx - ay = 0$ which has the solution form $f(x) = Ae^{ax}$. This suggests that a technique similar to that used in solving nth-order linear, homogeneous differential equations with constant coefficients might be applicable here.

By substituting the solution $f(x) = Ar^x$ into the homogeneous equation

$$a_n f(x + n) + a_{n-1} f(x + n - 1) + \cdots + a_1 f(x + 1) + a_0 f(x) = 0 \qquad (3\text{-}4)$$

there results

$$Ar^x(a_n r^n + a_{n-1} r^{n-1} + a_{n-2} r^{n-2} + \cdots + a_1 r + a_0) = 0 \qquad (3\text{-}5)$$

Obviously any value of r, say r_1, which is a root of the polynomial in the parentheses will make $f(x) = A_1 r_1{}^x$ a solution of the difference equation. If there are n distinct roots $r_1, r_2, \ldots, r_n$ of Eq. (3-5), then, since the equation is linear, the general solution of (3-4) must be of the form

$$f(x) = A_1 r_1{}^x + A_2 r_2{}^x + \cdots + A_n r_n{}^x \qquad (3\text{-}6)$$

where $A_1, A_2, \ldots, A_n$ are arbitrary constants.

Any root r of Eq. (3-5) can be expressed in complex form as $r = \rho(\cos \theta + i \sin \theta) = \rho e^{i\theta}$, where $\rho = |r|$ and where $i = \sqrt{-1}$, which shows that $r^x = \rho^x e^{ix\theta} = \rho^x(\cos \theta x + i \sin \theta x)$. Therefore, if r is real and positive, then $\theta = 0$ and so $A_1 r_1{}^x = A_1 \rho_1{}^x$, but if r is real and negative then $\theta = \pi$ so that $A_1 r_1{}^x = A_1 \rho_1{}^x e^{ix\pi} = A_1 \rho_1{}^x \cos \pi x$. Complex roots, of course, occur in pairs as $r_1 = \alpha + i\beta$ and $r_2 = \alpha - i\beta$ and thus $A_1 r_1{}^x + A_2 r_2{}^x$ becomes $\lambda^x(A_1' \cos \theta x + A_2' \sin \theta x)$, where $\lambda = \sqrt{\alpha^2 + \beta^2}$, $\theta = \tan^{-1} \beta/\alpha$, and A_1' and A_2' are again arbitrary constants since $A_1' = A_1 + A_2$ and $A_2' = i(A_1 - A_2)$.

If we consider the case where the roots of the polynomial are repeated, we find that the solution of Eq. (3-4) is analogous to that expected from our knowledge of differential equations. For example, if $A_1 r_1{}^x$ is a solution of the difference equation and r_1 is a root repeated k times, then $(A_1 + A_2 x + A_3 x^2 + \cdots + A_{k-1} x^{k-1}) r_1{}^x$ is that part of the general solution associated with this root.   The proof of this follows from the argument that this solution satisfies Eq. (3-4) provided the term in parentheses in Eq. (3-5) is zero for $r = r_1$ and also the first $k - 1$ derivatives of this same parentheses vanish for $r = r_1$, which is the case for the multiple root of order k.

A general linear difference equation with constant coefficients, but non-homogeneous, as Eq. (3-3), is solved by utilizing a procedure similar to that used in differential equations.   We let $g(x) = 0$ and then solve the resulting homogeneous difference equation.   Any particular solution of (3-3) is now obtained and added to the general solution of the homogeneous equation; this sum yields the general solution.   The problem of obtaining a particular solution is, in general, more complicated than in the nonhomogeneous differential equation, and the reader is referred to any text on the calculus of finite differences.   However, if $g(x)$ is a polynomial and the assumption is made that the particular solution $f_P(x)$ is a polynomial also, then $f_P(x + k)$ is a polynomial of the same degree as $f_P(x)$.   Thus it will usually be possible to determine the coefficients of the assumed polynomial solution.   For example, if $g(x) = 2x^2 + 3$ in Eq. (3-3) and we assume a solution $f_P(x)$ as $f_P(x) = ax^2 + bx + c$, then upon substitution in Eq. (3-3) we have

$$a_n[a(x + n)^2 + b(x + n) + c] + a_{n-1}[a(x + n - 1)^2$$
$$+ b(x + n - 1) + c] + \cdots + a_0(ax^2 + bx + c) = 2x^2 + 3$$

If the coefficients of x^2, x, 1 are equated to each other on the two sides of the equation, three algebraic equations in three unknowns a, b, c are generated which, in general, will have a unique solution.   If it so happens that an arbitrary constant is a solution of the homogeneous equation itself, then, as in differential equations, the particular solution would have to be of the form $f_P(x) = x(ax^2 + bx + c)$.

EXAMPLE 3-1   Find the general solution of the difference equation

$$f(x + 4) - 5f(x + 3) + 10f(x + 2) - 10f(x + 1) + 4f(x) = 2x$$

By setting the right-hand side of the equation equal to zero, the resulting homogeneous equation has a solution of the form Ar^x, which gives, upon substitution,

$$A(r^{x+4} - 5r^{x+3} + 10r^{x+2} - 10r^{x+1} + 4r^x) = 0$$

Since $r = 1$ is obviously a solution of the quartic $r^4 - 5r^3 + 10r^2 - 10r + 4 = 0$, we may divide by $r - 1$, which gives the resulting cubic $r^3 - 4r^2 + 6r - 4 = 0$.

Examination of this cubic equation indicates that $r = 2$ is a solution so that the original quartic is now easily factored as

$$(r - 1)(r - 2)(r + 1 + i)(r + 1 - i) = 0,$$

yielding the four roots $r = 1, 2, -1 \pm i$.

The solution of the homogeneous part of the original equation is thus

$$f_1(x) = A_1 1^x + A_2 2^x + A_3(-1 - i)^x + A_4(-1 + i)^x$$

or

$$f_1(x) = A_1 + A_2 2^x + (\sqrt{2})^x \left(A_5 \cos \frac{3\pi}{4} x + A_6 \sin \frac{3\pi}{4} x \right)$$

where A_1, A_2, A_5, and A_6 are arbitrary constants.

A particular solution $f_P(x)$ should be of the form $ax + b$ since $2x$ is a first-degree polynomial. However, a constant is already a solution of the homogeneous part $f_1(x)$ and so $ax + b$ must be multiplied by x, giving $f_P(x) = ax^2 + bx$ which can now be substituted in the original equation:

$$a(x + 4)^2 + b(x + 4) - 5[a(x + 3)^2 + b(x + 3)] + 10[a(x + 2)^2$$
$$+ b(x + 2)] - 10[a(x + 1)^2 + b(x + 1)] + 4(ax^2 + bx) = 2x$$

or

$$x^2(0) + x(-2a) + (a - b) = 2x$$

Equating coefficients gives $a = b = -1$ so that the final solution is

$$f(x) = f_1(x) + f_P(x) = A_1 + A_2 2^x + 2^{x/2} \left(A_5 \cos \frac{3\pi}{4} x + A_6 \sin \frac{3\pi}{4} x \right)$$
$$- x^2 - x \qquad\qquad ////$$

EXAMPLE 3-2 Fibonacci's numbers 0, 1, 1, 2, 3, 5, 8, 13, 21, etc., are generated by the rule that each member is the sum of the previous two elements. What is the general expression for $f(x)$ if we start with $f(0) = 0$? Obviously, the difference equation is

$$f(x + 2) = f(x + 1) + f(x)$$

and the substitution of Ar^x will yield $r^2 - r - 1 = 0$ which has the two roots $r = (\tfrac{1}{2})(1 \pm \sqrt{5})$. Hence

$$f(x) = A_1 \left(\frac{1 + \sqrt{5}}{2} \right)^x + A_2 \left(\frac{1 - \sqrt{5}}{2} \right)^x$$

Since $f(0) = 0$ and $f(1) = 1$, we have the two equations

$$0 = A_1 + A_2 \quad \text{and} \quad 1 = A_1 \frac{1 + \sqrt{5}}{2} + A_2 \frac{1 - \sqrt{5}}{2}$$

The solution of A_1 and A_2 yields $A_1 = -A_2 = 1/\sqrt{5}$ which now gives the solution of the difference equation with appropriate boundary conditions as

$$f(x) = \frac{1}{\sqrt{5}} \left(\frac{1 + \sqrt{5}}{2} \right)^x - \frac{1}{\sqrt{5}} \left(\frac{1 - \sqrt{5}}{2} \right)^x \qquad ////$$

(*Suggested Problems 3-1 to 3-6.*)

3-2   PROBABILITY GENERATING FUNCTIONS

A useful technique for solving homogeneous equations similar to (3-4) has as its basis the fact that an assumed solution as a function of α, $\phi(\alpha)$, can be uniquely determined in closed form if $\phi(\alpha)$ is originally in series form with the coefficients the unknown values of $f(x)$ as

$$\phi(\alpha) = f(0) + f(1)\alpha + f(2)\alpha^2 + \cdots = \sum_{k=0}^{\infty} f(k)\alpha^k \qquad (3\text{-}7)$$

Thus when $\phi(\alpha)$ is finally determined it can be expanded in a power series in α and the coefficient of α^k would now be $f(k)$. The domain of α is so chosen, $|\alpha| < \delta$, that the series converges.

Consider Eq. (3-4) again as

$$a_n f(x + n) + a_{n-1} f(a + n - 1) + \cdots + a_1 f(x + 1) + a_0 f(x) = 0 \qquad (3\text{-}4)$$

together with the expression

$$(a_n + a_{n-1}\alpha + a_{n-2}\alpha^2 + \cdots + a_1 \alpha^{n-1} + a_0 \alpha^n)\phi(\alpha) \qquad (3\text{-}8)$$

If expression (3-8) is expanded formally by substituting the series (3-7) for $\phi(\alpha)$, we have

$$(a_n + a_{n-1}\alpha + a_{n-2}\alpha^2 + \cdots + a_1 \alpha^{n-1} + a_0 \alpha^n)\phi(\alpha) = a_n f(0)$$
$$+ \alpha[a_n f(1) + a_{n-1} f(0)] + \alpha^2[a_n f(2) + a_{n-1} f(1) + a_{n-2} f(0)]$$
$$+ \cdots + \alpha^{n-1}[a_n f(n - 1) + a_{n-1} f(n - 2) + \cdots + a_1 f(0)]$$
$$+ \alpha^n[a_n f(n) + a_{n-1} f(n - 1) + \cdots + a_1 f(1) + a_0 f(0)] + \cdots$$

The coefficient of α^n is clearly equal to zero since it is Eq. (3-4) with $x = 0$. Similarly, the subsequent coefficients of α^{n+1}, α^{n+2}, ..., will also all be zero since again they are Eq. (3-4) with $x = 1, 2, \ldots$.   Therefore, if $f(0), f(1), \ldots$, $f(n - 1)$ are given, since they usually constitute the n pieces of information necessary to specify the arbitrary constants, $\phi(\alpha)$ is completely determined as the ratio of two polynomials with the polynomial in the numerator at least one degree less than that of the denominator as $\phi(\alpha) = A(\alpha)/B(\alpha)$.

If we assume for purposes of illustration that $B(\alpha)$ has m distinct roots (that is, $\alpha_1, \alpha_2, \ldots, \alpha_m$), then $\phi(\alpha)$ can be written as the following sum of fractions:

$$\phi(\alpha) = \frac{a_1}{\alpha_1 - \alpha} + \frac{a_2}{\alpha_2 - \alpha} + \cdots + \frac{a_k}{\alpha_k - \alpha} + \cdots + \frac{a_m}{\alpha_m - \alpha} \tag{3-9}$$

where $a_1, a_2, \ldots, a_m$ are constants.

By multiplying both sides of (3-9) by $\alpha_k - \alpha$ and finding the limit as $\alpha \to \alpha_k$, there results

$$a_k = \lim_{\alpha \to \alpha_k} (\alpha_k - \alpha)\phi(\alpha) = \lim_{\alpha \to \alpha_k} \frac{(\alpha_k - \alpha)A(\alpha)}{B(\alpha)} = -\frac{A(\alpha_k)}{B'(\alpha_k)}$$

by using L'Hospital's rule to evaluate the indeterminate form.

Since each term of Eq. (3-9) can be expanded in a series in powers of α as

$$\frac{a_k}{\alpha_k - \alpha} = \frac{a_k}{\alpha_k}\left[1 + \frac{\alpha}{\alpha_k} + \left(\frac{\alpha}{\alpha_k}\right)^2 + \cdots\right]$$

the coefficient of α^n in the sum of terms in Eq. (3-9) would be

$$\frac{a_1}{\alpha_1^{n+1}} + \frac{a_2}{\alpha_2^{n+1}} + \cdots + \frac{a_k}{\alpha_k^{n+1}} + \cdots + \frac{a_m}{\alpha_m^{n+1}} \tag{3-10}$$

It is apparent that as n becomes very large the dominant term in (3-10) is the one with the smallest magnitude in the denominator.   Thus the asymptotic value of $f(n)$ as n becomes large is determined from the root of the denominator of $\phi(\alpha)$ which is smallest in magnitude.

EXAMPLE 3-3   Given that $\phi(\alpha)$ was evaluated as

$$\phi(\alpha) = \frac{2(4 - \alpha^2)}{8 - 8\alpha + \alpha^3}$$

find the approximate value of $f(24)$ by computing the asymptotic value for $f(n)$. The roots of the cubic are $\alpha_1 = 2$, $\alpha_2 = -1 + \sqrt{5}$, and $\alpha_3 = -1 - \sqrt{5}$ so that, from the fact that

$$a_i = -\frac{2(4 - \alpha_i^2)}{3\alpha_i^2 - 8}$$

we immediately have $a_1 = 0$, $a_2 = (\sqrt{5} + 1)/\sqrt{5}$, and $a_3 = (\sqrt{5} - 1)/\sqrt{5}$. Since α_2 is numerically the smallest root, the asymptotic value of $f(n)$ is given by

$$f(n) \approx \frac{\sqrt{5} + 1}{\sqrt{5}} \left(\frac{1}{\sqrt{5} - 1} \right)^{n+1}$$

or

$$f(24) = \frac{\sqrt{5} + 1}{\sqrt{5}} \left(\frac{1}{\sqrt{5} - 1} \right)^{25} \approx .007236 \qquad ////$$

EXAMPLE 3-4 In Example 3-2 (Fibonacci's numbers), the difference equation was $f(x + 2) = f(x + 1) + f(x)$ with $f(0) = 0$, $f(1) = 1$. Solve this equation by means of a generating function satisfying the given initial conditions.

Assume the solution

$$\phi(\alpha) = f(0) + f(1)\alpha + f(2)\alpha^2 + \cdots$$

If the difference equation is written in the standard form, that is, $f(x + 2) - f(x + 1) - f(x) = 0$, then

$$(1 - \alpha - \alpha^2)\phi(\alpha) = f(0) + f(1)\alpha + f(2)\alpha^2 + f(3)\alpha^3 + \cdots$$

$$- \alpha [f(0) + f(1)\alpha + f(2)\alpha^2 + f(3)\alpha^3 + \cdots]$$

$$- \alpha^2 [f(0) + f(1)\alpha + f(2)\alpha^2 + \cdots]$$

$$= f(0) + \alpha[f(1) - f(0)] + \alpha^2[f(2) - f(1) - f(0)]$$

$$+ \alpha^3[f(3) - f(2) - f(1)] + \cdots$$

$$= 0 + \alpha[1 - 0] + \alpha^2[0] + \alpha^3[0] + \cdots$$

Hence

$$\phi(\alpha) = \frac{\alpha}{1 - \alpha - \alpha^2}$$

The roots of the denominator are $\alpha_1 = (-1 + \sqrt{5})/2$ and $\alpha_2 = (-1 - \sqrt{5})/2$, and the coefficients of the partial fractions are given by

$$a_i = \frac{-\alpha_i}{-1 - 2\alpha_i}$$

or

$$a_1 = \frac{\sqrt{5} - 1}{2\sqrt{5}} \qquad \text{and} \qquad a_2 = \frac{1 + \sqrt{5}}{2\sqrt{5}}$$

Thus the coefficient of α^n would be, from (3-10)

$$\frac{(\sqrt{5}-1)/2\sqrt{5}}{[(-1+\sqrt{5})/2]^{n+1}} + \frac{(1+\sqrt{5})/2\sqrt{5}}{[(-1-\sqrt{5})/2]^{n+1}}$$

or

$$\frac{1}{\sqrt{5}}\left(\frac{1+\sqrt{5}}{2}\right)^n - \frac{1}{\sqrt{5}}\left(\frac{1-\sqrt{5}}{2}\right)^n$$

which agrees with the original solution.   Thus the method is applicable whether the $f(n)$ are probabilities or not, provided the series converges for some $|\alpha| < \delta$.

$////$

3-3   SIMULTANEOUS DIFFERENCE EQUATIONS

As in total differential equations, it is possible to solve an nth-order difference equation of the type considered in the previous sections by solving a system of n first-order difference equations with n dependent variables.

The technique of forming such a system is straightforward.   We shall illustrate it by developing four first-order difference equations which have the same solution for $f(x)$ as a given fourth-order difference equation, namely:

$$f(x+4) = a_1 f(x+3) + a_2 f(x+2) + a_3 f(x+1) + a_4 f(x) + g(x) \qquad (3\text{-}11)$$

Introduce first the three dependent variables $u(x)$, $v(x)$, and $w(x)$ and then form the system of equations as

$$f(x+1) = a_1 f(x) + a_2 u(x) + a_3 v(x) + a_4 w(x) + g(x-3)$$
$$u(x+1) = f(x)$$
$$v(x+1) = u(x)$$
$$w(x+1) = v(x)$$

$$(3\text{-}12)$$

By advancing the value of x (i.e., replacing x by $x+1$, $x+2$, etc.) and multiplying through by the appropriate constants, the following seven equations may be generated:

$$f(x+4) = a_1 f(x+3) + a_2 u(x+3) + a_3 v(x+3) + a_4 w(x+3) + g(x)$$
$$a_2 u(x+3) = a_2 f(x+2) \qquad a_3 u(x+2) = a_3 f(x+1)$$
$$a_3 v(x+3) = a_3 u(x+2) \qquad a_4 v(x+2) = a_4 u(x+1)$$
$$a_4 w(x+3) = a_4 v(x+2) \qquad a_4 u(x+1) = a_4 f(x)$$

When these equations are added together, we obtain Eqs. (3-11) since the dependent variables $u(x)$, $v(x)$, and $w(x)$ cancel.   Therefore, it is necessary that the solution for $f(x)$ which is part of the solution for the entire system also be a solution of the fourth-order difference equation (3-11).

Obviously, the system (3-12) is just a very special case of the complete system

$$f_i(x + 1) = \sum_{j=1}^{4} a_{ij}f_j(x) + g_i(x) \qquad i = 1, 2, 3, 4 \qquad (3-13)$$

which is the mathematical model for some simple Markov processes discussed in a later chapter. The a_{ij} are constants.

Although, in general, this particular technique would not be used to solve a single difference equation, nevertheless it seems to us appropriate to analyze in some detail a system of two first-order difference equations since it enhances the reader's understanding of the solution characteristics of linear, constant-coefficient difference equations in general.

Consider two unknown functions $f_1(x)$ and $f_2(x)$ connected through the relationship

$$\begin{aligned} f_1(x + 1) &= a_1 f_1(x) + a_2 f_2(x) \\ f_2(x + 1) &= b_1 f_1(x) + b_2 f_2(x) \end{aligned} \qquad (3-14)$$

where a_1, a_2, b_1, and b_2 are constants such that

$$a_1 b_2 - a_2 b_1 \neq 0$$

Because of the form of the solution of the constant-coefficient difference equations in the previous section, it is reasonable to try solutions of the form $f_1(x) = Ar^x$ and $f_2(x) = Br^x$ in Eqs. (3-14). When these trial solutions are substituted in (3-14), there results, after dividing by r^x,

$$\begin{aligned} (a_1 - r)A + a_2 B &= 0 \\ b_1 A + (b_2 - r)B &= 0 \end{aligned} \qquad (3-15)$$

This pair of algebraic homogeneous equations has a solution for A and B (other than $A = B = 0$) only when the determinant of the coefficients is equal to zero or

$$\begin{vmatrix} a_1 - r & a_2 \\ b_1 & b_2 - r \end{vmatrix} = 0 \qquad (3-16)$$

The quadratic equation (3-16) always has two roots and commonly they are distinct, say r_1 and r_2. Since the substitution of r_1 in Eqs. (3-15) makes the two equations identical, they determine a single relationship between A_1 and B_1, the two proportionality constants assumed in the solution. This relation between the two corresponding constants for root r_1 may be obtained from either equation of the set.   A different relation between A_2 and B_2 associated with the second root r_2 is obtained in the same way by substituting r_2 in Eqs. (3-15).   The general solution of (3-14) would thus be

$$f_1(x) = A_1 r_1{}^x + A_2 r_2{}^x$$
$$f_2(x) = B_1 r_1{}^x + B_2 r_2{}^x \qquad (3\text{-}17)$$

where there are only two arbitrary constants, which is the necessary number for the general solution of two first-order equations.   If the two roots are equal (that is, $r_1 = r_2 = r$) and at the same time two arbitrary constants cannot be obtained from Eqs. (3-15), it is necessary to assume a solution of the form

$$f_1(x) = (A_1 + A_2 x)r^x \qquad \text{and} \qquad f_2(x) = (B_1 + B_2 x)r^x \qquad (3\text{-}18)$$

By substituting these values for $f_1(x)$ and $f_2(x)$ from (3-18) into either one of the pair of Eqs. (3-14) and equating like coefficients of r^x and xr^x on the two sides of that equation, the relationship between A_1 and B_1 and also A_2 and B_2 can be obtained through the two equations so generated.

EXAMPLE 3-5   Given the set of equations

$$f_1(x + 1) = f_1(x) - 4f_2(x)$$
$$f_2(x + 1) = -f_1(x) + f_2(x) \qquad (3\text{-}19)$$

find the general solution of the system.   Upon the substitution of $f_1(x) = Ar^x$ and $f_2(x) = Br^x$ in the pair of equations, there results

$$(1 - r)A - 4B = 0$$
$$-A + (1 - r)B = 0 \qquad (3\text{-}20)$$

These equations will have a solution only when the following equation is satisfied,

$$\begin{vmatrix} 1 - r & -4 \\ -1 & 1 - r \end{vmatrix} = r^2 - 2r - 3 = (r - 3)(r + 1) = 0$$

so that the roots are $r_1 = 3$, $r_2 = -1$.   If $r = r_1 = 3$ is substituted in either one of Eqs. (3-20) then $A_1 = -2B_1$ and, similarly, $A_2 = 2B_2$ upon the substitution of $r = r_2 = -1$.   Hence the general solution of the system (3-19) would be

$$f_1(x) = -2B_1 3^x + 2B_2(-1)^x = -2B_1 3^x + 2B_2 \cos \pi x$$
$$f_2(x) = B_1 3^x + B_2(-1)^x = B_1 3^x + B_2 \cos \pi x$$

where B_1 and B_2 are arbitrary constants which can be evaluated when specific initial conditions are imposed.   ////

EXAMPLE 3-6   Given the pair of difference equations

$$f_1(x + 1) = f_1(x) + 3f_2(x)$$
$$f_2(x + 1) = -f_1(x) + f_2(x) \qquad (3\text{-}21)$$

find the general solution.   Upon substitution of the same assumed solutions as before, the two equations now become

$$(1 - r)A + 3B = 0$$
$$-A + (1 - r)B = 0$$

so that the roots must satisfy $r^2 - 2r + 4 = 0$.   This gives the values of r as $1 \pm i\sqrt{3}$, which in turn show that $A_1 = -i\sqrt{3}\,B_1$ and $A_2 = i\sqrt{3}\,B_2$.   Therefore, the general solution of the system (3-21) becomes

$$f_1(x) = -i\sqrt{3}\,B_1\big(1 + i\sqrt{3}\big)^x + \sqrt{3}\,iB_2\big(1 - i\sqrt{3}\big)^x$$
$$f_2(x) = B_1\big(1 + i\sqrt{3}\big)^x + B_2\big(1 - i\sqrt{3}\big)^x$$

or, upon simplification,

$$f_1(x) = \sqrt{3}\,2^x(c_2 \cos x\theta + c_1 \sin x\theta)$$
$$f_2(x) = 2^x(c_1 \cos x\theta + c_2 \sin x\theta)$$

where the arbitrary constants c_1 and c_2 have the values $B_1 + B_2$ and $i(B_2 - B_1)$, respectively.   The value of θ is given by $\theta = \tan^{-1}\sqrt{3}$. $////$

EXAMPLE 3-7   Given the simultaneous equations

$$f_1(x + 1) = 3f_1(x) + f_2(x)$$
$$f_2(x + 1) = -f_1(x) + f_2(x) \qquad (3\text{-}22)$$

find the general solution.   Upon the substitution of the trial solutions in (3-22) we now have

$$(3 - r)A + B = 0$$
$$-A + (1 - r)B = 0 \qquad (3\text{-}23)$$

so that the roots are $r_1 = 2, r_2 = 2$.   Thus there exists only one possible relation between A and B, namely, $A = -B$ since both operations in (3-23) are now satisfied by this relationship between A and B.   It is therefore necessary to substitute $f_1(x) = (A_1 + A_2 x)2^x$ and $f_2(x) = (B_1 + B_2 x)2^x$ in one of the original equations as

$$[A_1 + A_2(x + 1)]2^{x+1} = 3(A_1 + A_2 x)2^x + (B_1 + B_2 x)2^x$$

Equating coefficients of 2^x and $x2^x$, we develop two equations:

$$2A_1 + 2A_2 = 3A_1 + B_1 \quad \text{and} \quad 2A_2 = 3A_2 + B_2$$

Therefore $B_2 = -A_2$ and $B_1 = 2A_2 - A_1$ so that the final solution of the system is

$$f_1(x) = (A_1 + A_2 x)2^x \quad \text{and} \quad f_2(x) = [(2A_2 - A_1) - A_2 x]2^x$$

with the two arbitrary constants now A_1 and A_2. ////

It is interesting to note that a double root does not always indicate that two arbitrary constants are impossible without resorting to the above special form for the solutions. For instance, the simultaneous pair of equations, $f_1(x + 1) = af_1(x)$ and $f_2(x + 1) = af_2(x)$, has the solution $f_1(x) = Aa^x$ and $f_2(x) = Ba^x$ and here both A and B are completely arbitrary, although the roots of the appropriate equation are $r = a, a$. This, of course, is a very special case for two simultaneous equations, but for a system with more equations with multiple roots, several other types of special cases arise.

EXAMPLE 3-8   Given the simultaneous equations

$$
\begin{aligned}
f_1(x + 1) &= 3f_1(x) + f_2(x) + x^2 \\
f_2(x + 1) &= -f_1(x) + f_2(x) + 2x + 3
\end{aligned}
\tag{3-24}
$$

find the general solution. This system is nonhomogeneous but still linear so that the general solution must consist of a complementary function which is a solution of the homogeneous part plus a particular solution. Since this pair's homogeneous part is the same as Example 3-7, we have

$$f_1(x) = (A_1 + A_2 x)2^x + f_{1p}(x)$$
$$f_2(x) = [(2A_2 - A_1) - A_2 x]2^x + f_{2p}(x)$$

where the particular solutions $f_{1p}(x)$ and $f_{2p}(x)$ must be polynomials of degree no higher than quadratic.

Thus if $f_{1p}(x) = a_1 x^2 + a_2 x + a_3$ and $f_{2p}(x) = b_1 x^2 + b_2 x + b_3$, where $a_1, a_2, \ldots, b_3$ are constants, these functions may now be substituted in the original system (3-24), giving

$$a_1(x + 1)^2 + a_2(x + 1) + a_3 = 3a_1 x^2 + 3a_2 x + 3a_3 + b_1 x^2 + b_2 x + b_3 + x^2$$

$$b_1(x + 1)^2 + b_2(x + 1) + b_3 = -a_1 x^2 - a_2 x - a_3 + b_1 x^2 + b_2 x + b_3 + 2x + 3$$

By equating similar powers of x on the two sides of each equation, six relations between the a's and b's are established which determine their unique value so that

$$f_{1p}(x) = 4x + 12$$
$$f_{2p}(x) = -x^2 - 8x - 20$$

which determines the general solution of the original system (3-24) when these particular solutions are added to the solutions of the homogeneous set.   ////

(*Suggested Problems 3-10 to 3-13.*)

3-4 APPLICATIONS

Utilizing a difference equation as a mathematical model to represent a probabilistic situation often constitutes a powerful and useful method of attack. This will be demonstrated particularly in the discussion of Markov processes in Chap. 9.   It is often difficult, however, to develop the required functional relationships since the probabilities of certain events are often dependent on several different antecedent conditions.   The difference equations are also often difficult to solve since these probabilities may be trial-dependent, which produces equations that are nonlinear and, even if linear, with nonconstant coefficients.

In this chapter, we shall consider only three types of problems which are amenable to the technique; they will, however, serve to illustrate the possibilities for this approach.

3-5 RANDOM WALKS

An interesting class of problems associated with discrete distributions is what is known as "random walks" and the allied problem as "first passage times." Most of these models are beyond the scope of the objectives of this text but we shall consider the original problem, known as the "gambler's ultimate ruin," from which the others were offshoots.

Two contestants are to play a series of games in which the probability of a particular player winning is p for each game ($q = 1 - p$ is the probability of his losing).   If the player loses, he gives one unit of money to his opponent and if he wins he receives one unit of money from the same opponent.   If this particular player starts with x units of money and his opponent with $s - x$ units, what is the probability of the player losing all his money?

FIGURE 3-1   0   1 x $s-1$ s

The game may be visualized geometrically by considering a line marked off in equal units from 0 to s, as indicated in Fig. 3-1. At the end of each play, the point x moves one unit to the right if our player wins or one unit to the left if he loses. The game is over when our player wins all his opponent's money and thus x moves to s, or he loses all his money and x moves to 0. This game is an example of a simple random walk with absorbing boundaries, since it is finished when either end point is reached.

Let us define $y(x)$ as the probability of the particular player losing all his money if he now has x units in his possession. The following relationship exists between the probability of his finally losing when he has x units of money expressed as a function of his finally losing when he has $x+1$ and $x-1$ units, respectively:

$$y(x) = py(x+1) + qy(x-1) \qquad (3\text{-}25)$$

This merely states that the original probability of ultimately losing the game with x dollars available must have not changed after one more game has been played. This game has only two mutually exclusive outcomes; thus the player wins with probability p or loses with probability $q(q = 1 - p)$ arriving at the point $x+1$ or $x-1$, respectively, on the scale of Fig. 3-1. It is clear that this relationship between the y's holds for values of x ranging from $x = 1$ up to $x = s-1$ and that the boundary conditions $y(0) = 1$ and $y(s) = 0$ must be satisfied if the solution is to satisfy the originally defined problem. Equation (3-25) is a second-order difference equation since the largest difference in the index of the y's is 2. It also is linear with constant coefficients as well as being homogeneous. If the solution r^x is substituted in Eq. (3-25), then $r^x = pr^{x+1} + qr^{x-1}$. This equation is quadratic in r and has two roots: one $r = 1$ and the other $r = q/p$. Accordingly, the general solution of the difference equation must be

$$y(x) = A + B\left(\frac{q}{p}\right)^x \qquad \text{for} \quad q \neq p \qquad (3\text{-}26)$$

where the constants A and B are obtained from the boundary conditions $y(0) = 1$ and $y(s) = 0$. These boundary values result in the two equations $A + B = 1$ and $A + B(q/p)^s = 0$. Solving for A and B from the simultaneous equations and substituting their values in Eq. (3-26), we have

$$y(x) = \frac{(q/p)^s - (q/p)^x}{(q/p)^s - 1} \qquad (3\text{-}27)$$

It is clear that if our player has an opponent who has a large amount of money so that s becomes very large then, in the limit, $y(x) = 1$, or $y(x) = (q/p)^x$, depending upon whether q is greater than p or less than p, respectively. If we replace q/p by r in Eq. (3-27) and use L'Hospital's rule to evaluate the limit of $y(x)$ as $r \to 1$ (i.e., as $p \to \frac{1}{2}$), we find that $y(x) = 1 - x/s$.

It is interesting to note that if the units of money held by each contestant are doubled (doubling the steps in Fig. 3-1) there results

$$y'(x) = \frac{(q/p)^{2s} - (q/p)^{2x}}{(q/p)^{2s} - 1} = y(x) \frac{(q/p)^s + (q/p)^x}{(q/p)^s + 1} \qquad (3\text{-}28)$$

where $y(x)$ is the probability of ultimately losing, computed on the basis of the original monetary units. Thus, here again, the effect of doubling the total units of money depends on the ratio of q/p so that $y'(x)$ can be either less or greater than $y(x)$.

3-6 GENERALIZED BINOMIAL DISTRIBUTION

An experiment is conducted which consists of a sequence of trials, but the probability of success on any one trial is a function of the trial number. An example might be a marksman shooting at a moving target which is either receding or approaching the marksman during his sequence of shots. What is the probability of obtaining x successes (bull's-eyes) in t trials where the probability of success on the kth trial is p_k (probability of failure $q_k = 1 - p_k$)?

Let $f(x, t)$ be the required probability of exactly x successes in the t trials. We observe that the following recurrence relation exists between the value of $f(x, t)$ itself and two mutually exclusive situations, which represent the possible states at $t - 1$. Thus

$$f(x, t) = p_t f(x - 1, t - 1) + q_t f(x, t - 1) \qquad (3\text{-}29)$$

This equation states that the probability of exactly x successes in t trials can result from either one of the situations: $x - 1$ successes in $t - 1$ trials followed by a success, or x successes in $t - 1$ trials followed by a failure. It should be observed that (3-29) is in reality a partial difference equation, since it contains two independent variables, x and t. However, it can be solved as an equation in one variable (x successes) since we can obtain its solution for a fixed number of trials t. There exist, of course, certain boundary conditions which must be satisfied, namely, that $f(x, 0) = 0$ for $x > 0$, and $f(0, t) = q_1 q_2 \cdots q_t$ for $t > 0$. We shall also arbitrarily set $f(0, 0) = 1$.

Let us assume a solution $\phi_t(\alpha)$ (which is a function of α for fixed t) of the following form, noting that the coefficient of α^k is the probability of exactly k successes in t trials:

$$\phi_t(\alpha) = f(0, t) + f(1, t)\alpha + f(2, t)\alpha^2 + f(3, t)\alpha^3 + \cdots$$

$$= \sum_{k=0}^{\infty} f(k, t)\alpha^k \tag{3-30}$$

If the difference equation (3-29) is written in the form

$$f(x, t) - p_t f(x - 1, t - 1) - q_t f(x, t - 1) = 0$$

it suggests that the following expression be evaluated in terms of the definition of $\phi_t(\alpha)$ [Eq. (3-30)]:

$$\phi_t(\alpha) - p_t \alpha \phi_{t-1}(\alpha) - q_t \phi_{t-1}(\alpha) \tag{3-31}$$

Note that, in anticipation of collecting terms in fixed powers of α, the second term is multiplied by α since its argument is $x - 1$, and x is now the variable, not t.

If (3-30) is substituted in (3-31), there results

$$\phi_t(\alpha) - p_t \alpha \phi_{t-1}(\alpha) - q_t \phi_{t-1}(\alpha) = [f(0, t) - q_t f(0, t - 1)]$$
$$+ \alpha[f(1, t) - p_t f(0, t - 1) - q_t f(1, t - 1)]$$
$$+ \alpha^2[f(2, t) - p_t f(1, t - 1) - q_t f(2, t - 1)] + \cdots$$

Since $f(0, t) - q_t f(0, t - 1) = 0$ from the boundary conditions and the coefficients of α, α^2, etc., are Eq. (3-29) with $x = 1, 2$, etc., expression (3-31) must equal zero so that

$$\phi_t(\alpha) = (q_t + p_t \alpha)\phi_{t-1}(\alpha) \tag{3-32}$$

Now

$$\phi_0(\alpha) = f(0, 0) + f(1, 0)\alpha + f(2, 0)\alpha^2 + \cdots = 1$$

from the boundary conditions so that

$$\phi_1(\alpha) = q_1 + p_1 \alpha \qquad \phi_2(\alpha) = (q_2 + p_2 \alpha)(q_1 + p_1 \alpha) \qquad \text{etc.}$$

and thus finally, using Eq. (3-32) recursively, we have

$$\phi_t(\alpha) = \prod_{i=1}^{t} (q_i + p_i \alpha) \tag{3-33}$$

The coefficient of α^x in the expansion of Eq. (3-33) will yield the probability of exactly x successes in t trials. In the special case where all the p_t are equal, the right-hand side of Eq. (3-33) is equal to $(q + p\alpha)^t$, which will give as the coefficient of α^x, upon expansion, the value for $f(x, t)$.   This value is, of course, the binomial distribution $f(x, t) = C(t, x)p^x q^{t-x}$.

3-7 SEQUENCE OF SUCCESSES

Assuming a constant probability p of success for each trial, there obviously exist 2^t possible sequences of successes and failures in t trials.   Certain of these sequences are easily defined and their probability obtained as the solution of a difference equation of the type we are considering.   As an illustration, what is the probability of a baseball player making at least a run of x hits at the next t times at bat, assuming a constant probability p of making a hit for each trip to the plate?   Thus $\overline{H}\,H\,H\,\overline{H}\,H\,H\,H\,H$ would have two runs of three successes and one run of four in seven times at bat.

Let $f(t)$ be the probability of obtaining at least one run of x successes (hits) in t trials.   There are two mutually exclusive ways of obtaining at least one run of x successes by the $(t + 1)$st trial.   Either the run has already occurred by the tth trial, or the $(t + 1)$st trial is needed to complete the run of x.   Thus,

$$f(t + 1) = f(t) + p^x q[1 - f(t - x)] \qquad (3\text{-}34)$$

The second term in (3-34) is the conditional probability that (1) we have not succeeded in obtaining a run of x by trial $t - x$, (2) the next trial $(t - x + 1)$ is a failure, and (3) this failure is followed by x successes in a row.   The boundary conditions for this problem must be

$$f(1) = f(2) = \cdots = f(x - 1) = 0 \qquad \text{and} \qquad f(x) = p^x \qquad (3\text{-}35)$$

The difference equation (3-34) is not homogeneous, but it can be made so by the substitution $g(t) = 1 - f(t)$.   This substitution amounts to considering the probability of not obtaining at least a run of x in t trials.   The boundary conditions, of course, change to conform to the new variable as

$$g(0) = g(1) = \cdots = g(x - 1) = 1 \qquad g(x) = 1 - p^x$$

The new homogeneous equation in terms of $g(t)$ is

$$g(t + 1) = g(t) - qp^x g(t - x) \qquad (3\text{-}36)$$

which now has the probability-generating function

$$\phi(\alpha) = g(0) + g(1)\alpha + g(2)\alpha^2 + \cdots = \sum_{k=0}^{\infty} g(k)\alpha^k$$

Remembering that the variable is t with x fixed, we have

$$(1 - \alpha + qp^x\alpha^{x+1})\phi(\alpha) = g(0) + [g(1) - g(0)]\alpha + [g(2) - g(1)]\alpha^2 + \cdots$$
$$+ [g(x) - g(x-1)]\alpha^x + [qp^xg(0) - g(x) + g(x+1)]\alpha^{x+1} + \cdots$$

Substituting the initial conditions and observing that the coefficient of α^{x+1} and all subsequent coefficients of powers of α are zero since they satisfy the difference equation, we have

$$\phi(\alpha) = \frac{1 - p^x\alpha^x}{1 - \alpha + qp^x\alpha^{x+1}}$$

Since $\phi(\alpha)$ must now be expanded in powers of α in order to obtain the coefficients of α^k which are $g(k)$ and $f(k) = 1 - g(k)$, it is apparent that this is not feasible except for small k. However, if a numerical value for p and x are given, the smallest root can be found for the polynomial in the denominator; this immediately gives the asymptotic value for $f(k)$. As a practical alternative, $f(t + 1)$ can be computed recursively from Eq. (3-34) to obtain a numerical solution for specified p and x.

It is to be noted that very simple sequence problems do not always lend themselves to this particular technique for their solution, as the difference equation as a model turns out to be either nonlinear or not to have constant coefficients. This fact is illustrated in the following simple sample where the solution, however, is easily obtained through another approach.

EXAMPLE 3-9 An unbiased coin is tossed successively until two heads in a row occur, at which time the game is over. What is the probability that this event will occur on the nth toss of the coin? Let P_n be the desired probability, and let us examine its value for small n to see if the series of probabilities is recognizable. If $n = 2$, the possible sequence is HH and $P_2 = 1/2^2$. If $n = 3$, the possible sequence is THH and $P_3 = 1/2^3$. If $n = 4$, the possible sequences are HTHH, TTHH and thus $P_4 = 2/2^4$. If $n = 5$, the possible sequences are THTHH, HTTHH, TTTHH and thus $P_5 = 3/2^5$. Finally, if $n = 6$, the possible sequences are HTHTHH, TTHTHH, THTTHH, HTTTHH, TTTTHH and thus $P_6 = 5/2^6$. It is now apparent that $P_7 = 8/2^7$ since all the sequences for $n = 6$ can have a T in front of them, making a total of five, plus three more which also can have an H in front since they start with a T. The latter three come from the sequence of three for $n = 5$. Thus the number of possible sequences for any n is the sum of the possible sequences in $n - 1$ and $n - 2$

which is the model for Fibonacci's numbers in Example 3-2. Therefore, $P_n = f(n-1)/2^n$, where

$$f(x) = \frac{1}{\sqrt{5}}\left[\left(\frac{1+\sqrt{5}}{2}\right)^x - \left(\frac{1-\sqrt{5}}{2}\right)^x\right]$$

so that

$$P_n = \frac{1}{\sqrt{5}}\frac{[(1+\sqrt{5})/2]^{n-1} - [(1-\sqrt{5})/2]^{n-1}}{2^n}$$

If one sets up the difference equation from first principles, there would result the first-order equation

$$P_{n+1} = \frac{f(n)}{2f(n-1)}\,P_n$$

with the boundary condition $P_2 = \frac{1}{4}$. Since the coefficient of P_n is a function of n, our techniques for solving difference equations with constant coefficients would not apply.   ////

(*Suggested Problems 3-14 to 3-22.*)

PROBLEMS

In Probs. 3-1 to 3-6, find the general solution of the difference equations.

3-1 (a) $f(x+2) - f(x) = 0$
    (b) $f(x+2) - f(x) = 2x$ *Ans.* (b) $A + B\cos\pi x + \frac{1}{2}x^2 - x$.

3-2 (a) $f(x+2) + f(x) = 0$
    (b) $f(x+2) + f(x) = x + 3$
                Ans. (b) $A\cos(\pi/2)x + B\sin(\pi/2)x + \frac{1}{2}x + 1$.

3-3 $f(x+3) + f(x+2) - f(x+1) - f(x) = 0$

3-4 $f(x+3) - 3f(x+2) + 4f(x) = 2x + 4$
                Ans. $A2^x + Bx2^x + C\cos\pi x + x + \frac{7}{2}$.

3-5 $f(x+4) - 2f(x+2) + f(x) = 0$

3-6 $f(x+3) - f(x+2) = 5$ *Ans.* $A + 5x$.

3-7 Given that the probability of the occurrence of an event on the nth trial is p_n, which satisfies the difference equation

$$p_n = p_{n-1} - (\tfrac{5}{16})p_{n-2} + (\tfrac{1}{32})p_{n-3}$$

where $p_0 = p_1 = p_2 = 1$.
    (a) Find the generating function $\phi(\alpha)$ for the p_n.
    (b) Using the asymptotic approximation, find p_{18}.   *Ans.* (b) $9/2^{19}$.

3-8   The probability of the occurrence of an event on the nth trial is given by

$$p_n = \tfrac{7}{10} p_{n-1} - \tfrac{1}{10} p_{n-2}$$

where $p_0 = 1$ and $p_1 = 1$.   Find the asymptotic expression for p_n.

3-9   The generating function of a certain probability distribution is given by the expression

$$\phi(\alpha) = \exp [\lambda(\alpha - 1)]$$

By expanding $\phi(\alpha)$ in powers of α, identify the distribution.

Find the general solution of the system of difference equations in Probs. 3-10 to 3-13.

3-10   (a) $f_1(x + 1) = f_1(x) + f_2(x)$
$f_2(x + 1) = 3f_1(x) - f_2(x)$

(b) $f_1(x + 1) = f_1(x) + f_2(x) + 2$
$f_2(x + 1) = 3f_1(x) - f_2(x) + 4$

Ans.   (b) $f_1(x) = A_1 2^x + A_2 2^x \cos \pi x - \tfrac{8}{3}$.
$f_2(x) = A_1 2^x - 3A_2 2^x \cos \pi x - 2$.

3-11 $f_1(x + 1) = 2f_1(x) - 4f_2(x)$
$f_2(x + 1) = f_1(x) + 2f_2(x)$

3-12 $f_1(x + 1) = -2f_1(x) - f_2(x)$
$f_2(x + 1) = f_1(x) - 4f_2(x)$

Ans, $f_1(x) = A_1 3^x \cos \pi x + A_2 x 3^x \cos \pi x$
$f_2(x) = (A_1 + 3A_2)3^x \cos \pi x + A_2 x 3^x \cos \pi x$.

3-13 $f_1(x + 1) = 3f_1(x) + f_2(x) + x$ Ans. $f_1(x) = A2^x - 2x - 2$.
$f_2(x + 1) = -3f_1(x) - f_2(x) + 1$ $f_2(x) = -A2^x + 3x + 2$.

3-14   The probability of an event occurring on exactly the xth trial is the average of the probabilities for the previous two trials.   If $f(0) = 0$ and $f(1) = 1$, what is the value approached by $f(x)$ as $x \to \infty$?   Ans. $\tfrac{2}{3}$.

3-15   Write the difference equation for the geometric distribution with the appropriate initial condition.   Find the generating function for the probabilities and then expand it in powers of α to check your analysis.

3-16   A biased coin (probability of a head equals p) is tossed until either two heads or two tails occur consecutively.   If $f(x)$ is the probability that the experiment will be completed on the xth trial, set up the difference equation and solve completely with the appropriate initial conditions.   What happens when $p = \tfrac{1}{2}$?

3-17   The probability of success for a particular event is $\tfrac{1}{4}$ on each trial, and therefore the probability of at least one success on n trials is $P_n = 1 - (\tfrac{3}{4})^n$.   Obtain this result by setting up the appropriate difference equation connecting successive P_n's and solving this equation formally.   Ans. $P_{n+1} = P_n + \tfrac{1}{4}(1 - P_n)$.

3-18   Each of n urns contains five white and seven black balls.   One ball chosen at random is transferred from the first urn to the second, and afterwards a random ball is transferred from the second urn to the third.   This procedure is followed until finally a random ball is transferred from the $(n-1)$st to the nth urn.   A ball is now drawn at random from the nth urn.   Calculate the probability that it is white, knowing that the ball transferred from the first to the second urn was white, by setting up the appropriate difference equation.

$$\textit{Ans.} \quad (^{91}\!/_{12})(^{1}\!/_{13})^n + ^5\!/_{12}.$$

3-19 In the random-walk problem in Sec. 3-5, suppose that the possibility of a tie exists on each play so that the probability of winning is p_1, of losing, p_2, and of tying, p_3, with $p_1 + p_2 + p_3 = 1$. Find the probability of ultimate ruin for the assumption.　　　　$\textit{Ans.}$ $\textit{Same as Sec. 3-5 with } p = p_1/(p_1 + p_2)$.

3-20 In the random-walk problem in Sec. 3-5, suppose that either winner is willing to consider the possibility of the contest continuing by returning one unit of money to the loser. Thus the contest will continue, if it continues, with one player having one unit of money and the other with $s - 1$ units of money. If the probability of our player winning is $\frac{1}{3}$, and the continuance or noncontinuance determined by a toss of an unbiased die [1, 2, 3, 4, 5 stop; 6 continue], what is the probability of ultimate ruin? Assume that both contestants can take advantage of this possible continuance only once.

3-21 It can be shown that if the probability of not obtaining the sequence HTH in x consecutive tosses of an unbiased coin is $f(x)$, then the difference equation satisfied by consecutive values of $f(x)$ would be

$$f(x) = f(x-1) - \tfrac{1}{4}f(x-2) + \tfrac{1}{8}f(x-3) \quad \text{where } f(0) = f(1) = f(2) = 1$$

Show that the generating function, $\phi(\alpha)$, for the probabilities would be

$$\phi(\alpha) = \frac{8 + 2\alpha^2}{8 - 8\alpha + 2\alpha^2 - \alpha^3}$$

and that the asymptotic value for $f(x)$ would be

$$f(x) \approx \frac{1.44}{(1.14)^{n+1}}$$

(The reader should observe that the derivations of the difference equations both here and in Prob. 3-22 are more difficult than the text examples since the probabilities $f(x), f(x-1)$, etc., are not total probabilities.)

3-22 It can be shown that if $f(x)$ is the probability that the combination HHH occurs for the first time on the last three tosses of an unbiased coin [that is, $f(x)$, $f(x-1), f(x-2)$] and if at the same time the sequence HTH never occurred, then the difference equation for $f(x)$ is

$$f(x) = \tfrac{1}{2}f(x-1) + \tfrac{1}{8}f(x-3) + \tfrac{1}{16}f(x-4)$$

with $f(0) = f(1) = f(2) = 0$ and $f(3) = \frac{1}{8}$.

Find the generating function $\phi(\alpha)$ and show that $f(x)$ for $x \to \infty$ is approximately

$$f(x) \approx \frac{.153}{(1.24)^{n+1}}$$

Find the probability of the sequence HHH occurring before the sequence HTH by setting $\alpha = 1$ in $\phi(\alpha)$. Why is this true?

4

CONTINUOUS RANDOM VARIABLES

4-1 DISTRIBUTION FUNCTIONS

Ordinarily, the specification of the admissible values of a continuous random variable is perfectly straightforward and can be done simply by stating the range or ranges of continuous variation, as $A \leq x \leq B$; but a valid association with probability is less direct. No one-to-one correspondence can be set up between admissible values of a continuous variate and any infinite ordinal series of events E_1, E_2, E_3, ... because the points on a continuum are not subject to enumeration. Therefore, the problem of defining the distribution of a continuous random variable calls for some ingenuity. Fortunately, however, an elementary idea proves powerful enough to cope with the most general situation.

Choosing an arbitrary number a, let us define two mutually exclusive and complementary classes of values: (1) $x \leq a$, and (2) $x > a$. With class 1 we associate the event E_a in the sense that any value of the variate belonging to class 1 corresponds to the occurrence of E_a; similarly, we associate the event $\bar{E}_a$ with class 2. Thus $P(E_a)$ and $P(\bar{E}_a)$ are rigorously definable, but since they are complementary, it is sufficient to know $P(E_a)$. This construction can be carried out for any value of a whatever, and the result will obviously be a function of a.

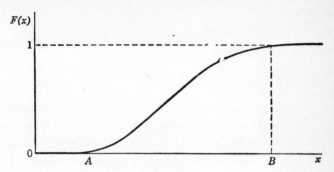

FIGURE 4-1
Typical distribution function of a continuous variate.

Hence we could replace $P(E_a)$ by the equivalent functional notation

$$F(a) \equiv P(E_a) \equiv P(x \le a) \qquad (4\text{-}1)$$

but since we are going to allow a to take on all possible values and thus serve much the same purpose as the symbol x, it is more sensible to adopt the notation $F(x)$ rather than $F(a)$. That is, $F(x)$ denotes the probability that the random variable in question assumes any value less than or equal to x and is thus again the *distribution function* of x.

In a rigorous development of this subject, a distinction is made between the random variable itself and the real variable which includes all its possible values; thus X is used for the former and x for the latter. However, it must be understood that the variable x in the mathematical expression for $F(x)$ is a pure number; for instance, if the actual random variable assumes the value 15 volts, then x is the pure number 15.

In contrast to the distribution function of a discrete variate, which is necessarily a step function, the distribution function of a continuous variate is commonly continuous throughout its range of definition. Although discontinuities do sometimes occur, they are usually confined to a few isolated points. A typical distribution function of a continuous variate x, where the admissible values include all points in the range $-\infty < x < \infty$, is illustrated in Fig. 4-1.

4-2 USE OF DISTRIBUTION FUNCTIONS IN DEFINING DISTRIBUTIONS

We have previously defined the distribution of a discrete variate by stating the equation of its point-probability function $f(x)$ together with the specification

FIGURE 4-2

of admissible values.   For reasons soon to become apparent, a point-probability function is meaningless when the distribution function is continuous, although we are going to develop an analogous function in terms of point density.   In lieu of the probability function, however, we may define the distribution of a continuous variate (or a discrete variate, too, for that matter) by stating the equation of its distribution function $F(x)$.   This notion is best elucidated through specific illustrations.

EXAMPLE 4-1   From geometric probability, we draw an example of a continuous variate.   The permissible region, shown in Fig. 4-2, consists of the line segment OL.   If points are picked at random on OL, their distances from O will generate a continuous random variable, the admissible values of which will range from 0 to L inclusive.   Assuming all points equally likely, let us find the distribution of this variate.   Let X denote an arbitrary point on OL and x its distance from O.   The probability that any point X' picked at random will fall somewhere on the segment OX is

$$P(X' \text{ on } OX) = \frac{OX}{OL} = \frac{x}{L}$$

But having X' fall on OX is precisely the condition that the distance OX' be less than or equal to x.   Hence by definition of the distribution function, we have

$$F(x) = \frac{x}{L}$$

an equation which holds everywhere in the admissible range of x, namely, $0 \le x \le L$. *Adopting the convention that the distribution function is identically zero for all values of x below the minimum admissible value of the variate and identically* 1 *for all values of x above the maximum admissible value of the variate,* we may express the distribution of x in one statement by combining the range of definition with the equation of $F(x)$.   Thus the distribution of x is

$$F(x) = \frac{x}{L} \qquad 0 \le x \le L$$

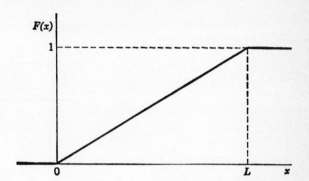

FIGURE 4-3

The graph of the distribution function is shown in Fig. 4-3. Although this distribution function is linear over the allowable range of x, geometric problems readily give rise to curved distribution functions, as illustrated by the next example. ////

EXAMPLE 4-2   The permissible region, indicated by heavy lines in Fig. 4-4, is bounded below by the X axis and above by the parabola $y = 2x - x^2$. If all points within the region are equally likely, let us find the distribution of the distances of random points to the Y axis. At any distance $x \leq 2$ to the right of the Y axis, pass the straight line L parallel to OY. The distance of a random point R from OY is equal to its abscissa r, and the probability that $r \leq x$ is the same as the probability that R lies on or to the left of L. Hence

$$F(x) = P(r \leq x) = \frac{\text{area to left of } L}{\text{total area}} = \frac{\int_0^x y \, dx}{\int_0^2 y \, dx} = \frac{3x^2 - x^3}{4}$$

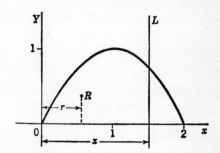

FIGURE 4-4

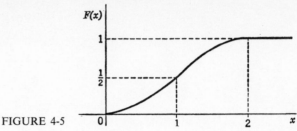

FIGURE 4-5

Thus the distribution of the random variable x is

$$F(x) = \frac{3x^2 - x^3}{4} \qquad 0 \leq x \leq 2$$

The graph of $F(x)$ is shown in Fig. 4-5.   ////

4-3 COMPUTATION OF PROBABILITIES

The probability that x lies in any particular range $a < x \leq b$ may be computed from the distribution function of x in the following manner. Since the event $x \leq b$ can be decomposed into two mutually exclusive events $x \leq a$ and $a < x \leq b$, it follows from the law of total probability that

$$P(x \leq b) = P(x \leq a) + P(a < x \leq b) \qquad (4\text{-}2)$$

whence
$$P(a < x \leq b) = P(x \leq b) - P(x \leq a)$$

$$= F(b) - F(a) \qquad (4\text{-}3)$$

Thus, in Example 2-1, the probability that x exceeds 0 but does not exceed 2 is

$$P(0 < x \leq 2) = F(2) - F(0) = \tfrac{26}{27} - \tfrac{8}{27} = \tfrac{18}{27} = f(1) + f(2)$$

In passing, we note that Eq. (4-3) implies that $F(b) \geq F(a)$ whenever $b > a$, for the left-hand side of this equation can never be negative, since it is the probability of a definable event.

Suppose now that $F(x)$ is continuous, and let $a = \xi - \varepsilon$, $b = \xi + \varepsilon$, where ξ is any value of x and ε is an arbitrarily small, positive number (see Fig. 4-6). Any point which lies in the interval $\xi \leq x \leq \xi + \varepsilon$ necessarily lies also in the interval $\xi - \varepsilon < x \leq \xi + \varepsilon$ because the former interval is entirely contained in the latter. Therefore, it follows that $P(\xi \leq x \leq \xi + \varepsilon) \leq P(\xi - \varepsilon < x \leq \xi + \varepsilon)$, and so, applying Eq. (4-3), we have

$$P(\xi \leq x \leq \xi + \varepsilon) \leq P(\xi - \varepsilon < x \leq \xi + \varepsilon) = F(\xi + \varepsilon) - F(\xi - \varepsilon) \qquad (4\text{-}4)$$

FIGURE 4-6

In the limit as $\varepsilon \to 0$, the extreme left-hand side of Eq. (4-4) may be interpreted as $P(x = \xi)$, whereas on the right-hand side, the hypothesized continuity of $F(x)$ requires that $F(\xi - \varepsilon) \to F(\xi)$ and $F(\xi + \varepsilon) \to F(\xi)$, whence $F(\xi + \varepsilon) - F(\xi - \varepsilon) \to 0$. Hence, for any variate x having a continuous distribution function, it follows that

$$P(x = \xi) = 0 \qquad \text{for all } \xi \qquad (4\text{-}5)$$

In words: The probability that any continuous variate x which has a continuous distribution function $F(x)$ will take on a value exactly equal to any preassigned number ξ is vanishingly small.   Consequently, if $F(x)$ is continuous, any interval may be so defined as to include or exclude one or both end points without affecting the probability that a random point will belong to it; that is,

$$P(a < x \le b) = P(a \le x \le b) = P(a \le x < b) = P(a < x < b)$$
$$= F(b) - F(a) \qquad (4\text{-}6)$$

Thus, in Example 4-2 the probability that a random value of distance lies in the open interval $1 < x < 2$ and the probability that it lies in the closed interval $1 \le x \le 2$ are both given by the same subtraction $F(2) - F(1) = 1 - \frac{1}{2} = \frac{1}{2}$.

As explained in Chap. 1, the zero probability given by Eq. (4-5) for $x = \xi$ does not mean that the value $x = \xi$ is impossible but merely that it fades into immeasurable insignificance as compared with the total, inexhaustible field of distinct possibilities.

4-4   DENSITY FUNCTIONS

The concept of density, which has proved fruitful and convenient in physics, also aids in the study of random variables.   Consider the physical notion of density as applied to a long, thin wire.   For a homogeneous wire, the linear mass density is constant and is simply the mass per unit length, but for a nonhomogeneous wire, it varies as a function of the distance x measured from the origin, taken at one end of the wire.   Hence if $M(x)$ represents the mass of the segment of wire from the origin to any point x, the linear mass density $\rho(x)$ is defined as the limit of the ratio of the mass increment $\Delta M(x)$ to the length increment Δx as $\Delta x \to 0$.   Thus $\rho(x)$ is the derivative of the mass function $M(x)$

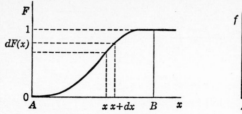 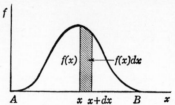

FIGURE 4-7
Typical distribution function and corresponding density function.

and, in turn, $M(x) = \int_0^x \rho(\xi)\, d\xi$. (NOTE: When x is a limit of integration, it is conventional to use some other letter, say ξ, to represent the variable with respect to which the integration is performed.)

Provided that the distribution function of a variate is differentiable (hence certainly continuous), we may define probability density in a similar fashion. The probability that a random value r of the variate will lie between x and $x + \Delta x$ may be interpreted as the probability increment at the point x and (for simplicity) represented by ΔP.

$$\Delta P = P(x \le r \le x + \Delta x) = F(x + \Delta x) - F(x) = \Delta F(x) \qquad (4\text{-}7)$$

At the same time, the average "concentration" of probability in the neighborhood of x will be given by the relative increment $\Delta P/\Delta x = \Delta F(x)/\Delta x$ and under the assumption that $F(x)$ is differentiable, this ratio will approach a limit as $\Delta x \to 0$. This limit, which will be the derivative of the distribution function at the point x, will be called the *probability density* at x and will be denoted by $f(x)$. That is,

$$f(x) = \frac{dF(x)}{dx} \qquad (4\text{-}8)$$

Correspondingly, the infinitesimal element of probability $dF(x)$ is given by

$$dF(x) = f(x)\, dx \qquad (4\text{-}9)$$

and represents the probability that the variate will assume a value lying somewhere in the range x to $x + dx$. The geometric connection between $F(x)$ and $f(x)$ is shown schematically in Fig. 4-7.

Although we are using the notation $f(x)$ both for the probability that a discrete variate assumes a particular admissible value x and for the probability density of a continuous variate at the point x, this usage need not cause confusion, for the defining equation of the function will always be accompanied by a statement of admissible values. In the discrete case, these will be indicated as a series, but in the continuous case, by a range of continuous variation.

4-5 PROPERTIES OF DENSITY FUNCTIONS

From the fact that $F(x + \Delta x) \geq F(x)$, and hence $\Delta F(x) \geq 0$ for all $\Delta x \geq 0$, it follows at once that the density function $f(x)$, being the derivative of the distribution function, is inherently nonnegative. If A and B represent, respectively, the minimum and maximum admissible values of any variate x, then $F(A) \equiv P(x \leq A) = P(x = A)$ and $F(B) \equiv P(x \leq B) = 1$. If the variate possesses a density function, its distribution function is certainly continuous, because continuity is a necessary condition for the existence of the derivative; therefore, by virtue of Eq. (4-5), $P(x = A) = 0$, whence $F(A) = 0$. Consequently, we arrive at the following general conclusions with regard to density functions:

(1)
$$f(x) \geq 0 \qquad \text{for all } x$$

(2)
$$\int_A^{x'} f(x)\, dx = F(x') - F(A) = F(x')$$

(3)
$$\int_A^B f(x)\, dx = F(B) = 1 \tag{4-10}$$

(4)
$$P(a \leq x \leq b) = \int_a^b f(x)\, dx = P(a < x < b)$$

In words: The probability density function $f(x)$ of a variate x is a nonnegative function having unit area when integrated over the admissible range of x. When integrated from the lower limit of x up to any point x', it yields the corresponding value of the distribution function at $x = x'$, and when integrated between any two limits (a, b) it yields the probability that x lies in the interval (a, b).

4-6 EXAMPLES OF DENSITY FUNCTIONS

Examples 4-1 and 4-2 serve as illustrations of the density function. The distribution function found in Example 4-1 was $F(x) = x/L$ in the range $0 \leq x \leq L$. Therefore, in the same range,

$$f(x) = \frac{dF(x)}{dx} = \frac{1}{L}$$

and $f(x) \equiv 0$ for $x < 0$ and $x > L$. Again, the distribution function found in Example 4-2 was $F(x) = (3x^2 - x^3)/4$ in the range $0 \leq x \leq 2$. Therefore, in the same range,

$$f(x) = \frac{dF(x)}{dx} = \frac{6x - 3x^2}{4} = \tfrac{3}{4}x(2 - x)$$

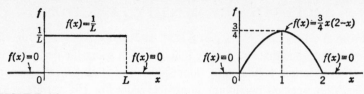

FIGURE 4-8
Density functions corresponding to the variates defined in Examples 4-1 and 4-2.

and $f(x) \equiv 0$ for $x < 0$ and $x > 2$. These two density functions are exhibited in Fig. 4-8.

Both of these density functions admit of simple geometric interpretations. In Example 4-1, for instance, the probability that a random point X' falls in any segment of length S is equal to S/L. Hence if x' is the abscissa of X', we have

$$f(x)\,dx = P(x \le x' \le x + dx) = \frac{dx}{L}$$

whence $f(x) = 1/L$. In Example 4-2 (see Fig. 4-9) let M and M' denote two straight lines parallel to OY and cutting the x axis at x and $x + dx$, respectively. The probability that a random point R will fall between M and M' is equal to the ratio of the area included between these lines to the total area. Hence if r denotes the abscissa of the point R, we have

$$f(x)\,dx \equiv P(x \le r \le x + dx) = \frac{\text{area between } M \text{ and } M'}{\text{total area}}$$

$$= \frac{y\,dx}{\int_0^2 y\,dx} = \frac{(2x - x^2)\,dx}{\frac{4}{3}} = \tfrac{3}{4}x(2 - x)\,dx$$

whence $f(x) = \tfrac{3}{4}x(2 - x)$.

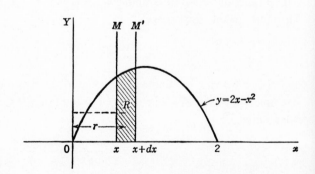

FIGURE 4-9

4-7 USE OF DENSITY FUNCTIONS IN DEFINING DISTRIBUTIONS

When the density function exists, its equation, supplemented by the specification of the admissible range, can be used to define the distribution of the random variable. If the range is $A \leq x \leq B$, which includes both limits, we shall adopt the abbreviated notation (A, B); but if the admissible range excludes one or both of the limits, we shall write the specification in full, for example $(A < x \leq B)$, $(A < x < B)$, etc. Thus the distributions of the variates considered in Examples 4-1 and 4-2 may be expressed as follows:

Example 4-1:

$$f(x) = \frac{1}{L} \qquad (0, L)$$

Example 4-2:

$$f(x) = \tfrac{3}{4}x(2 - x) \qquad (0, 2)$$

Outside of the *expressed range* of definition, it shall be taken for granted that the *density function* is *identically* zero.

In defining a distribution by either the distribution function or the density function, the specification of the admissible range must always be included. Otherwise the definition of the distribution is incomplete, and substitution of inadmissible values into the formulas could lead to absurdities, such as negative probabilities or probabilities greater than 1.

4-8 EXCEPTIONAL SITUATIONS

From the elementary standpoint, the density function is more convenient to work with and more appealing to the intuition than the distribution function, but whereas the latter always exists for any definable distribution, the density function exists only when the distribution function is differentiable. This condition, however, is not very restrictive in practice, inasmuch as most applications of probability theory are confined to variates which do have differentiable distribution functions. Even the few exceptions to this rule are usually amenable to rather obvious modifications of the ordinary analysis. A case in point is the poundage of brass rod shipped daily by a dealer in nonferrous metals. Since the minimum value, zero, has a finite probability of occurrence, the distribution function takes a finite jump at $x = 0$ as we cross from the left to the right of the origin; but thereafter it behaves in a perfectly regular manner.

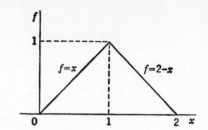

FIGURE 4-10

Therefore, a density function could be defined in the range $x > 0$.   A similar situation is presented by the amount of rainfall over a particular city in a 24-hour period.   Here there is a finite probability of a zero amount and also of a detectable but nonmeasurable amount called a "trace," but in the measurable range beyond, a density function is definable.

Later on, in the solution of some fairly simple problems, we shall encounter a few instances of another type of irregularity, such that the density function must be defined by two or more equations, each of which holds for a specified interval.   Strictly speaking, the distribution function might not be differentiable at boundary points between adjacent intervals, and so the density function would be undefined there.   Nevertheless, because of the continuity of the distribution function, it is not necessary to distinguish between the integral of the density function over the rigorously correct open interval which excludes the boundary points and over the more convenient closed interval which includes them.   Therefore, it is permissible to define the density function at a boundary point arbitrarily by the equation which holds for the interior of either interval.   This convention applies also when the distribution function becomes nondifferentiable at the extremes of the admissible region.   Let us consider two examples.

EXAMPLE 4-3   (See Fig. 4-10.)

$$f(x) = x \qquad (0, 1)$$
$$f(x) = 2 - x \qquad (1, 2)$$

Here both equations yield the same value, $f = 1$, at the boundary point $x = 1$; hence the distribution function is differentiable at $x = 1$, and the density function is rigorously defined as its derivative.   To put it another way, we note that for any value $0 < a < 1$, the distribution function is given by

$$F(a) = \int_0^a x \, dx = \frac{a^2}{2}$$

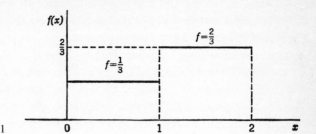

FIGURE 4-11

whereas for any value $1 \le b \le 2$ it is given by

$$F(b) = \int_0^1 x \, dx + \int_1^b (2 - x) \, dx = 2b - \frac{b^2}{2} - 1$$

Both formulas yield the same value at their common point, that is, $F(1) = \frac{1}{2}$, and both have the same derivative, $f = 1$.   ////

EXAMPLE 4-4   (See Fig. 4-11.)

$$f(x) = \tfrac{1}{3} \quad (0, 1) \qquad f(x) = \tfrac{2}{3} \quad 1 < x \le 2$$

Here we exclude $x = 1$ from the definition of $f(x)$ in the second interval in order to avoid a contradiction of the definition in the first interval.   However, in the integrals by which probabilities are computed, it is permissible to include $x = 1$ in both formulas.   For if the lower limit of the second one were taken as $1 + \varepsilon$, the ultimate value of the integral as $\varepsilon \to 0$ would equal that obtained by taking the lower limit as 1 to begin with.   For any value $0 \le a \le 1$, the distribution function is given by

$$F(a) = \int_0^a \tfrac{1}{3} \, dx = \frac{a}{3}$$

and for any value $1 \le b \le 2$ it is given by

$$F(b) = \int_0^1 \tfrac{1}{3} \, dx + \int_1^b \tfrac{2}{3} \, dx = \frac{2b}{3} - \tfrac{1}{3}$$

both of which yield $F(1) = \frac{1}{3}$.   However, the two functions yield different derivatives at $x = 1$, and so the density function is not defined at this point except by the stated arbitrary convention.   In this example also, the strict definition of the density function as a derivative breaks down at the extremes of the admissible region $(0, 2)$.   At the lower extreme, we have $F(x) = 0$ if

$x \leq 0$ but $F(x) = x/3$ if $0 \leq x \leq 1$; and although $F(0) = 0$ in either case, the two functions have different derivatives at $x = 0$. At the upper extreme, $F(x) = (2x - 1)/3$ if $1 \leq x \leq 2$ but $F(x) = 1$ if $x \geq 2$; and as before, the functions agree at their common point, yielding $F(2) = 1$, but their derivatives are different. Consequently, the definition of the density function at $x = 0$ and at $x = 2$ must rest upon convention. Finally, we note the same situation in the density function associated with Example 4-1. Here the equation $f(x) = 1/L$ represents the derivative of $F(x)$ only in the range $0 < x < L$, but no inconsistency is caused by setting $f(0) = 1/L = f(L)$, thereby extending the definition of the density function to the end points of the range. So much for this type of irregularity—hereafter we shall take it in stride.   ////

(*Suggested Problems 4-1 to 4-16*)

4-9   THE GAMMA DISTRIBUTION[1]

The rest of this chapter is devoted to the study of three families of continuous distributions which are useful in formulating mathematical models of random processes. These three are known, respectively, as the gamma, the beta, and the normal distributions, and it so happens that in each case the common forms of their density functions contain two parameters. We discuss the gamma distribution at some length because of the simple introduction it offers to certain concepts and methods of approach having wide application in the field of random variables.

The ordinary form of the gamma distribution pertains to a continuous variate x which may assume any nonnegative value $0 \leq x < \infty$ and is defined in terms of its density function as follows:

Gamma distribution:

$$f(x) = \frac{1}{\beta^{\alpha+1}\Gamma(\alpha + 1)} x^{\alpha}e^{-x/\beta} \qquad \alpha > -1; \beta > 0; 0 \leq x < \infty \qquad (4\text{-}11)$$

This distribution is skewed to the right for all values of the parameters α, β but as α increases, the skewness becomes less pronounced. We shall see that the conventional practice of writing the coefficient of x in the exponential factor as $1/\beta$ simplifies the physical and geometrical interpretation of this parameter.

The general effect of each parameter on the shape of the curve can be investigated by the usual method of examining ordinates and derivatives for

---

[1] See Appendix for discussion of beta and gamma functions.

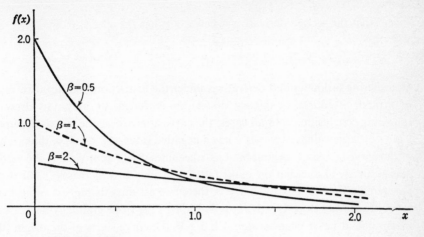

FIGURE 4-12
Exponential distribution for various selections of β.

points of interest. We readily find that β operates mainly as a scale factor, whereas α produces qualitatively different effects, depending on its magnitude.

The special case with $\alpha = 0$ is important enough to have a name of its own and is known as the *exponential distribution*. Thus, the exponential distribution is defined in terms of its density function as,

Exponential distribution:

$$f(x) = \frac{1}{\beta} e^{-x/\beta} \qquad \beta > 0; 0 \leq x < \infty \qquad (4\text{-}12)$$

The first two derivatives in the exponential case are

$$f'(x) = -\frac{1}{\beta^2} e^{-x/\beta} \qquad f''(x) = \frac{1}{\beta^3} e^{-x/\beta} \qquad (4\text{-}13)$$

Figure 4-12 displays the exponential distribution for three values of β ($\beta = .5, 1, 2$).

Letting $k = 1/\beta^{\alpha+1}\Gamma(\alpha + 1)$ we may write the first two derivatives for values of α other than zero as follows:

$$f'(x) = \frac{k}{\beta} x^{\alpha-1} e^{-x/\beta}(\alpha\beta - x) \qquad \text{root: } x = \alpha\beta \qquad (4\text{-}14)$$

$$f''(x) = \frac{k}{\beta^2} x^{\alpha-2} e^{-x/\beta}[\alpha(\alpha - 1)\beta^2 - 2\alpha\beta x + x^2] \qquad \text{roots: } x = \beta(\alpha \pm \sqrt{\alpha})$$

$$(4\text{-}15)$$

except that the second derivative, when $\alpha = 1$, has the special form

$$f''(x) = \frac{1}{\beta^4} e^{-x/\beta}(x - 2\beta) \qquad \text{root: } x = 2\beta \qquad (4\text{-}16)$$

Considering ordinates and derivatives under the restriction that negative values of x (and, of course, imaginary values) are excluded, we may now draw the following conclusions. In all cases, the curve approaches the X axis asymptotically for large values of x. If α has a negative value $-1 < \alpha < 0$, the ordinate of the curve increases without limit as x becomes small, approaching the vertical axis as an asymptote, but the curve falls off steadily as x increases. The ordinates remain finite if $\alpha \geq 0$. If $\alpha = 0$, the curve assumes its highest value $1/\beta$ at $x = 0$; the initial slope is $-1/\beta^2$, and although the slope gradually decreases in magnitude, it never changes sign. If $\alpha > 0$, the curve starts at the origin [that is, $f(0) = 0$], rises to a maximum at $x = \alpha\beta$, and then falls off. The curve is tangent to the vertical axis at the origin if α lies in the range $0 < \alpha < 1$, it has an initial slope equal to $1/\beta^2$ if $\alpha = 1$ (just opposite to the slope if $\alpha = 0$), and it is tangent to the X axis at the origin if $\alpha > 1$. There is one point of inflection, located at a distance $\beta\sqrt{\alpha}$ to the right of the maximum, if α lies in the range $0 < \alpha \leq 1$; but if $\alpha > 1$, there are two points of inflection, symmetrically placed at a distance $\beta\sqrt{\alpha}$ to either side of the maximum.

Curves which rise to an asymptote or to a cusp at one extreme but fall off steadily in the opposite direction are called J-*shaped*, although the "J" is often backwards. Thus the gamma distribution is J-shaped if $\alpha \leq 0$. Curves with a single rounded peak are called *unimodal*, and the value of x (the abscissa) at which the peak occurs is called the *mode*. Accordingly, the gamma distribution is unimodal when $\alpha > 0$. By an extension of the term, the mode is also understood as the point of highest density or, in the discrete case, the point of highest probability. However, it is not customary to refer to J-shaped curves as unimodal, even though we may be willing to call the high end of the curve the "mode." Curves with two, three, or more peaks are called *bimodal*, *trimodal*, or *multimodal*, respectively.

The gamma distribution is exhibited in the following figures. The effect of α is illustrated in Fig. 4-13, where β is held constant ($\beta = 1$) and α is given a series of values ($\alpha = -.5, 0, .5, 1, 2, 5$). The effect of β is displayed in Fig. 4-14 where α is held constant ($\alpha = 1$) and β is given a series of values ($\beta = .5, 1, 2$). The combined effect of α and β is indicated in Fig. 4-15 by means of three pairs of values ($\alpha = 5$, $\beta = .4$; $\alpha = 10$, $\beta = .2$; $\alpha = 20$, $\beta = .1$) which yield the same mode ($\alpha\beta = 2$).

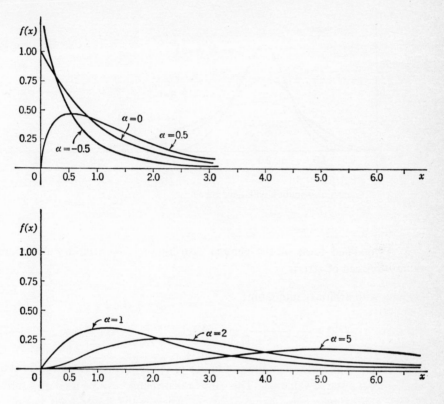

FIGURE 4-13
Gamma distributions with unit β but different values of α.

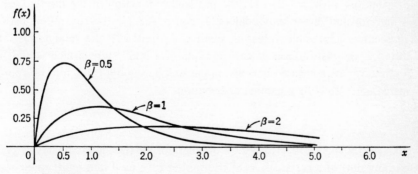

FIGURE 4-14
Gamma distributions with unit α but different values of β.

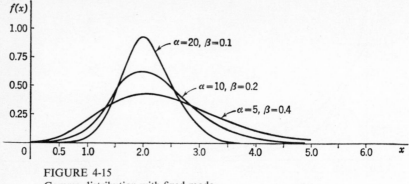

FIGURE 4-15
Gamma distribution with fixed mode.

A modified form of the gamma distribution with arbitrary minimum value A instead of zero is

Gamma with arbitrary minimum:

$$f(x) = \frac{1}{\beta^{\alpha+1}\Gamma(\alpha + 1)} (x - A)^{\alpha} e^{-(x-A)/\beta} \qquad A \leq x < \infty \qquad (4\text{-}17)$$

This form is applicable when the conditions of the problem exclude values of x smaller than a stated value A. The gamma distribution can be used as a satisfactory approximation in many situations wherein the upper limit is finite, because the exponential factor $e^{-x/\beta}$ can be so adjusted as to render negligibly small the probability of exceeding the actual physical limit. Moreover, the gamma distribution often furnishes an excellent approximation to discrete distributions provided that (1) the probability function of the discrete variate is smooth and either unimodal or J-shaped and (2) the separation between consecutive admissible values is small compared with the total range. For instance, the distribution of cash receipts of a retail store, which would be a discrete variate measured in cents, might well be smooth enough to be approximated adequately by a gamma distribution.

EXAMPLE 4-5   The distribution of length of life (i.e., hours of service before burning out) of a certain type of transmitter tube used in aircraft radar sets is exponential with $\beta = 180$. What is the probability that a tube will last less than 90 hours? More than 360 hours? If three tubes are picked at random for life testing, what is the probability that one will last 90 hours or less, another

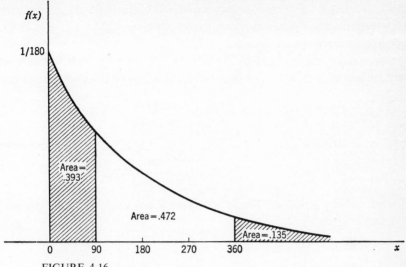

FIGURE 4-16
Exponential distribution with $\beta = 180$.

between 90 and 360 hours, and another more than 360 hours? (See Fig. 4-16.)
Here, the distribution is given by

$$f(x) = (\tfrac{1}{180})e^{-x/180} \qquad 0 \le x < \infty$$

In general, the exponential-distribution function with arbitrary β is

$$F(x) = 1 - e^{-x/\beta} \qquad (4\text{-}18)$$

and in this case, the distribution function is $F(x) = 1 - e^{-x/180}$. The respective
values of this distribution function at the points 90 and 360 are

$$F(90) = 1 - e^{-.5} = .393 \qquad F(360) = 1 - e^{-2} = .865$$

Hence the probability that a tube will last less than 90 hours is

$$P(x < 90) = F(90) = .393$$

The probability that a tube will last between 90 and 360 hours is

$$P(90 \le x \le 360) = F(360) - F(90) = .472$$

and the probability that a tube will last more than 360 hours is

$$P(x > 360) = 1 - F(360) = .135$$

The probability that a random sample of three will contain one instance of each
of these three ranges is given by the multinomial law of probability as

$$\frac{3!}{1!1!1!}(.393)(.472)(.135) = .150 \qquad ////$$

EXAMPLE 4-6   Expressed as excess over $500, annual family income in dollars in a certain city follows the gamma distribution with $\alpha = 1$, $\beta = 2,000$. If advertising circulars are sent out to 1,000 families picked at random, what is the probability that more than 15 circulars will be sent to families with annual income less than $800? The distribution, in terms of its density function, is

$$f(x) = \frac{1}{(2,000)^2}(x - 500)e^{-(x-500)/2,000} \qquad 500 \leq x < \infty$$

and the distribution is

$$F(x) = \int_{500}^{x} f(\xi)\, d\xi = 1 - \frac{1,500 + x}{2,000} e^{-(x-500)/2,000}$$

The probability p that any family with income less than $800 will be selected at random is given by

$$p = P(x < 800) = F(800) = 1 - (1.15)e^{-.15} = 1 - .9898 \approx .01$$

Strictly speaking, the distribution of the number of low-income families selected in a (conditionally) random sample of 1,000 is hypergeometric, but assuming that sampling depletions are negligible, we may regard the distribution as binomial with $n = 1,000$, $p = .01$. Again, since n is large and p is small, we may make the further simplification that the binomial distribution, in turn, may be replaced by the Poisson with $\mu = np = 10$. Hence, letting c denote the number of circulars sent to families with income less than $800, we obtain the required probability from Molina's tables. This is

$$P(c > 15) = .049 \approx .05 \qquad ////$$

The density function of the gamma distribution can be integrated exactly if α is an integer, but the work becomes inordinately laborious for hand computation if α is at all large; for an arbitrary value of α not an integer, numerical integration is necessary. For this reason, suitable tables are a practical necessity. Extensive tables of the gamma distribution with $\beta = 1$, entitled "Tables of the Incomplete Γ-Function," were published in 1922 by the office of *Biometrika*, under the editorship of Karl Pearson. These tables may be applied to arbitrary $\beta > 0$ by the change of variable $v = x/\beta$. For

$$F(x) = \frac{1}{\beta^{\alpha+1}\Gamma(\alpha + 1)}\int_0^x y^\alpha e^{-y/\beta}\, dy = \frac{1}{\Gamma(\alpha + 1)}\int_0^{x/\beta} v^\alpha e^{-v}\, dv \qquad (4\text{-}19)$$

In Pearson's tables, a further change of scale $u = v/\sqrt{\alpha + 1}$ was made in order to facilitate tabulation. Hence, to evaluate $F(x)$ one enters the tables under the appropriate α with the figure $u = x/(\beta\sqrt{\alpha + 1})$. Selected percentage points of the gamma distribution, obtained by adapting Thompson's tables of chi-square,[1] are given in Table 4-1. This table gives values of x which correspond to fixed values of $F(x)$. For such a value of x, corresponding to a stated fraction of the total area, Hald[2] has proposed the convenient term *fractile*. Thus a value of x, say x_p, is a fractile of level p if $F(x_p) = p$.

Table 4-1 SELECTED PERCENTAGE POINTS OF THE GAMMA DISTRIBUTION: VALUES OF x/β CORRESPONDING TO STATED VALUES OF $F(x)$†

α \ F	.01	.05	.10	.25	.50	.75	.90	.95	.99
−.5	.0000785	.00197	.00790	.0508	.227	.662	1.353	1.921	3.317
0	.01005	.0513	.105	.288	.693	1.386	2.303	2.996	4.605
.5	.0574	.176	.292	.606	1.183	2.054	3.126	3.907	5.672
1.0	.149	.355	.532	.961	1.678	2.693	3.890	4.744	6.638
1.5	.277	.573	.805	1.337	2.176	3.313	4.618	5.535	7.543
2.0	.436	.818	1.102	1.727	2.674	3.920	5.322	6.296	8.406
2.5	.620	1.084	1.417	2.127	3.173	4.519	6.008	7.034	9.238
3.0	.823	1.366	1.745	2.535	3.672	5.109	6.681	7.754	10.045
3.5	1.044	1.663	2.084	2.949	4.171	5.694	7.342	8.460	10.833
4.0	1.279	1.970	2.433	3.369	4.671	6.274	7.994	9.154	11.605
4.5	1.527	2.287	2.789	3.792	5.170	6.850	8.638	9.838	12.362
5.0	1.785	2.613	3.152	4.219	5.670	7.423	9.275	10.513	13.108
5.5	2.053	2.946	3.521	4.650	6.170	7.992	9.906	11.181	13.844
6.0	2.330	3.285	3.895	5.083	6.670	8.558	10.532	11.842	14.571
6.5	2.615	3.630	4.273	5.518	7.169	9.123	11.154	12.498	15.289
7.0	2.906	3.981	4.656	5.956	7.669	9.684	11.771	13.148	16.000
7.5	3.204	4.336	5.043	6.396	8.169	10.244	12.384	13.794	16.704
8.0	3.507	4.695	5.432	6.838	8.669	10.802	12.995	14.435	17.403
8.5	3.816	5.058	5.825	7.281	9.169	11.359	13.602	15.072	18.095
9.0	4.130	5.425	6.221	7.726	9.669	11.914	14.206	15.705	18.783
9.5	4.449	5.796	6.620	8.172	10.169	12.467	14.808	16.335	19.466
10.0	4.771	6.169	7.021	8.620	10.668	13.020	15.407	16.962	20.145
11.0	5.428	6.924	7.829	9.519	11.668	14.121	16.598	18.208	21.490
12.0	6.099	7.690	8.646	10.422	12.668	15.217	17.782	19.443	22.821
13.0	6.782	8.464	9.470	11.329	13.668	16.310	18.958	20.669	24.139
14.0	7.477	9.246	10.300	12.239	14.668	17.400	20.128	21.886	25.446
15.0	8.180	10.035	11.135	13.152	15.668	18.487	21.293	23.098	26.744
20.0	11.825	14.072	15.382	17.755	20.668	23.883	27.045	29.062	33.104
25.0	15.623	18.218	19.717	22.404	25.667	29.234	32.711	34.916	39.308
30.0	19.532	22.444	24.113	27.085	30.667	34.552	38.315	40.691	45.401

†Adapted from Catherine M. Thompson, Tables of Percentage Points of the Chi-square Distribution, *Biometrika*, vol. 32, pt. II, p. 187, October, 1941.

[1] Catherine M. Thompson, Tables of Percentage Points of the Chi-square Distribution, *Biometrika*, vol. 32, pt. II, p. 187, October 1941.

[2] A. Hald, "Statistical Theory with Engineering Applications," p. 66, John Wiley & Sons, Inc., New York, 1952.

The problem of estimating the parameters of a distribution from empirical data comes under the heading of *statistical estimation*, a branch of statistics for which a well-developed theory exists. However, in areas to which the application of probability represents an innovation or a comparatively recent approach, one must be prepared to face the obstacle of extreme paucity of data. Often the most that can be elicited from available records or from experienced workers in the field is a reasonable guess about certain features of the distribution—for instance, whether the curve is J-shaped or unimodal, and if the latter, the approximate location of the mode or, in either case, perhaps a pair of percentage points. Under economic or military pressure, the incentive to attack the problem may be compelling, whereas the procurement of substantially more detailed information might be out of the question. (This is true particularly of data which require years to accumulate.) The gamma distribution, because of the variety of shapes it can assume with only two parameters, is often hypothesized as the form of the probability distribution, and the next step is to decide upon plausible values of its parameters. With the aid of Table 4-2 (or a similar one especially suited to the information at hand) these parameters can be evaluated from estimates of 2 percentage points or, as the case may be, the mode and 1 percentage point. The percentage points chosen for this table are *deciles*. The ith decile D_i is defined as that value of x for which $F(x) = i(.1)$. Thus, $F(D_1) = .10$; $F(D_5) = .50$; $F(D_9) = .90$. A more common name for the fifth decile D_5 is the *median*.

EXAMPLE 4-7 A city undertaking a study of corrosion of water pipes has records of total footage of each type of pipe laid yearly since the inauguration of public water supply, but breakage data for only the last 5 years. Hypothesizing a gamma distribution for the length of life of a foot of pipe, estimate α, β if the median is 30 years and the ninth decile is 60 years; again, if the mode is 22 years and the first decile is 12 years. From the median and ninth decile we compute the ratio $D_9/D_5 = {}^{60}\!/_{30} = 2$. Referring to Table 4-2, we find that the nearest value given is $D_9/D_5 = 1.99$ at $\alpha = 2$. Hence we take α as equal to 2. To determine β we refer to Table 4-1 and find that $D_5/\beta = 2.674$ when $\alpha = 2$. Hence

$$\beta = \frac{D_5}{2.674} = \frac{30}{2.674} = 11.2 \approx 11$$

As a tentative hypothesis, then, we put

$$f(x) = \frac{1}{2,662} x^2 e^{-x/11}$$

Alternatively, using the estimates of the mode and D_1, we have $D_1/M = {}^{12}\!/_{22} = .545$. Again, the nearest value in Table 4-2 corresponds to $\alpha = 2$, and since the mode is $M = \alpha\beta$, we obtain $\beta = M/\alpha = {}^{22}\!/_2 = 11$, as before. ////

When α is an integer, an interesting connection exists between the upper tail area of the gamma distribution and the Poisson-distribution function. Considering the gamma distribution with α an integer and $\beta = 1$, let $G(\mu)$

Table 4-2 RATIOS FACILITATING THE ESTIMATION OF THE PARAMETERS α, β OF THE GAMMA DISTRIBUTION

α	D_1/M	D_5/M	D_9/M	D_1/D_5	D_9/D_5	D_1/D_9
$-.5$	Curve J-shaped			.0348	5.960	.0058
0	Curve J-shaped			.152	3.323	.0455
.5	†	2.366	6.252	.247	2.642	.0934
1.0	.532	1.678	3.890	.317	2.318	.137
1.5	.537	1.451	3.079	.370	2.122	.174
2.0	.551	1.337	2.661	.412	1.990	.207
2.5	.567	1.269	2.403	.447	1.893	.236
3.0	.582	1.224	2.227	.475	1.819	.261
3.5	.595	1.192	2.098	.500	1.760	.284
4.0	.608	1.168	1.999	.521	1.711	.304
4.5	.620	1.149	1.920	.539	1.671	.323
5.0	.630	1.134	1.855	.556	1.636	.340
5.5	.640	1.122	1.801	.571	1.606	.355
6.0	.649	1.112	1.755	.584	1.579	.370
6.5	.657	1.103	1.716	.596	1.556	.383
7.0	.665	1.096	1.682	.607	1.535	.396
7.5	.672	1.089	1.651	.617	1.516	.407
8.0	.679	1.084	1.624	.627	1.499	.418
8.5	.685	1.079	1.600	.635	1.483	.428
9.0	.691	1.074	1.578	.643	1.469	.438
9.5	.697	1.070	1.559	.651	1.456	.447
10.0	.702	1.067	1.541	.658	1.444	.456
11.0	.712	1.061	1.509	.671	1.423	.472
12.0	.720	1.056	1.482	.682	1.404	.486
13.0	.728	1.051	1.458	.693	1.387	.500
14.0	.736	1.048	1.438	.702	1.372	.512
15.0	.742	1.045	1.420	.711	1.359	.523
20.0	.769	1.033	1.352	.744	1.309	.569
25.0	.789	1.027	1.308	.768	1.274	.603
30.0	.804	1.022	1.277	.786	1.249	.629

† Mode to left of D_i.
KEY: $D_i = i$th decile; $M =$ mode.

denote the probability $P(x \geq \mu)$ that x assumes a value equal to or greater than an arbitrary positive quantity μ:

$$G(\mu) = \frac{1}{\Gamma(\alpha + 1)} \int_{\mu}^{\infty} x^{\alpha} e^{-x} \, dx \qquad (4\text{-}20)$$

Integrating by parts, we obtain

$$G(\mu) = \frac{1}{\Gamma(\alpha + 1)} \mu^{\alpha} e^{-\mu} + \frac{1}{\Gamma(\alpha)} \int_{\mu}^{\infty} x^{\alpha - 1} e^{-x} \, dx \qquad (4\text{-}21)$$

Now if α is an integer, the exponent of x can be reduced one unit at a time to zero; moreover, the gamma functions become factorials of integers. Consequently

$$G(\mu) = e^{-\mu} \left[\frac{\mu^{\alpha}}{\alpha!} + \frac{\mu^{\alpha-1}}{(\alpha-1)!} + \cdots + \frac{\mu^2}{2!} + \mu + 1 \right] \qquad (4\text{-}22)$$

which we recognize as the distribution function of some discrete variate, call it y, having the Poisson distribution with parameter μ. Thus we conclude that

$$P(y \leq \alpha) = \sum_{y=0}^{\alpha} e^{-\mu} \frac{\mu^{y}}{y!} = \frac{1}{\Gamma(\alpha + 1)} \int_{\mu}^{\infty} x^{\alpha} e^{-x} \, dx = G(\mu) \qquad (4\text{-}23)$$

Therefore, the Poisson-distribution function can be evaluated from the incomplete gamma function and vice versa. This result was first published by E. C. Molina in 1915.

EXAMPLE 4-8   In 1920, Rutherford, Chadwick, and Ellis studied the emission of alpha particles from a radioactive substance. It was found empirically that the distribution of the number of particles emitted during a time interval of 7.5 seconds was adequately represented by a Poisson distribution with $\mu = 3.87$, or more generally that the number of particles emitted in t seconds was a Poisson variate with $\mu = \rho t$, where $\rho = 3.87/7.5 = .516$. Instrumentally, it is customary to clock the length of time taken to register a fixed number of particles rather than to count the number of particles emitted during a fixed interval of time. Let us find the distribution of the length of time required to register 40 particles. In general, let $G(t)$ denote the probability that it will take more than t seconds for the emission of $n + 1$ particles. Clearly, this can happen in $n + 1$ mutually exclusive ways, namely, by the emission of any smaller number of particles 0, 1, 2, ..., n in t seconds. Hence the corresponding total probability, obtained by summing the separate probabilities of each number of particles as given by the Poisson distribution, is

$$G(t) = \sum_{y=0}^{n} e^{-\mu} \frac{\mu^{y}}{y!}$$

But by Eq. (4-23), the latter sum can be expressed as an integral, which yields

$$G(t) = \frac{1}{\Gamma(n+1)} \int_{\mu}^{\infty} x^n e^{-x} \, dx$$

Now the distribution function $F(t)$ is the complement of $G(t)$, whence

$$F(t) = 1 - G(t) = \frac{1}{\Gamma(n+1)} \int_{0}^{\mu} x^n e^{-x} \, dx$$

and the density function $f(t)$ is given by

$$f(t) = \frac{dF(t)}{dt} = \frac{dF(\mu)}{d\mu} \frac{d\mu}{dt} = \frac{\rho^{n+1}}{\Gamma(n+1)} t^n e^{-\rho t}$$

Therefore, the distribution of waiting time t for the emission of $n+1$ particles is

Distribution of waiting time for emission of $n+1$ alpha particles:

$$f(t) = \frac{\rho^{n+1}}{\Gamma(n+1)} t^n e^{-\rho t} \qquad (0, \infty) \qquad (4\text{-}24)$$

This is a gamma distribution with $\alpha = n$, $\beta = 1/\rho$. In particular, for the problem at hand, $n = 39$, $\rho = .516$ and the distribution is

$$f(t) = \frac{(.516)^{40}}{39!} t^{39} e^{-(.516)t} \qquad 0 \leq t < \infty \qquad \text{////}$$

Further interesting deductions follow readily from this result. Setting $n = 0$ in Eq. (4-24) we find that the distribution of waiting time for one particle is given by the exponential distribution

Waiting time for one particle:

$$f(t) = \rho e^{-\rho t} \qquad 0 \leq t < \infty \qquad (4\text{-}25)$$

Now it is obvious that the length of time required for the emission of a total of $n+1$ particles is equal to the sum of the waiting times for $n+1$ individual particles. Moreover, since these are emitted independently, their waiting times are independent. Therefore we arrive at the following theorem:

Theorem 4-1 If $t_1, t_2, \ldots, t_k$ are independent variates all governed by the same exponential distribution,

$$h(t_i) = \frac{1}{\beta} e^{-t_i/\beta} \qquad (0, \infty) \qquad (4\text{-}26)$$

then their sum $t = t_1 + t_2 + \cdots + t_k$ is a gamma variate with the same β and with $\alpha = k - 1$:

$$f(t) = \frac{1}{\beta^k \Gamma(k)} t^{k-1} e^{-t/\beta} \qquad (0, \infty) \qquad (4\text{-}27)$$

////

More generally, the length of time for any number of particles can be regarded as the sum of times for sets of smaller numbers that add up to the same total. Hence we may state:

Theorem 4-2   If x_1, x_2, ..., x_k are independent gamma variates with common β and integral values of α, namely α_1, α_2, ..., α_k, then their sum $x = x_1 + x_2 + \cdots + x_k$ is a gamma variate with the same β and with $\alpha = \alpha_1 + \alpha_2 + \cdots + \alpha_k + k - 1$.   ////

In a later chapter we shall see that this result holds even when the α's are not integers, but we are not yet prepared for the proof.

EXAMPLE 4-9   The distribution of length of telephone calls in minutes was found by Molina to be exponential with $\beta = 2.26$ for calls between a pair of adjacent cities in 1925.   Let us suppose (for the sake of illustration) that the same applies today for any random selection of local calls.   A man goes to a public telephone booth and finds that five people are ahead of him, the first having just stepped into the booth.   Neglecting dialing time, what is the probability that this man will have to wait more than 18 minutes for his turn? Assuming that the five people ahead of him constitute a random sample, the distribution of their total length of call will be gamma with $\alpha = 5 - 1 = 4$ and $\beta = 2.26$.   Here $x/\beta = 18/2.26 = 7.96$.   Referring to Table 4-1 under $\alpha = 4$, we find that the probability of exceeding this value of x/β is just over .10. Hence the probability is about .10 that he will have to wait more than 18 minutes for his turn.   ////

We have previously noted that exact values of the binomial-distribution function can be obtained from the incomplete beta function and approximate values for large n either from the Poisson or from the normal distribution (Sec. 4-11), depending on the circumstances.   The gamma distribution also can be used as an approximation, filling the gap for intermediate values of n beyond existing tables of the binomial but not large enough to yield good approx-

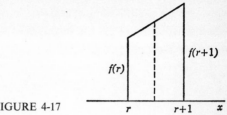

FIGURE 4-17

imations by the normal. Even within the tabulated range of n, the gamma distribution is also useful in situations wherein a continuous variate is easier to handle.

The problem of adjusting the parameters of a curve so that it will pass through or close to a given set of points comes under the general heading of *curve fitting*, and several techniques are available for this purpose. Whereas most of these rest on principles which we have not yet discussed, the method we are about to employ is based on simple geometry.

Letting r stand for any value of the binomial variate and $f(r)$ the corresponding value of the probability function of that variate, consider the line segment joining the tops of the two ordinates $f(r)$, $f(r + 1)$ (see Fig. 4-17). The slope of this line segment is

$$\frac{f(r + 1) - f(r)}{(r + 1) - r} = f(r + 1) - f(r)$$

and the mid-height (i.e., the height at the midpoint) is $[f(r + 1) + f(r)]/2$. Since the recursion formula for the binomial-probability function gives $f(r + 1)$ in terms of $f(r)$ as

$$f(r + 1) = \frac{(n - r)p}{(r + 1)(1 - p)} f(r)$$

it is evident that both the slope and mid-height contain $f(r)$ as a factor. Hence $f(r)$ will cancel if we form the ratio of slope to mid-height, and the simplified expression becomes

$$\frac{\text{Slope}}{\text{Mid-height}} = 2 \frac{(n + 1)p - r - 1}{(n - 1)p + r(1 - 2p) + 1} \qquad (4\text{-}28)$$

Let us now determine a smooth curve $g(x)$ which will yield this same ratio of slope to ordinate when $x = r + \frac{1}{2}$. Substituting $r = x - \frac{1}{2}$ in Eq. (4-28) we obtain

$$\frac{1}{g(x)} \frac{dg(x)}{dx} = 2 \frac{(n + 1)p - x - \frac{1}{2}}{np + \frac{1}{2} + x(1 - 2p)} = \frac{-2(x - a)}{bx + c} \qquad (4\text{-}29)$$

where $\qquad a = (n + 1)p - \frac{1}{2} \qquad b = 1 - 2p \qquad c = np + \frac{1}{2}$

Hereafter we shall assume that $0 < p < .5$ so that b will be positive. A curve which satisfies Eq. (4-29) for all values of x will certainly satisfy it at the particular points $x = r + \frac{1}{2}$ $(r = 0, 1, 2, \ldots, n)$. Hence we may integrate, obtaining

$$\ln g(x) = \ln K_1 + \frac{2(ab + c)}{b^2} \ln(bx + c) - \frac{2x}{b}$$

whence

$$g(x) = K_2(x + A)^\alpha e^{-x/\beta} = k(x + A)^\alpha e^{-(x+A)/\beta}$$

where

$$A = \frac{c}{b} = \frac{2np + 1}{2(1 - 2p)}$$

$$\alpha = \frac{2(ab + c)}{b^2} = \frac{4(1 + n)p(1 - p)}{(1 - 2p)^2} \qquad (4\text{-}30)$$

$$\beta = \frac{b}{2} = \frac{1 - 2p}{2}$$

If we now choose k so that the curve has unit area over the range $-A \le x < \infty$ we arrive at a gamma distribution:

Gamma approximation to binomial distribution with $0 < p < .5$:

$$g(x) = \frac{1}{\beta^{\alpha+1}\Gamma(\alpha + 1)}(x + A)^\alpha e^{-(x+A)/\beta} \qquad -A \le x < \infty \qquad (4\text{-}31)$$

Table 4-3   COMPARISON OF EXACT BINOMIAL-DISTRIBUTION FUNCTION (F) AND APPROXIMATION (Γ) USING GAMMA DISTRIBUTION

$n = 9, p = .2$			$n = 16, p = .2$		
r	F	Γ	r	F	Γ
0	.13	.15	0	.03	.04
1	.44	.43	1	.14	.16
2	.74	.72	2	.352	.357
3	.91	.89	3	.598	.595
4	.98	.97	4	.798	.791
5	.997	.993	5	.918	.909
6	.9997	.9985	6	.973	.967
7	.99998	.99975	7	.993	.990
			8	.999	.997
			9	.9998	.9993
			10	.99997	.99985

where the parameters A, α, β are defined by Eq. (4-30). In applying this approximation, we must remember to evaluate the gamma distribution function at $x = r + \frac{1}{2}$ in order to estimate the binomial-distribution function at r. Two sample cases using $n = 9$, $p = .2$ and $n = 16$, $p = .2$ are shown in Table 4-3. The approximations are good enough for most practical or even theoretical purposes.

(Suggested Problems 4-17 to 4-24)

4-10   THE BETA DISTRIBUTION

In its standard form, the beta distribution applies to a continuous variate, the admissible values of which lie between 0 and 1 with probability density:

Beta distribution:

$$f(x) = \frac{1}{\beta(\alpha + 1, \lambda + 1)} x^{\alpha}(1 - x)^{\lambda} \qquad \alpha, \lambda > -1; 0 \leq x \leq 1 \qquad (4\text{-}32)$$

This distribution is often a suitable model for the random behavior of percentages.

The effects of the parameters α, λ on the shapes of the curves are indicated in Fig. 4-18. The special case with both parameters equal to zero is called the *rectangular* distribution, for the density function reduces to the constant $f = 1$, and the figure resembles a rectangle, except that the two ends are not actually part of the density function. Another special case, called the *triangular* distribution, arises when one parameter is zero and the other unity. The curve then becomes a straight line with slope $+2$ if $\alpha = 1$, or -2 if $\lambda = 1$. The curve is U-shaped if both parameters are negative, J-shaped if only one is negative, and unimodal if both are positive. When both parameters have the same sign the curve has one turning point, which occurs at $x = \alpha/(\alpha + \lambda)$; this is the mode if the parameters are positive, but the lowest point if they are negative. If both α, λ exceed unity, there are two points of inflection, located at a distance $\sqrt{\alpha\lambda(\alpha + \lambda - 1)}/(\alpha + \lambda)(\alpha + \lambda - 1)$ to either side of the mode. If one parameter equals unity and the other exceeds unity, there is one point of inflection; if $\alpha = 1$, $\lambda > 1$ the inflection point is located at $x = 2/(1 + \lambda)$; if $\alpha > 1$, $\lambda = 1$ the inflection point occurs at $x = (\alpha - 1)/(\alpha + 1)$.

In either case the location is given as a special case of the general formula, except that there is only one true root. The curve is symmetrical if $\alpha = \lambda$, skewed to the right if $\alpha < \lambda$, and skewed to the left if $\alpha > \lambda$. An interchange

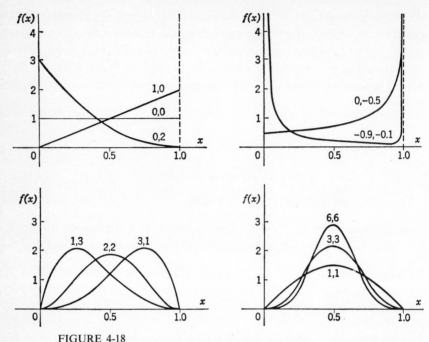

FIGURE 4-18
Beta distributions with various selections of parameters α, λ. The symbol a, b means $\alpha = a$, $\lambda = b$.

of parameters yields the mirror image of the initial curve. If the mode is held fixed, the concentration in the neighborhood of the mode increases as the parameters increase.

EXAMPLE 4-10   Ceramic parts for various electrical devices are manufactured to order, and as a concession to the manufacturer, the customer placing an order for a given number of items N is willing to accept at the same unit price up to 10 percent more than the number ordered. However, the manufacturer must supply at least the number ordered. Because of cracking, chipping, and other defects, the number of pieces S of standard quality is a fraction x ($0 \leq x \leq 1$) of the initial number I put into the molds. For a particular type of part, the distribution of this fraction is beta with $\alpha = 8$, $\lambda = 2$; that is,

$$f(x) = 495x^8(1 - x)^2 \qquad (0, 1)$$

If the initial number is chosen as a multiple m of the number of items ordered ($I = mN$), determine as a function of m the probability (*a*) that the number of standard pieces will fall short of the number ordered, (*b*) that the number of

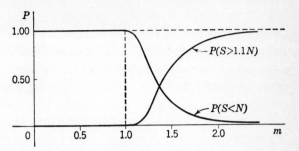

FIGURE 4-19
Respective probabilities of falling short of ordered number and of exceeding the acceptable number.

standard pieces will exceed the maximum the customer will accept at the same unit price.   Since $x = S/I$ we have

$$S = xI = xmN$$

Hence $S < N$ whenever $xm < 1$, which implies that $x < 1/m$; again, $S > 1.1N$ whenever $xm > 1.1$, or $x > 1.1/m$.   Consequently,

$$P(S < N) = P\left(x < \frac{1}{m}\right) = F\left(\frac{1}{m}\right)$$

and

$$P(S > 1.1N) = P\left(x > \frac{1.1}{m}\right) = 1 - F\left(\frac{1.1}{m}\right)$$

Now in the range $0 \le x \le 1$ the distribution function of x is defined as

$$F(x) = \int_0^x f(\xi)\, d\xi = x^9(55 - 99x + 45x^2)$$

but $F(x) \equiv 1$ if $x > 1$.   Hence, provided $m \ge 1$, the probability of falling short is given by

$$P(S < N) = P\left(x < \frac{1}{m}\right) = \frac{55m^2 - 99m + 45}{m^{11}}$$

However, if $m < 1$, this means that the initial number of items I is less than the number ordered, and so it is obvious that $P(S < N) \equiv 1$ if $m < 1$.   In the same way, it is impossible to exceed the number $1.1N$ which the customer has agreed to buy unless $I > 1.1N$ whence $m > 1.1$.   Thus $P(S > 1.1N) \equiv 0$ if $m \le 1.1N$, and otherwise it is given by

$$P(S > 1.1N) = P\left(x > \frac{1.1}{m}\right) = 1 - (1.1)^9 \frac{55m^2 - 99(1.1)m + 45(1.1)^2}{m^{11}}$$

These two functions are displayed in Fig. 4-19.

In general, it is costly to fall short of the number ordered, because the manufacturing process has to be repeated, and this entails setup time, schedule delays, tying up equipment, and increased costs of labor and power for the same item. The risk of falling short can be fixed at any desired level >0 by choosing I sufficiently large. However, playing it safe in this respect incurs its own penalty in occupying equipment and labor, which could be more profitably employed, and in wasting power and raw materials. The aims of ensuring the required number of items and of making profitable use of limited resources are in conflict. The problem of striking an optimum balance involves relative costs, and its solution entails the concept of mathematical expectation, which we treat in Chap. 7. $////$

The beta density function cannot be integrated formally from 0 to x unless α and λ are both integral multiples of $\frac{1}{2}$. Unlike the gamma distribution, which permits the simplification of one-parameter tabulation because all gamma integrals with arbitrary parameter β can be obtained from the integral with $\beta = 1$, the beta distribution requires tabulation with respect to both parameters. Pearson's "Tables of the Incomplete Beta-function," to which we referred in Chap. 2, stands as the most comprehensive work thus far produced on this distribution. Very often, however, it is convenient to have a table of fractiles, that is, one giving the values of x which correspond to selected levels of probability. A good table of this type has been constructed by Catherine M. Thompson.[1]

The scope of the beta distribution can be broadened by allowing the variate to range between any pair of finite limits A, B. The modified distribution then becomes

Beta with arbitrary limits:

$$f(x) = \frac{1}{C}(x - A)^\alpha (B - x)^\lambda \qquad (A, B) \qquad (4\text{-}33)$$

where
$$C = (B - A)^{\alpha + \lambda + 1}\beta(\alpha + 1, \lambda + 1)$$

Fortunately the two extra parameters A, B do not necessitate additional tables, for if we substitute $u = (x - A)/(B - A)$, we obtain an integral in standard form. Thus

$$\int_A^\xi f(x)\,dx = \frac{1}{\beta(\alpha + 1, \lambda + 1)} \int_0^{\xi'} u^\alpha (1 - u)^\lambda\,du \qquad (4\text{-}34)$$

[1] Catherine M. Thompson, Percentage Points of the Incomplete Beta-function, *Biometrika*, vol. 32, pt. II, pp. 168–181, October, 1941.

where $\xi' = (\xi - A)/(B - A)$. From the equation for u, we see that the effect of A and B is to shift the origin and change the scale of the conventional beta variate, but otherwise the density curves are similar.

4-11 THE NORMAL DISTRIBUTION

The normal distribution was discovered by De Moivre, whose published works as early as 1733 contained a derivation of it as the limiting form of the binomial. It was also known to Laplace no later than 1774, but through historical error, it has been attributed to Gauss, whose earliest published reference to it appeared in 1809. Nevertheless, the term "gaussian distribution" is an accepted synonym for "normal distribution." During the eighteenth and nineteenth centuries, various attempts were made to establish this distribution as the underlying law governing all continuous variates—hence the name "normal." Although these attempts failed, being based on false premises, the normal distribution rightly occupies a preeminent place in the field of probability; for besides having exceptionally convenient properties of its own, it also serves as a useful approximation to many other distributions which are less tractable. In particular, the averages of n observations taken at random from almost any population tend to become normally distributed as n increases.

A variate x is said to be normally distributed if its density function $f(x)$ is given by an expression of the form

Normal distribution:

$$f(x) = \frac{1}{\sigma\sqrt{2\pi}} e^{-(x-\mu)^2/2\sigma^2} \qquad (-\infty, \infty) \qquad (4\text{-}35)$$

where μ can be any real number and σ any real number which is greater than zero. The distribution is unimodal, the mode being $x = \mu$, and there are two points of inflection, located at a distance σ to either side of the mode. By inspection, it is evident that the density curve is symmetrical about $x = \mu$ and falls off rapidly as the magnitude of $x - \mu$ increases. Although the normal distribution has an infinite range, the probability of very large deviations from μ is small enough to be neglected for most practical purposes; hence this distribution is capable of approximating others for which the true range is finite. Figure 4-20 shows the distribution for three values of μ and a constant value of σ; in Fig. 4-21, μ is constant and three values of σ are used. Clearly, a shift in μ displaces the curve as a whole, whereas a change in σ alters its relative proportions with reference to a fixed scale. In view of the symmetry of the

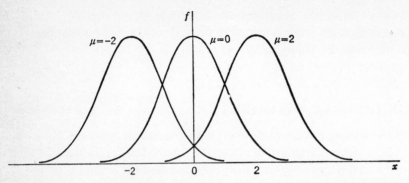

FIGURE 4-20
Normal distributions with fixed $\sigma(\sigma = 1)$ and selected values of μ.

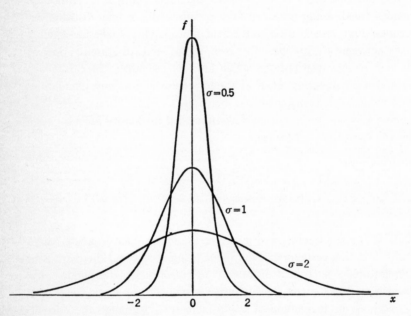

FIGURE 4-21
Normal distributions with fixed $\mu(\mu = 0)$ and selected values of σ.

curve about the mode, it is an accepted idiom to say that x is "normally distributed about λ" when we mean that the distribution of x is normal with $\mu = \lambda$.

A normal distribution with $\mu = 0$ and $\sigma^2 = 1 = \sigma$ is termed a *unit normal* distribution. Accordingly, this distribution is as follows,

Unit normal distribution:

$$f(x) = \frac{1}{\sqrt{2\pi}} e^{-x^2/2} \qquad (-\infty, \infty) \qquad (4\text{-}36)$$

Since the density function of the normal distribution cannot be integrated explicitly, the evaluation of probabilities would entail numerical integration were it not for extensive tables already available. These tables have been constructed for the unit normal distribution only but can be adapted to any other normal distribution by the substitution $u = (x - \mu)/\sigma$. For example, if x is a normal variate with arbitrary parameters, then

$$P(a \le x \le b) = \int_a^b \frac{1}{\sigma\sqrt{2\pi}} e^{-(x-\mu)^2/2\sigma^2} \, dx = \int_{a'}^{b'} \frac{1}{\sqrt{2\pi}} e^{-u^2/2} \, du \qquad (4\text{-}37)$$

where
$$a' = \frac{a - \mu}{\sigma} \qquad b' = \frac{b - \mu}{\sigma}$$

A brief tabulation of ordinates, areas, and fractiles of the unit normal distribution is given in Table 4-4, where only positive values of x are recorded, inasmuch as $f(-x) = f(x)$ and $F(-x) = 1 - F(x)$. We have used the notation $R(x)$ to denote the area to the right of x, and the notation $W(x)$ to denote the area within $-x$ and x. Thus if ξ denotes a random value of a unit normal variate, and $x \ge 0$, then

$$P(\xi \le x) = F(x) = \int_{-\infty}^x \frac{1}{\sqrt{2\pi}} e^{-y^2/2} \, dy = \int_{-\infty}^x f(y) \, dy$$

$$P(\xi \ge x) = R(x) = \int_x^\infty f(y) \, dy = 1 - F(x)$$

$$P(\xi \le -x) = F(-x) = \int_{-\infty}^{-x} f(y) \, dy = R(x) = 1 - F(x) \qquad (4\text{-}38)$$

$$P(|\xi| \ge x) = \int_{-\infty}^{-x} f(y) \, dy + \int_x^\infty f(y) \, dy = F(-x) + R(x) = 2R(x)$$

$$P(-x \le \xi \le x) = W(x) = \int_{-x}^x f(y) \, dy$$

Table 4-4 UNIT NORMAL DISTRIBUTION

x	$f(x)$	$F(x)$	$R(x)$	$2R(x)$	$W(x)$
.0	.3989	.5000	.5000	1.0000	0
.1	.3970	.5398	.4602	.9203	.0797
.2	.3910	.5793	.4207	.8415	.1585
.3	.3814	.6179	.3821	.7642	.2358
.4	.3683	.6554	.3446	.6892	.3108
.5	.3521	.6915	.3085	.6171	.3829
.6	.3332	.7257	.2743	.5485	.4515
.7	.3123	.7580	.2420	.4839	.5161
.8	.2897	.7881	.2119	.4237	.5763
.9	.2661	.8159	.1841	.3681	.6319
1.0	.2420	.8413	.1587	.3173	.6827
1.1	.2179	.8643	.1357	.2713	.7287
1.2	.1942	.8849	.1151	.2301	.7699
1.3	.1714	.9032	.0968	.1936	.8064
1.4	.1497	.9192	.0808	.1615	.8385
1.5	.1295	.9332	.0668	.1336	.8664
1.6	.1109	.9452	.0548	.1096	.8904
1.7	.0940	.9554	.0446	.0891	.9109
1.8	.0790	.9641	.0359	.0719	.9281
1.9	.0656	.9713	.0287	.0574	.9426
2.0	.0540	.9772	.0228	.0455	.9545
2.1	.0440	.9821	.0179	.0357	.9643
2.2	.0355	.9861	.0139	.0278	.9722
2.3	.0283	.9893	.0107	.0214	.9786
2.4	.0224	.9918	.0082	.0164	.9836
2.5	.0175	.9938	.0062	.0124	.9876
2.6	.0136	.9953	.0047	.0093	.9907
2.7	.0104	.9965	.0035	.0069	.9931
2.8	.0079	.9974	.0026	.0051	.9949
2.9	.0060	.9981	.0019	.0037	.9963
3.0	.0044	.9987	.0013	.0027	.9973

Fractiles

1.2816	.1755	.9000	.1000	.2000	.8000
1.6449	.1031	.9500	.0500	.1000	.9000
1.9600	.0584	.9750	.0250	.0500	.9500
2.0537	.0484	.9800	.0200	.0400	.9600
2.3263	.0267	.9900	.0100	.0200	.9800
2.5758	.0145	.9950	.0050	.0100	.9900

EXAMPLE 4-11   By route A, the travel time in minutes from hotel to airport is normal with $\mu = 27$, $\sigma = 5$; by route B, the distribution is normal with $\mu = 30$, $\sigma = 2$.   Which route is the better choice if (a) one has 30 minutes, (b) 34 minutes?   In either situation, we shall choose the route for which the probability of exceeding the allotted travel time is the smaller.   Let t denote any value of travel time.   In case a we have,

Route A:
$$P(t > 30) = R\left(\frac{30 - 27}{5}\right) = R(.6) = .27$$

Route B:
$$P(t > 30) = R\left(\frac{30 - 30}{2}\right) = R(0) = .50$$

Hence in case a, route A is the better risk.   In case b we have,

Route A:
$$P(t > 34) = R\left(\frac{34 - 27}{5}\right) = R(1.4) = .08$$

Route B:
$$P(t > 34) = R\left(\frac{34 - 30}{2}\right) = R(2.0) = .02$$

Hence in case b, route B is the better risk.   ////

Suppose n independent observations $x_1, x_2, \ldots, x_n$ of the same variate x are taken at random, and the sample average $\bar{x}$ is computed in the usual way, i.e., by adding the numbers and then dividing by n.   Every time this experiment is performed, a particular value of $\bar{x}$ will result from the computation, and if we imagine the experiment repeated indefinitely with the same number of observations n each time, the computed averages will range over a series of values according to some definite law of probability.   A theorem of far-reaching importance and known as the *central limit theorem* asserts that the sample averages from almost any population likely to be encountered in practice tend to become normally distributed as the sample size increases, despite the fact that no value of $\bar{x}$ can possibly be smaller than the smallest admissible value of x or larger than the largest admissible value.   The reason is that the parameter σ which figures in the normal approximation to the distribution of $\bar{x}$ decreases with increasing n (actually σ is proportional to $1/\sqrt{n}$) at such a rate that the total probability associated with the region which lies outside the true physical limits of variation is negligibly small for sufficiently large n.

EXAMPLE 4-12 To a good approximation, the individual values of the average monthly temperatures for a given locality are normally distributed about the climatological mean temperature at that locality for the month in question. Toward the end of December, a meteorologist predicts that at two localities A and B the average temperatures for the following January will be at least $5°$ above the corresponding climatological means. These forecasts prove to be correct. If σ for the distribution of the average January temperature is $2°$ at A and $4°$ at B, what are the respective probabilities of obtaining average temperatures within the ranges forecast? Which forecast is the more indicative of skill? Let $\overline{T}_A$ denote the average temperature for any January and μ_A the climatological mean at locality A. According to the forecast, $\overline{T}_A \geq \mu_A + 5°$, whence $\overline{T}_A - \mu_A \geq 5°$. The probability is

$$P(\overline{T}_A \geq \mu_A + 5°) = R(\tfrac{5}{2}) = R(2.5) = .006$$

Similarly, for locality B, the forecast asserts that $\overline{T}_B \geq \mu_B + 5°$, and the probability of such an occurrence is

$$P(\overline{T}_B \geq \mu_B + 5°) = R(\tfrac{5}{4}) = R(1.25) = .106$$

The forecast for locality A is the more indicative of skill, because it correctly calls for a much rarer event. ////

The normal distribution is the limiting form of a great many distributions; in particular, it is the limiting form approached by the binomial for large n. Thus, if x is binomially distributed,

$$f(x) = C(n, x)p^x q^{n-x} \qquad x = 0, 1, 2, \ldots, n$$

then as n increases,

Normal approximation to binomial:

$$f(x) \to \frac{1}{\sqrt{npq}\sqrt{2\pi}} e^{-(x-np)^2/2npq} \qquad (-\infty, \infty) \qquad (4\text{-}39)$$

which is of normal form with $\mu = np$, $\sigma^2 = npq$. For moderately large values of n, the binomial distribution is adequately approximated for most purposes by the normal with $\mu = np$, $\sigma^2 = npq$, provided that $np \geq 5$. However, in applying the normal approximation, a compensation is made for the fact that the variate is actually discrete and n is finite. The conventional correction consists of changing the numerical deviation from μ by one-half unit before dividing by σ, on the grounds that the discrete value $x = a$ should correspond on a continuous scale to the interval $a - \frac{1}{2} \le x \le a + \frac{1}{2}$. Thus to estimate the probability

$$P(x \ge a) = \sum_{x=a}^{n} C(n, x)p^x q^{n-x}$$

where $a \ge \mu + \frac{1}{2}$ we set

$$y = \frac{a - \frac{1}{2} - \mu}{\sigma}$$

and look up $R(y)$ in the unit normal table. Again, to estimate the probability

$$P(x \le a) = \sum_{x=0}^{a} C(n, x)p^x q^{n-x}$$

where $a \le \mu - \frac{1}{2}$ we set

$$y' = \frac{\mu - (a + \frac{1}{2})}{\sigma}$$

and look up $R(y')$. Finally, to estimate $P(a \le x \le b)$ we extend the interval one-half unit to either side before dividing by σ.

EXAMPLE 4-13   In the manufacture of 1-inch wood screws at a certain plant, it occasionally happens that a screw is not slotted. This occurs at random, and the compound probability of its taking place and escaping notice on inspection is .02. If a retail hardware dealer purchases a batch of 2,500 screws, what is the probability that 64 or more will lack slots? 36 or less? Between 36 and 64 inclusive? The true distribution, which is binomial with $n = 2,500$, $p = .02$, was given in Table 2-12. The respective probabilities are

$$P(x \ge 64) = 1 - F(63) = .031$$
$$P(x \le 36) = F(36) = .023$$
$$P(36 \le x \le 64) = F(64) - F(35) = .962$$

For the normal approximation we have $\mu = np = 50$, $\sigma^2 = npq = 49$, $\sigma = 7$, and the corresponding probabilities are

$$P(x \ge 64) \approx R\left(\frac{13.5}{7}\right) = R(1.93) = .027$$

$$P(x \le 36) \approx R\left(\frac{13.5}{7}\right) = R(1.93) = .027$$

$$P(36 \le x \le 64) \approx W\left(\frac{14.5}{7}\right) = W(2.07) = .962$$

A comparison between the normal and Poisson approximations to this binomial is given in Table 4-5.   The Poisson is closer for this small value of p, but the normal is adequate for most purposes, and for larger values of p the normal approximation improves.   ////

As suggested by Table 4-5, the fact that the Poisson distribution with $\mu = np$ approximates the binomial for sufficiently small p, whereas the normal approximates the binomial for large n provided that np is not too small, implies that the normal approximates the Poisson when μ is sufficiently large.   Here the normal approximation uses the same μ as the Poisson, and σ is taken as $\sqrt{\mu}$.

The gamma distribution also approaches the normal as α increases.   In this case, we take $\mu = \beta(1 + \alpha)$ and $\sigma = \beta\sqrt{1 + \alpha}$.   R. A. Fisher showed that a more nearly normal variate y can be derived from a gamma variate x with parameters α, β by putting $y = 2\sqrt{x/\beta}$.   When α exceeds 14, the distribution of y becomes approximately normal with $\mu = \sqrt{4\alpha + 3}$, $\sigma = 1$.   A still more rapid approach to normality may be obtained by another transformation of a gamma variate due to Wilson and Hilferty.   Here if we put $w = [x/\beta(\alpha + 1)]^{1/3}$ it turns out that the distribution of w is approximately normal with $\mu = (9\alpha + 8)/(9\alpha + 9)$, $\sigma = 1/(3\sqrt{\alpha + 1})$.   A comparison between the exact values of $F(x)$

Table 4-5   COMPARISON OF NORMAL AND POISSON APPROXIMATIONS TO THE BINOMIAL PROBABILITY FUNCTION $f(x)$ WITH $n = 2,500$, $p = .02$

x	Binomial	Normal	Poisson
25	.0000	.0001	.0000
30	.0006	.0010	.0007
35	.0052	.0057	.0054
40	.0212	.0205	.0215
45	.0460	.0442	.0458
50	.0569	.0570	.0563
55	.0424	.0442	.0422
60	.0199	.0205	.0201
65	.0061	.0057	.0063
70	.0013	.0010	.0014
75	.0002	.0001	.0002

for the gamma distribution with $\alpha = 15$, $\beta = 1$ and the approximate values, first treating the gamma itself as normal and then treating the two functions y and w as normal, is presented in Table 4-6. Even the first approximation, treating the gamma itself as normal, is close enough for many qualitative uses; Fisher's transformation, which is fairly easy to use, is in error only in the third decimal; and the Wilson-Hilferty transformation yields results which differ from the exact values only in the fourth decimal. Since the parameter β can be eliminated in computing probabilities, equally good results will be obtained with any value of β; moreover, the approximations improve with increasing values of α.

The foregoing transformations were originally developed in connection with a certain variate known as *chi-square*, the distribution of which is gamma with $\beta = 2$, $\alpha = (n/2) - 1$, where n is a positive integer. The effectiveness of the transformations illustrates the general rule that if x is a nonnegative variate with a skewed distribution, the distributions of $\sqrt{x}$ and $x^{1/3}$ tend to be more nearly symmetrical and more nearly normal than the distribution of x. The same is true of $\log x$ provided that $f(0) = 0$. Other normalizing transformations have also been devised for special uses. Besides the obvious advantage that existing probability tables may be applied, normalizing transformations serve the purpose of reducing many difficult analytical problems to solvable forms.

(Suggested Problems 4-25 to 4-30)

Table 4-6 COMPARISON BETWEEN EXACT AND APPROXIMATE VALUES OF $F(x)$ FOR GAMMA DISTRIBUTION WITH $\alpha = 15$, $\beta = 1$

x	$F(x)$	Direct normal	Fisher's transformation	Wilson-Hilferty
8.180	.01	.025	.013	.0101
10.035	.05	.068	.055	.0497
11.135	.10	.112	.103	.0999
13.152	.25	.238	.247	.2498
15.668	.50	.467	.492	.5000
18.487	.75	.733	.746	.7501
21.293	.90	.907	.902	.9001
23.098	.95	.962	.953	.9501
26.744	.99	.996	.992	.9900

4-12 CONFIDENCE LIMITS (INTERVAL ESTIMATION)

In this chapter we have considered continuous density functions $f(x)$. This function may contain unknown parameters which one would like to estimate, even though the general form of the function is known. Therefore, in this section we shall consider how to bracket the true value of a single parameter by an interval called a "confidence interval." It should be kept in mind in all this discussion that the parameter itself has a definite value and therefore, in general, has no probability distribution. Although there are several methods of determining confidence limits, we shall use the one which seems to have the widest application, inasmuch as it is a direct approach to the problem.

Let us denote this probability density function by $f(x, \theta)$ and therefore the probability that x will lie between x and $x + dx$ as $f(x, \theta)\,dx$. (Actually, the variable x may be a function of other random variables such as the average of n observations drawn from a particular distribution.) Since we do not know the true value of θ in a particular instance, we must allow for the possibility of its assuming any value which is logically admissible, even though its actual value is unique. Thus both x and θ are defined jointly over a range of possible values, which may or may not be infinite in extent. The limits on x may well be functions of θ itself, such as $a(\theta) \leq x \leq b(\theta)$.

If we fix the parameter θ at a specific value, then a value of x may be chosen, say x_1, so that

$$\int_{a(\theta)}^{x_1} f(x, \theta)\,dx = p_1$$

and a value of x_2 chosen so that

$$\int_{x_2}^{b(\theta)} f(x, \theta)\,dx = p_2$$

where p_1 and p_2 are constants whose sum is less than unity. It is clear, then, that the probability that $x_1 \leq x \leq x_2$ is $1 - p_1 - p_2$. These two integral equations actually give x_1 as a function of θ and p_1 and x_2 as a function of θ and p_2. If we denote these relationships as $x_1 = w_1(\theta, p_1)$ and $x_2 = w_2(\theta, p_2)$ it is now possible to visualize graphically the behavior of these two monotonically increasing functions of the variable θ as in Fig. 4-22. The area between the two curves shown represents what is termed the *confidence belt*, in that no matter what the value of θ is during a single experiment, the probability is $1 - p_1 - p_2$ that a random observation (from that particular population determined by θ) will yield a value of x inside that region.

Now let us start from the opposite direction and assume we have a value of x given and now wish to say something about the likelihood of certain values

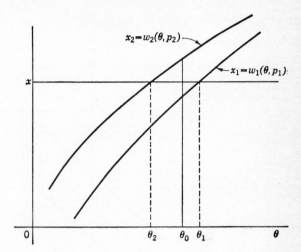

FIGURE 4-22

of θ.   A horizontal line through the given observed point x cuts each of the two curves $x_2 = w_2(\theta, p_2)$ and $x_1 = w_1(\theta, p_1)$ at one point because of the monotonic and continuous character of the curves.   Let these two points have the θ coordinate of θ_2 and θ_1, respectively.   If θ_0 is the true value of the parameter, the interval $\theta_1 - \theta_2$ will include θ_0 provided x lies between $x_1 = w_1(\theta_0, p_1)$ and $x_2 = w_2(\theta_0, p_2)$.   This, however, will occur exactly $1 - p_1 - p_2$ fraction of the time. Thus, if θ_0 is the true value of the parameter, the probability that $\theta_1 - \theta_2$ will overlap θ_0 is also $1 - p_1 - p_2$.   This same probability, however, equally well applies to any value of the parameter which may be the true one that we are estimating.   In the long run, therefore, considered over a large number of experiments in which the value of x is calculated from observed data, and the interval $\theta_1 - \theta_2$ computed according to the above scheme, $1 - p_1 - p_2$ fraction of the time the interval will cover the true value of the parameter which gave rise to the calculated x.   It is important to realize that this statement holds *only* in the *long run*, inasmuch as in a particular instance the point either is covered or not.   In the first instance, the probability is 1 and in the second, zero. There are obviously an infinite number of values of p_1 and p_2 which will give a series of confidence intervals based on the same confidence coefficient $(1 - p_1 - p_2)$.   Intuitively one would feel that the value of p_1 and p_2 which made the length of the confidence intervals as small as possible should be the one preferred.   If the function $f(x, \theta)$ is reasonably symmetrical for a fixed θ, a fairly good approximation to the intervals of shortest length would be obtained by setting $p_1 = p_2$ which is customarily chosen.

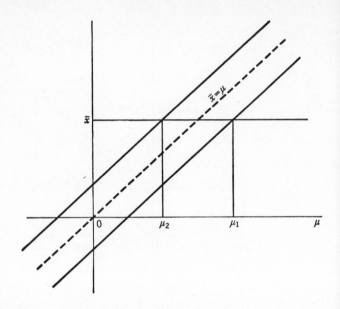

FIGURE 4-23

EXAMPLE 4-14   Given that the distribution of the mean of a sample of size n is normally distributed as $(\sqrt{n}/\sigma\sqrt{2\pi})\exp[-n(\bar{x}-\mu)^2/2\sigma^2]\,(-\infty,\infty)$, what are the 95-percent confidence limits for $\bar{x}$ if σ is known?

For a fixed value of μ, $\bar{x}$ will lie within the limits $\mu \pm (1.96/\sqrt{n})\sigma$ about 95 percent of the time.   Since both σ and n are constant, the functions w_1 and w_2 are two lines parallel to the line $\bar{x} = \mu$ as in Fig. 4-23.   Therefore, for a fixed value of $\bar{x}$ we have the two limits of a confidence interval as $\mu_2 = \bar{x} - 1.96\sigma/\sqrt{n}$ and $\mu_1 = \bar{x} + 1.96\sigma/\sqrt{n}$.   Thus 95 percent of the time, irrespective of the true value of μ, this parameter will be overlapped by this confidence interval calculated on the basis of the observed value of $\bar{x}$, or $P(\mu_2 \le \mu \le \mu_1) = .95$.

////

EXAMPLE 4-15   The percentage of defects is observed in a random sample of size n taken from a large lot.   Assuming that for a fixed value of the parameter p (percentage of defects for the entire lot) the number of defects is binomially distributed in the sample, what are the 50-percent confidence limits in estimating the lot parameter p?

Since for a fixed value of this parameter p a binomial distribution is hypothesized for the sample, the observed p' will lie 50 percent of the time with-

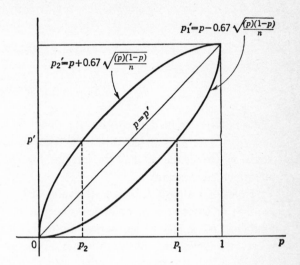

FIGURE 4-24

in the limits $p' = p \pm .67\sqrt{pq/n}$ approximately.   Therefore, $p'_2 = w_2(p, .25) = p + .67\sqrt{(p)(1-p)/n}$ and $p'_1 = w_1(p,.25) = p - .67\sqrt{(p)(1-p)/n}$.   Since n is fixed, these two equations in p and p'_2 and p and p'_1 form the top and bottom of an ellipse (see Fig. 4-24).   The p_1 and p_2 which form the upper and lower confidence limits are thus the two roots of the above quadratic equations and are as follows:

$$\frac{n}{n + (.67)^2}\left[p' + \frac{(.67)^2}{2n} \pm .67\sqrt{\frac{p'q'}{n} + \frac{(.67)^2}{4n^2}}\right] \qquad ////$$

(*Suggested Problems 4-31 to 4-33*)

APPENDIX TO CHAPTER 4

Beta and Gamma Functions

4-13   GAMMA FUNCTIONS

The value of the integral $\int_0^\infty x^{n-1}e^{-x}\,dx$ is dependent upon the value of n and represents a finite quantity for values of n greater than zero.   This particular function of n is called the *gamma function*, written $\Gamma(n)$, and has two common forms:

$$\Gamma(n) = \int_0^\infty x^{n-1} e^{-x} \, dx \qquad n > 0 \qquad (4\text{-}40)$$

$$\Gamma(n) = 2 \int_0^\infty y^{2n-1} e^{-y^2} \, dy \qquad n > 0 \qquad (4\text{-}41)$$

Equation (4-41) can be derived from Eq. (4-40) by making the change of variable $x = y^2$. Actually it makes no difference whether the variable in (4-41) is called x or y, since the value of the integral, for a given n, depends only on the limits which are constant.

Unless n is a positive integer, the value of $\Gamma(n)$ must be obtained by series expansion or some type of numerical integration. For this reason, tables of the gamma function have been worked out for various values of n. Commonly, however, the tables are confined to values of n between 1 and 2, inasmuch as the function can be determined for other values of n by means of a recurrence formula, which we shall now derive. If Eq. (4-40) is integrated by parts, we have

$$\Gamma(n) = \int_0^\infty x^{n-1} e^{-x} \, dx = \left[-x^{n-1} e^{-x} \right]_0^\infty + (n-1) \int_0^\infty x^{n-2} e^{-x} \, ax$$
$$= (n-1)\Gamma(n-1)$$

since the integrated part vanishes at both limits for $n > 1$. Therefore if the gamma function is tabulated for values of n between two consecutive integers, say 1 and 2, the function can be computed for values outside of the table by repeated use of the recurrence formula.

On the other hand, if n is a positive integer, the gamma function of n reduces to the ordinary factorial of $n - 1$ as defined in algebra. To begin with, we determine $\Gamma(1)$ from the defining integral, that is,

$$\Gamma(1) = \int_0^\infty e^{-x} \, dx = 1$$

Then from the recurrence formula we obtain $\Gamma(2)$, $\Gamma(3)$, ..., $\Gamma(n)$ successively as follows:

$$\Gamma(2) = (1)\Gamma(1) = 1 \qquad \Gamma(3) = (2)\Gamma(2) = 2 \times 1$$
$$\Gamma(4) = (3)\Gamma(3) = 3 \times 2 \times 1 \qquad \text{etc.}$$

and thus

$$\Gamma(n) = (n-1)(n-2) \cdots 3 \times 2 \times 1 = (n-1)! \qquad n = 1, 2, 3, \ldots$$

where by definition $0! = 1$.

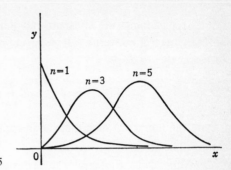

FIGURE 4-25

For any positive value of n, we may interpret the gamma function as the area under the curve $y = x^{n-1}e^{-x}$ from $x = 0$ to $x = \infty$. A plot of these curves for various n's is shown in Fig. 4-25. The areas under these curves are given in Fig. 4-26 and thus show the plot of $\Gamma(n)$ as a function of n.

An important point in Fig. 4-26 corresponds to $n = \frac{1}{2}$. To find its value, consider

$$\Gamma(\tfrac{1}{2}) = 2 \int_0^\infty e^{-x^2}\, dx \qquad \text{or} \qquad \Gamma(\tfrac{1}{2}) = 2 \int_0^\infty e^{-y^2}\, dy$$

where the definition (4-41) for the gamma function is being utilized. By multiplying these two values of $\Gamma(\frac{1}{2})$ together, we obtain

$$[\Gamma(\tfrac{1}{2})]^2 = \left(2 \int_0^\infty e^{-x^2}\, dx\right)\left(2 \int_0^\infty e^{-y^2}\, dy\right)$$
$$= 4 \int_0^\infty dx \int_0^\infty e^{-(x^2 + y^2)}\, dy$$

This last expression may be considered a double integral to be evaluated over the first quadrant in the XOY plane. The above procedure is permitted since

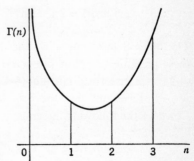

FIGURE 4-26

both sets of limits are constant and both integrals as well as the product or double integral converge. If polar coordinates are now introduced, we have, since $dx\,dy = r\,d\theta\,dr$,

$$[\Gamma(\tfrac{1}{2})]^2 = 4 \int_0^{\pi/2} d\theta \int_0^\infty e^{-r^2} r\,dr = 4\,\frac{\pi}{2}\left[\frac{e^{-r^2}}{-2}\right]_0^\infty = \pi$$

where the limits are such that the integral is still evaluated over the first quadrant. Therefore $\Gamma(\tfrac{1}{2}) = \sqrt{\pi}$.

Many definite integrals can be evaluated by means of the gamma function either because they are of the gamma form or because they can be transformed into that form. A few illustrations follow with the appropriate transformation

$$\int_0^\infty x^{4/5} e^{-x}\,dx = \Gamma(\tfrac{9}{5})$$

$$\int_0^\infty x^2 e^{-2x}\,dx = \tfrac{1}{8}\int_0^\infty y^2 e^{-y}\,dy = \tfrac{1}{8}\Gamma(3) = \tfrac{1}{4} \qquad 2x = y$$

$$\int_0^\infty x^9 e^{-x^3}\,dx = \tfrac{1}{3}\int_0^\infty y^{7/3} e^{-y}\,dy = \tfrac{1}{3}\Gamma(\tfrac{10}{3}) = {}^{28}\!\!\big/\!_{27}\Gamma(\tfrac{4}{3}) \qquad x^3 = y$$

$$-\int_0^1 (\ln x)^{1/5}\,dx = -\int_\infty^0 y^{1/5} e^{-y}\,dy = \int_0^\infty y^{1/5} e^{-y}\,dy = \Gamma(\tfrac{6}{5}) \qquad \ln x = -y$$

4-14   BETA FUNCTIONS

A second useful type of integral in the field of operations research and statistics involves two parameters and is called the *beta function*. It can be defined as follows:

$$\beta(m, n) = \int_0^1 x^{m-1}(1-x)^{n-1}\,dx \qquad\qquad m > 0, n > 0 \qquad (4\text{-}42)$$

$$\beta(m, n) = 2 \int_0^{\pi/2} \sin^{2m-1}\theta\,\cos^{2n-1}\theta\,d\theta \qquad m > 0, n > 0 \qquad (4\text{-}43)$$

The integral has a specific value for any given choice of m and n, provided these quantities are greater than zero; otherwise the integral diverges. Equation (4-42) is transformed into Eq. (4-43) by the change of variable $x = \sin^2\theta$.

It turns out that the beta function is related to the gamma function. Consider the following integrals:

$$\Gamma(m) = 2\int_0^\infty y^{2m-1} e^{-y^2}\,dy \qquad \Gamma(n) = 2\int_0^\infty x^{2n-1} e^{-x^2}\,dx$$

If these two integrals defining gamma functions of m and n are multiplied together, we have, since the appropriate conditions are satisfied,

$$\Gamma(m)\Gamma(n) = 4 \int_0^\infty dx \int_0^\infty x^{2n-1} y^{2m-1} e^{-(x^2+y^2)} \, dy$$

As before, we may introduce polar coordinates and again evaluate the integral over the first quadrant:

$$\Gamma(m)\Gamma(n) = \left(2 \int_0^{\pi/2} \cos^{2n-1}\theta \sin^{2m-1}\theta \, d\theta\right)\left(2 \int_0^\infty r^{2(m+n)-1} e^{-r^2} \, dr\right)$$

Utilizing our definition of both beta and gamma functions,

$$\Gamma(m)\Gamma(n) = \beta(m, n)\Gamma(m + n)$$

and therefore

$$\beta(m, n) = \frac{\Gamma(m)\Gamma(n)}{\Gamma(m + n)} \qquad (4\text{-}44)$$

From this it is obvious that the beta function is symmetrical in the parameters m and n, that is,

$$\beta(m, n) = \beta(n, m)$$

The beta function is the area under the curve $y = x^{m-1}(1 - x)^{n-1}$ between the values of $x = 0$ and $x = 1$. Not only is this total area known, but tables are available for the percentage area included between 0 and x, where $0 < x < 1$, and these computational aids make the curves very useful for the representation of simple mathematical models considered in later chapters. The effect of the parameters m and n on the shapes of these curves may be studied by dividing the ordinates by $\beta(m, n)$, thus making the total area equal to unity in all cases. The order of contact of the curves to the x axis at 0 and 1 is determined by the value of m and n and increases with increasing values of these parameters. When m and n are equal, the curves are symmetrical about $x = \frac{1}{2}$. The effect of assigning a value of m between 0 and 1 is to cause the curve to rise to infinity with $x = 0$ as an asymptote, although the area under the curve is still finite. A similar situation exists at $x = 1$ if n is allowed to take on a value between 0 and 1. Some of these effects are shown in Fig. 4-27a and b, where m and n are varied for purposes of illustration, but the total area is kept equal to unity in each case. A wide variety of shapes can thus be obtained with these curves, provided only one maximum is desired. When two or more maxima are required by the mathematical models being considered, several of these curves can be added together in some predetermined proportion in order to give the desired effect.

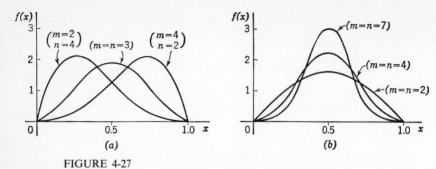

FIGURE 4-27

Graphs of $f(x) = \dfrac{1}{B(m,n)} x^{m-1}(1-x)^{n-1}$.

It should be observed from the following illustrative examples that the range for x from 0 to 1 is not essential, since a simple linear transformation changes the origin and the scale in such a manner that the curve will vanish at any other two points in lieu of 0 and 1. The following examples illustrate the use of the beta function in evaluating definite integrals. Where substitutions are needed, they are shown in parentheses.

$$\int_0^1 x^3(1-x)^4 \, dx = B(4,5) = \frac{\Gamma(4)\Gamma(5)}{\Gamma(9)} = \frac{1}{280}$$

$$\int_0^{\pi/2} \tan^{1/2} x \, dx = \int_0^{\pi/2} \sin^{1/2}x \cos^{-1/2} x \, dx = \tfrac{1}{2}B(\tfrac{3}{4}, \tfrac{1}{4})$$

$$= \frac{1}{2} \frac{\Gamma(\tfrac{3}{4})\Gamma(\tfrac{1}{4})}{\Gamma(1)} = \frac{8}{3} \Gamma(\tfrac{7}{4})\Gamma(\tfrac{5}{4})$$

$$\int_0^1 \frac{x^2 \, dx}{\sqrt{1-x^4}} = \tfrac{1}{4} \int_0^1 y^{-1/4}(1-y)^{-1/2} \, dy \qquad (x^4 = y)$$

$$= \tfrac{1}{4}B(\tfrac{3}{4}, \tfrac{1}{2}) = \frac{1}{4} \frac{\Gamma(\tfrac{3}{4})\Gamma(\tfrac{1}{2})}{\Gamma(\tfrac{5}{4})} = \frac{\sqrt{\pi}}{3} \frac{\Gamma(\tfrac{7}{4})}{\Gamma(\tfrac{5}{4})}$$

$$\int_{-2}^3 (2+x)^2(x-3)^3 \, dx = \int_0^5 u^2(u-5)^3 \, du \qquad (u = 2+x)$$

$$= 5^6 \int_0^1 y^2(y-1)^3 \, dy \qquad (u = 5y)$$

$$= -5^6 B(3,4) = \frac{-5^6}{60} = \frac{-15,625}{60} \approx -260.4$$

PROBLEMS

4-1 Points are uniformly distributed inside a circle of radius R. (*a*) Find the distribution function $F(x)$ for the distribution of the distance of interior points from the center. (*b*) For what value of x are the chances even that a random interior point will be nearer to the center than that distance or farther away?

4-2 Points are uniformly distributed along the periphery of a semicircle of radius R. (*a*) Find the distribution function $F(x)$ for the distribution of the shortest distance x from one end of the diameter to random points of the periphery. (*b*) Find $P(x \le R)$. (*Hint:* What angle is inscribed in a semicircle?) *Ans.* (*b*) $P = \frac{1}{3}$.

4-3 Points are equally dense along the line $y = x - 1$ from the point $(1, 0)$ to $(3, 2)$. What is the probability density of the distance z from an arbitrary point on that segment to the y axis? *Ans.* $f(z) = \frac{1}{2}$ $(1, 3)$.

4-4 Points are uniformly distributed inside the area bounded by $y = 1$, $y = 2x$, and $y = -x$. What are the distribution function and the density function of the shortest distance from the line $x = 1$ to a randomly chosen point in the interior of this area? *Ans.* $F(z) = \frac{4}{3}(z - \frac{1}{2})^2$ $(\frac{1}{2}, 1)$, $F(z) = 1 - \frac{2}{3}(z - 2)^2$ $(1, 2)$.

4-5 Given that $f(x) = 3x^2$ $(0 \le x \le 1)$, what is the probability that $\frac{1}{2} \le x \le \frac{3}{4}$?

4-6 A density function is said to be symmetrical about zero when $f(-x) = f(x)$. If a is any positive value of x, derive the following results, using geometric intuition freely:

(*a*) $F(0) = .5$

(*b*) $P(x > a) = P(x \ge a) = .5 - \int_0^a f(x)\,dx$

(*c*) $F(-a) + F(a) = 1$

(*d*) $P(-a < x < a) \equiv P(-a \le x \le a) = 2F(a) - 1$

(*e*) $P(|x| > a) \equiv P(|x| \ge a) = 2F(-a) = 2[1 - F(a)]$

4-7 A distribution is said to be unimodal if its density function has only one turning point, and this is a maximum value of the ordinate; the abscissa of the maximum is called the *mode*. In Probs. 4-7a to 4-7e, show that the distributions are unimodal and find the mode in each case. Also determine k in each case so that the total area will be unity.

(*a*) $f(x) = kx^4 e^{-x}$ $(0, \infty)$ *Ans.* $k = \frac{1}{24}$, mode $= 4$.

(*b*) $f(x) = kxe^{-ax}$ $a > 0, 0 \le x < \infty$ *Ans.* $k = a^2$, mode $= 1/a$.

(*c*) $f(x) = kxe^{-x^2}$ $(0, \infty)$ *Ans.* $k = 2$, mode $= 1/\sqrt{2}$.

(*d*) $f(x) = kx(a - x)^2$ $(0, a)$ *Ans.* $k = 12/a^4$, mode $= a/3$.

(*e*) $f(x) = k(ax + x^2)e^{-bx}$ $a, b > 0, 0 \le x < \infty$

4-8 Given that X is a random variate with probability density

$$f(x) = \begin{cases} \frac{1}{3} & 2 \le x \le 4 \\ \frac{1}{12} & 8 \le x \le 12 \\ 0 & \text{otherwise} \end{cases}$$

What is the probability that the smallest observation in a total of four drawn independently at random from this population is less than 9? *Ans.* $1 - (\frac{1}{4})^4$.

4-9   The probability density function of a random variable x is $f(x) = 1$ $(0 \leq x \leq 1)$. The probability density function of a random variable y is $g(y) = \frac{1}{2}$ $(0 \leq y \leq 2)$. An observation is drawn from each density function simultaneously. What is the probability that the smaller of the two observations is less than $\frac{1}{2}$?

Ans. $\frac{5}{8}$.

4-10 The *median* of a variate x which has a continuous distribution function $F(x)$ is defined as that value of c such that $F(c) = \frac{1}{2}$. If x has a density function $f(x)$ and the lower and upper limits of x are A and B, respectively, show that the median satisfies the following equation:

$$\int_A^c f(x)\, dx = \int_c^B f(x)\, dx$$

4-11 The variate x is distributed as follows:

$$f(x) = \frac{1}{\sigma\sqrt{2\pi}}\, e^{-(x-\mu)^2/2\sigma^2} \qquad \sigma > 0,\ -\infty < x < \infty$$

(a) Show that the total area is a unity. (b) Determine the points of inflection. (c) Locate the median and mode by inspection. (d) Sketch the curve.

4-12 The variate x is distributed as follows:

$$f(x) = \frac{k}{a^2 + (x - b)^2} \qquad (-\infty,\ \infty)$$

(a) Determine k so that the total area will be unity. (b) Sketch the curve. (c) Locate the median and mode by inspection, checking the former by integration and the latter by differentiation.

4-13 A random observation x_1 is made of the variate x, the distribution of which is

$$f(x) = a^2 x e^{-ax} \qquad a > 0;\ 0 \leq x < \infty$$

(a) What is the probability that x_1 will be less than the mode? (b) More than twice the mode? *Ans.* (a) .264. (b) .406.

4-14 The random variable x is distributed as follows:

$$f(x) = \frac{6x}{(1 + x)^4} \qquad (0,\ \infty)$$

(a) Show that the total area is unity and find the distribution function $F(x)$. (b) Find $P(x \leq \frac{1}{4})$, (c) $P(x \geq 4)$, (d) $P(1 \leq x \leq 2)$.

4-15 Given

$$P(x \geq a) = \frac{1}{(1 + a/k)^k} \qquad k > 0;\ 0 \leq a < \infty$$

(a) Find $F(x)$ and $f(x)$. (b) Plot $F(x)$ and $f(x)$ if $k = 2$. (c) If three observations are taken at random and $k = 2$, what is the probability that all three will exceed 4? (d) What is the probability that one will lie between 0 and 1, another between 2 and 3, and another between 4 and 5? *Ans.* (c) $\frac{1}{729}$, (d) $\frac{13}{1470}$.

4-16 To an adequate approximation, the distribution of length of life (period of satisfactory service) of a certain manufactured product is given by $f(x) = 2xe^{-x^2}$ $(0 \leq x < \infty)$. (a) To three decimals, find the value of a such that $P(x \geq a) = .99$. (b) If 1,000 articles are sold, what is the probability that the life of at least one will be less than this value a? *Ans.* (a) .100, (b) .999956.

4-17 Assume that annual income is a gamma variate. If the average family income is $10,000 and 10 percent of all families have an income of less than $4,120, how many have incomes greater than $20,000? *Ans.* About 10%.

4-18 The rush-hour traffic demand (in thousands) along a freeway is a gamma variate with $\alpha = 6, \beta = 3$. If a demand of 31.5 thousand cars causes a traffic jam, what are the chances of being caught in one this afternoon? *Ans.* About .10.

4-19 In months, the distribution of business life of a certain type of enterprise is exponential, the risk of failure in the early stages being fairly great. (a) What is the value of β if the first decile is 7.56? (b) If a man rents buildings to two such companies, and independence may be assumed, what is the median value of the total rent-months where, from the revenue standpoint, concurrent months are regarded as distinct? *Ans.* (a) 72, (b) 120.82.

4-20 The distribution of x is known to be gamma with $\beta = 1$ but unknown α. A random observation of x turns out to have the value 7.5. (a) For what value of α would this value of x correspond to a fractile of .99? (b) Of .01? These two values of α are called "98-percent confidence limits."
Ans. (a) 1.5, (b) 14.0.

4-21 Under the same conditions as Prob. 4-20, suppose that, instead of one observation, three are drawn at random, and their sum S is computed. (a) If $S = 19.5$ what are the 98-percent confidence limits for α in the distribution of S? (b) What are the corresponding limits for α in the distribution of x? (Assume validity of Theorem 4-2 for nonintegral α's.)

4-22 Given an exponential distribution, show that the distribution of x such that $x \geq a \geq 0$ is also exponential with the same value of β; that is show that $F(x + a)$, given that $x \geq a$, is equal to $F(x)$, given that $x \geq 0$. What does this imply concerning replacement policy for tubes with an exponential decay rate?

4-23 Customers use a multiparty telephone line at random. The length of any call is an exponentially distributed random variable with density function $f(x) = \frac{1}{2}e^{-x/2}$. The time between the start of one call and the time when another customer wishes to place a call is another random variable and has the density function $g(y) = \frac{1}{4}e^{-y/4}$. What is the probability that a person making a call will not complete his conversation before another customer picks up his phone to place another call? *Ans.* $\frac{2}{3}$.

4-24 Using the Wilson-Hilferty transformation, estimate the ninth decile of the gamma distribution with $\alpha = 24, \beta = 1$. Check by linear interpolation in Table 4-1.

4-25 The average number of years at school, $\bar{x}$, of a random sample of n people from a certain company is normally distributed with $\mu = 12$, $\sigma = 3/\sqrt{n}$. How large a sample should we draw in order that the probability is at least .90 that $\bar{x}$ is between 11 and 13? *Ans.* $n = 25$.

4-26 Air-conditioning units in an office building will operate without malfunction for x months without servicing, where x is normally distributed with $\mu = 13.5$, $\sigma = 1.5$. All n units will be cleaned and serviced at a unit cost of c at the same time each year; those that fail before this annual servicing will be repaired at a unit cost of $10c$. (*a*) How many units will fail between servicings? (*b*) Would you recommend lengthening or shortening the period between servicings? What savings would result from a one-month change in service time?

4-27 An event has a probability of success of $p = .2$ in 50 trials at this event. What is the probability of 15 or 16 successes; (*a*) binomial (arithmetic unnecessary); (*b*) by Poisson approximation (arithmetic unnecessary); (*c*) by normal approximation? *Ans.* (*c*) About .05.

4-28 The height of high-school-age males in a city is normally distributed with $\mu = 72$, $\sigma = 2$. What is the probability that the centers on the city's two rival basketball teams will each be six-foot-six or taller? One school has 385 boys; the other has 310. (Assume that the tallest boy in each school is the center.)

4-29 Suppose $p = .2$ is the probability of success in a single trial. Estimate the total probability of obtaining less than 6 or more than 15 successes in 50 trials.

4-30 Given a Poisson distribution with $\mu = 25$, estimate $P(x > 35)$. Use $\sigma = 5$.

4-31 In Chap. 6 it is proved that if x is normally distributed with parameters (μ, σ) then $\bar{x}$, the average of n independent observations of x, is also normally distributed with corresponding parameters $(\mu, \sigma/\sqrt{n})$. This result is not merely approximate in this case, but exact. Given a normal population with $\sigma = 1$, plot the distance between the lower and upper 98-percent confidence limits for μ as a function of the sample size n on which the sample means $\bar{x}$ is based.

4-32 The breaking strength (in pounds) of string manufactured by an established process is normally distributed with $\mu = 15$, $\sigma = 3$. A new process is tried, and a random sample of nine pieces of standard length have an average breaking strength of 18. Regarding the new breaking strengths as belonging to another normal population, with σ the same as before, find 98-percent confidence limits for μ. *Ans.* 15.7 *to* 20.3.

4-33 A certain probability distribution is a function of a single parameter a. If the value of a is fixed and an extremely large number of observations are made on this population, 95 percent of them will always lie in the region $a \pm (a - 1)$ $(3 - a)$. It is known that all the possible values of the parameter a are between $1 < a < 3$. Construct the 95-percent confidence belt. If an observation is made on the population experimentally, it is found that $a' = 1.5$. What is the confidence interval in this case and exactly what is meant by that figure? *Ans.* $(5 - \sqrt{7})/2$ to $(3 + \sqrt{3})/2$.

JOINT DISTRIBUTIONS

5-1 FUNDAMENTAL CONCEPTS

So far we have been concerned with distributions of single variates. These are called *univariate* distributions to distinguish them from *bivariate* distributions, which involve two variates simultaneously, or *multivariate* distributions, which are generally understood to involve three or more variates simultaneously. The term "multivariate" may also be used broadly to designate an arbitrary number of variates and in that sense would include univariate and bivariate distributions as special cases. Since natural phenomena are characterized by the coexistence of many attributes and dimensions, it is often necessary to consider multivariate distributions in representing the relevant conditions of practical problems. Once the conceptual step is taken from univariate to bivariate distributions, the transition to higher dimensions involves nothing essentially new. Therefore, let us consider some concrete examples of bivariate distributions.

EXAMPLE 5-1 The team for a competitive game consists of two players, each of whom can score 0, 1, or 2 points. Player 1 has the probability .2 of scoring 0, .5 of scoring 1, and .3 of scoring 2, and for player 2 the corresponding probabilities are .1, .4, .5.   The problem is to derive the joint distribution of the scores of these two teammates, assuming their performances are independent. Let x_1 denote any possible score of player 1 and $f_1(x_1)$ his probability of gaining that score; also let x_2 denote any possible score of player 2 and $f_2(x_2)$ the corresponding probability.   This much is univariate material; we now introduce the bivariate aspect by letting $f(x_1, x_2)$ denote the probability of the compound event that player 1 scores x_1 points and in the same game player 2 scores x_2 points.   By the assumption of independence we have

$$f(x_1, x_2) = f_1(x_1)f_2(x_2)$$

Thus, the probability that both players score 0 is

$$f(0, 0) = f_1(0)f_2(0) = (.2)(.1) = .02$$

Again, the probability that player 1 scores 0 and player 2 scores 1 is

$$f(0, 1) = f_1(0)f_2(1) = (.2)(.4) = .08$$

and so on.   Table 5-1 presents the complete bivariate distribution, comprising nine distinct possibilities together with their associated probabilities.   The margins of the table contain the sums of the tabular entries $f(x_1, x_2)$ by rows (horizontal) and columns (vertical).   Each row total yields precisely the probability $f_1(x_1)$ that player 1 scores x_1 points, and each column total yields the probability that player 2 scores x_2 points.   Because of the fact that the single distributions are thus displayed in the margins of the bivariate table, single distributions have come to be called *marginal* distributions.

Supposing now that the score for the team is the sum of the scores gained by the two players, let us find the distribution of the team score t.   Clearly the possible values of t are 0, 1, 2, 3, 4 and the probability $p(t)$ of any particular

Table 5-1   BIVARIATE DISTRIBUTION OF SCORES

x_1 \ x_2	0	1	2	Sum $= f_1(x_1)$
0	.02	.08	.10	.20
1	.05	.20	.25	.50
2	.03	.12	.15	.30
Sum $= f_2(x_2)$	.10	.40	.50	1.00

value is equal to the sum of the probabilities of the distinct ways in which that combined score can be obtained. Hence

$$p(0) = f(0, 0) = .02$$

$$p(1) = f(0, 1) + f(1, 0) = .08 + .05 = .13$$

$$p(2) = f(0, 2) + f(1, 1) + f(2, 0) = .10 + .20 + .03 = .33$$

$$p(3) = f(1, 2) + f(2, 1) = .25 + .12 = .37$$

$$p(4) = f(2, 2) = .15$$

In this way, we arrive at the distribution of the team score t as shown in Table 5-2. ////

A joint distribution can be synthesized from the marginal distributions only if the variates are independent. The joint distribution of dependent variates can be derived only when there is sufficient mathematical information to determine the compound probabilities either directly or by analytic deduction; without such information, one must resort to empirical estimates based on relative frequencies of compound events. A bivariate distribution with dependent coordinates is illustrated in the next example.

EXAMPLE 5-2 Companies A and B control, respectively, 50 and 20 percent of the market for a certain commodity. The problem is to derive the bivariate distribution of the numbers of customers patronizing each company in random samples of two buyers. The symbol (x_1, x_2) will represent the event that x_1 of the two are customers of A, and x_2 are customers of B; also $f(x_1, x_2)$ will represent the probability of this event. Since the two companies together sell to only 70 percent of the market, the remaining 30 percent goes to neither. Letting N denote neither A nor B, the possible allocations of two buyers are shown in Table 5-3, wherein compound probabilities are computed on the

Table 5-2 DISTRIBUTION OF TEAM SCORE

t	0	1	2	3	4	Sum
$p(t)$	.02	.13	.33	.37	.15	1.00

Table 5-3

Allocation Buyer 1	Buyer 2	Compound probability of this allocation	Corresponding bivariate event	Total probability of bivariate event
N	N	$(.3)(.3) = .09$	$(0, 0)$	$f(0, 0) = .09$
N	B	$(.3)(.2) = .06$	$(0, 1)$	
B	N	$(.2)(.3) = .06$	$(0, 1)$	$f(0, 1) = .06 + .06$ $= .12$
B	B	$(.2)(.2) = .04$	$(0, 2)$	$f(0, 2) = .04$
N	A	$(.3)(.5) = .15$	$(1, 0)$	
A	N	$(.5)(.3) = .15$	$(1, 0)$	$f(1, 0) = .15 + .15$ $= .30$
B	A	$(.2)(.5) = .10$	$(1, 1)$	
A	B	$(.5)(.2) = .10$	$(1, 1)$	$f(1, 1) = .10 + .10$ $= .20$
A	A	$(.5)(.5) = .25$	$(2, 0)$	$f(2, 0) = .25$

assumption that sampling depletions may be ignored.   The bivariate distribution itself is presented compactly in Table 5-4.   As in the previous example, the row and column totals yield the marginal distributions, for even though the variates are not independent, the separate combinations along a row or column represent the exhaustive and mutually exclusive ways in which a fixed value of one of the variates can occur.   ////

In Chap. 1 the conditional probability of an event B, given the occurrence of another event A, was obtained as the ratio of the compound probability of $(A \cap B)$ to the unconditional probability of A; that is, $P(B|A) = P(A \cap B)/P(A)$. Similarly, the conditional probability that a discrete variate x_2 takes on a certain

Table 5-4   BIVARIATE DISTRIBUTION OF CUSTOMERS

x_1 \ x_2	0	1	2	Sum $= f_1(x_1)$
0	.09	.12	.04	.25
1	.30	.20	0	.50
2	.25	0	0	.25
Sum $= f_2(x_2)$	.64	.32	.04	1.00

admissible value b when another discrete variate x_1 takes on a specific value a is given by

$$\phi(b\,|\,a) = \frac{f(a, b)}{f_1(a)}$$

Holding x_1 fixed at the value a but letting x_2 vary over all of its (conditionally) admissible values, we obtain what is known as the *conditional distribution* of x_2, given that $x_1 = a$. It should be noted that the conditional distribution summed for all possible values of x_2 always equals unity. Inasmuch as a, in turn, can represent any admissible value of x_1, we represent the conditional-probability function of x_2 by the symbol $\phi(x_2\,|\,x_1)$ for general reference, or more specifically by $\phi(x_2\,|\,x_1 = a)$ when a particular value a of x_1 is intended. The conditional distributions of x_2 for successive choices of x_1 in Example 5-2 are shown in Table 5-5. For example, $\phi(1\,|\,x = 1) = .20/.50 = .40$. An advantage of the conditional distribution is the comparative ease with which it often can be derived by elementary changes of variable or deduced from physical reasoning.

Summarizing the fundamental concepts: The joint distribution of *independent* variates can be synthesized from the respective marginal distributions; but for *dependent* variates, either the joint or conditional distribution must be known in advance or derivable from the given information. For discrete variates, the marginal probabilities may be obtained by summing the appropriate joint probabilities, and conditional probabilities may be obtained by dividing joint probabilities by associated marginal probabilities; in turn, the joint distribution can be constructed if the marginal and conditional distributions are known.

Geometric language is helpful in discussing multivariate distributions. The simultaneously admissible values of n variates may be thought of as cutting out a region in n-dimensional space, the coordinate axes of which represent the respective variates. We shall develop this idea as we proceed.

(*Suggested Problems 5-1 to 5-9*)

Table 5-5 CONDITIONAL DISTRIBUTIONS

	Value of x_2			
	0	1	2	Sum
$\phi(x_2\,|\,x_1 = 0)$	.36	.48	.16	1.00
$\phi(x_2\,|\,x_1 = 1)$	.60	.40	0	1.00
$\phi(x_2\,|\,x_1 = 2)$	1.00	0	0	1.00

5-2   MULTIVARIATE-DISTRIBUTION FUNCTIONS AND DENSITY FUNCTIONS

In the joint distribution of several random variables x_1, x_2, x_3, ... the joint-distribution function $F(a, b, c, ...)$ yields the probability that the inequalities $x_1 \leq a$, $x_2 \leq b$, $x_3 \leq c$, ... are simultaneously satisfied.   As suggested by the univariate case, the general notation employed for the joint-distribution function is $F(x_1, x_2, x_3, ...)$ inasmuch as the symbols $a, b, c, ...$ merely represent arbitrary values of the given variates.   Because of its definition, $F(x_1, x_2, x_3, ...)$ must be a monotonic function in all the variables with $F(-\infty, -\infty, ..., -\infty) = 0$ and $F(\infty, \infty, ..., \infty) = 1$.

Several random variables are *mutually independent* provided that their joint-distribution function factors identically into the product of their separate (i.e., marginal) distribution functions, as

$$F(x_1, x_2, ..., x_n) = F_1(x_1)F_2(x_2) \cdots F_n(x_n) \qquad (5\text{-}1)$$

We shall take this equation as the general definition of independence for random variables.   It is, of course, compatible with the definition previously laid down for qualitative events, but we shall not go through the algebraic manipulation necessary to establish the connection.

The density concept introduced for continuous univariate distributions can be extended to any number of dimensions, and although the joint-distribution function has the mathematical advantage of complete generality, the joint density function, when it exists, is much easier to work with from the elementary viewpoint.   If the joint-distribution function is continuous and differentiable, the joint density function may be defined as its nth partial derivative wherein each variate in turn is taken once:

$$f(x_1, x_2, ..., x_n) = \frac{\partial^n}{\partial x_1 \, \partial x_2 \cdots \partial x_n} [F(x_1, x_2, ..., x_n)] \qquad (5\text{-}2)$$

The computation of the density function from the distribution function $F(x, y)$ may be visualized in two dimensions by reference to Fig. 5-1.   The probability that x lies between x and $x + \Delta x$ at the same time that y lies between y and $y + \Delta y$ is the change in $F(x, y)$ across the shaded rectangle which has the value

$$F(x + \Delta x, y + \Delta y) - F(x + \Delta x, y) - F(x, y + \Delta y) + F(x, y)$$

We observe that $F(x, y)$ is added in as a positive quantity, as the probability that the random variable is less than x and y simultaneously was subtracted out twice, since it was contained in both the second and third terms of the above.   If this probability is now divided by $\Delta x \, \Delta y$, we have the probability

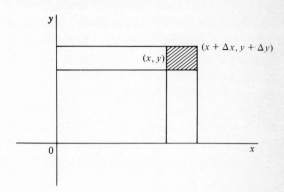

FIGURE 5-1

per unit area, which is the probability density if we take the limit as $\Delta x \to 0$ and $\Delta y \to 0$ independently. If we start with $F(x, y)$ and calculate the change in $F(x, y)$ by increasing x by Δx and dividing the result by Δx, we have by definition

$$\frac{\partial F}{\partial x} = \frac{F(x + \Delta x, y) - F(x, y)}{\Delta x}$$

$$\Delta x \to 0$$

If the change in the numerator of this expression is computed (before the limit is taken) by increasing y by Δy and dividing this change by Δy, we have the definition of the $\partial^2 F / \partial x \, \partial y$ as Δx and Δy both approach zero independently; this expression is identical with that for the probability per unit area.

Although Eq. (5-2) constitutes a formal definition, in this book the joint density function itself will be taken as primary information or else derived under stated hypotheses, without first explicitly solving for the corresponding distribution function. The meaning of the joint density function is analogous to that of the univariate density function; that is, the expression

$$f(x_1, x_2, \ldots, x_n) \, dx_1 \, dx_2, \ldots, dx_n$$

represents the probability that a random point $x_1', x_2', \ldots, x_n'$ will fall within an infinitesimal region such that $x_1 < x_1' < x_1 + dx_1$; $x_2 < x_2' < x_2 + dx_2$; $\ldots$; $x_n < x_n' < x_n + dx_n$. The total integral of any joint density function over the entire region for which it is defined is necessarily equal to unity.

By taking the appropriate nth partial derivatives of both sides of (5-1), we see at once that if the variates are independent, their joint density function factors identically into the product of their marginal density functions:

$$f(x_1, x_2, \ldots, x_n) = f_1(x_1) f_2(x_2) \cdots f_n(x_n) \qquad (5\text{-}3)$$

From this fact it follows that the joint density function of independent variates can be constructed by taking the product of their respective marginal density functions, just as the joint-probability function of independent discrete variates can be analogously constructed. Equation (5-3) will serve as an alternative in the case of independence, applicable to continuous variates for which the joint density function exists and also formally applicable to discrete variates with the understanding that f then represents the discrete-probability function.

From now on, for the sake of simplicity of exposition, we shall concentrate on bivariate distributions, although general results will be stated where feasible. The reader can easily work out the n-dimensional analogies if he understands the bivariate case, particularly with the aid of Chap. 1.

5-3 ADMISSIBLE REGIONS

The mathematical specification of a joint distribution can be expressed either in terms of the joint-distribution function or the joint density function, but the region of definition must always be stated to make the specification complete. One way of defining the admissible region is to describe its geometric boundaries. For example, we might say: "The admissible region of x and y consists of the triangle (Fig. 5-2) bounded by the lines $y = 0$, $y = x$, and $x = 1$." Or in another case we might say: "The admissible region of u and v (Fig. 5-3) lies in the first quadrant and consists of that infinite wedge bounded below by the hyperbola $v = 1/u$ and above by the straight line $v = u$." Verbal descriptions such as these are sometimes necessary for clarity, but if the geometric figure is simple enough, it often happens that the admissible region can be defined concisely by stating the extreme limits of one variate and then stating the functional limits of the other for an arbitrary value of the first. For instance, we may represent the triangular region of Fig. 5-2 by the analytic specification

$$0 \le x \le 1 \qquad 0 \le y \le x$$

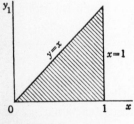

FIGURE 5-2

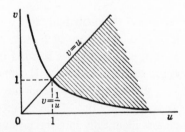

FIGURE 5-3

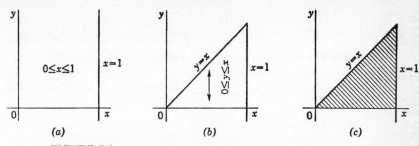

FIGURE 5-4
Steps in the visualization of an admissible region. (a) Outside limits; (b) complementary limits; (c) admissible region.

This compact specification is interpreted to mean that the outside limits of the figure are the lines $x = 0$ and $x = 1$; within these limits, the figure is further bounded by the lines $y = 0$ and $y = x$. The separate steps in the interpretation are indicated graphically in Fig. 5-4. For the same region, we could have written, equally well, the alternative specification

$$0 \leq y \leq 1 \qquad y \leq x \leq 1$$

for which the steps in the corresponding graphical interpretation are indicated in Fig. 5-5.

The infinite-wedge-shaped region of Fig. 5-3 can be defined by the concise specification

$$1 \leq u < \infty \qquad \frac{1}{u} \leq v \leq u$$

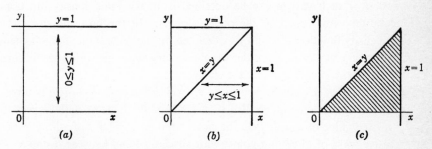

FIGURE 5-5
Visualization of alternative construction. (a) Outside limits; (b) complementary limits; (c) admissible region.

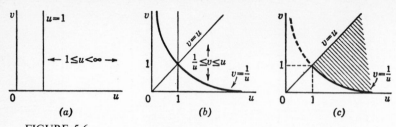

FIGURE 5-6
(*a*) Outside limits; (*b*) complementary limits; (*c*) admissible region.

for which the graphical interpretation is shown in Fig. 5-6. Here, however, in order to interchange the roles of u and v, as we did with x and y in the previous example, we would need two statements:

$$0 \le v \le 1; \frac{1}{v} \le u < \infty \quad \text{and} \quad 1 \le v < \infty; v \le u < \infty$$

Thus the specification may be simpler from one viewpoint than another, and we shall ordinarily choose the simpler one. Where geometrically feasible, we shall describe the admissible region by stating constant limits for one variate, followed by corresponding limits for the second when the first assumes an arbitrary value within those constant limits; it is always understood that the density is identically zero outside of the admissible region. With a little practice, the reader will rapidly acquire facility in the interpretation of compact representations of the admissible region.

5-4   CONTINUOUS BIVARIATE DISTRIBUTIONS

As would be expected from analogy with discrete variates, the *marginal* density function of each variate can be obtained from the joint density function by integration with respect to the other variates. In the bivariate case, let the joint density function of x and y be denoted by $f(x, y)$ and the marginal density function of x by $f_1(x)$. Consider the elementary rectangle which has its corners located at the points (x, y), $(x + dx, y)$, $(x, y + dy)$, $(x + dx, y + dy)$. The probability that a random point (x', y') will fall in this rectangle is given by $f(x, y)\, dx\, dy$, and the integral of such probability elements with respect to y, for a fixed value of x, will yield the sum of the probabilities of all the mutually exclusive ways of obtaining points with abscissas lying between x and $x + dx$. Now, unless $f(x, y)$ is defined for the entire plane, the admissible region will be bounded by lines or curves, and the integral with respect to y will be taken from

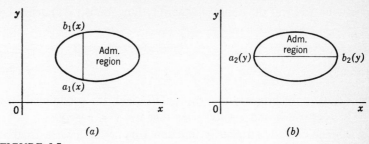

FIGURE 5-7
(a) Path of integration for marginal density of x; (b) path of integration for marginal density of y.

the bottom to the top of the vertical strip (Fig. 5-7a) corresponding to the fixed value of x. The lower limit of y, ordinarily a point on a curve, will be expressed in terms of x as $a_1(x)$ and the upper limit expressed as $b_1(x)$. Consequently,

$$P(x \leq x' \leq x + dx) = \left[\int_{a_1(x)}^{b_1(x)} f(x, y) \, dy \right] dx = f_1(x) \, dx \qquad (5\text{-}4)$$

whence, equating coefficients of dx,

$$f_1(x) = \int_{a_1(x)}^{b_1(x)} f(x, y) \, dy \qquad (5\text{-}5)$$

By similar reasoning, the marginal density function of y is given by

$$f_2(y) = \int_{a_2(y)}^{b_2(y)} f(x, y) \, dx \qquad (5\text{-}6)$$

where $a_2(y)$ and $b_2(y)$ are the functional values of x in terms of y at the two ends of the horizontal strip (Fig. 5-7b) corresponding to a fixed value of y.

Given the information that the ordinate y' of a random point lies between y and $y + dy$, the conditional probability that its abscissa x' will lie between x and $x + dx$ can be obtained in the usual way from the joint and unconditional probabilities:

$$P[(x \leq x' \leq x + dx) | (y \leq y' \leq y + dy)]$$

$$= \frac{P[(x \leq x' \leq x + dx) \cap (y \leq y' \leq y + dy)]}{P(y \leq y' \leq y + dy)}$$

$$= \frac{f(x, y) \, dx \, dy}{f_2(y) \, dy} = \frac{f(x, y)}{f_2(y)} \, dx \qquad (5\text{-}7)$$

We thus arrive at the *conditional* density function $\phi_1(x|y)$ defined by the equation

$$\phi_1(x|y) \equiv \frac{f(x, y)}{f_2(y)} \qquad (5\text{-}8)$$

and having the property that $\phi_1(x|y)\,dx$ yields the conditional probability that x' lies between x and $x + dx$, given that $y \leq y' \leq y + dy$.

The conditional density function $\phi_2(y|x)$ of y for a fixed value of x is correspondingly defined by the equation

$$\phi_2(y|x) \equiv \frac{f(x, y)}{f_1(x)} \qquad (5\text{-}9)$$

and has a similar interpretation.   Reciprocally, the joint density function can be constructed as the product of the marginal density function of one variate and the conditional density function of the other:

$$f(x, y) = f_1(x)\phi_2(y|x) = f_2(y)\phi_1(x|y) \qquad (5\text{-}10)$$

By including the range of definition with the equation of the conditional density function, we obtain the complete specification of the conditional distribution. Thus, referring to Fig. 5-7b, the conditional distribution of x for a fixed value of y would be defined by the equation for $\phi_1(x|y)$ together with the admissible range, $a_2(y) \leq x \leq b_2(y)$; similarly (Fig. 5-7a), the conditional distribution of y for a fixed value of x would be given by the equation for $\phi_2(y|x)$ together with the statement that $a_1(x) \leq y \leq b_1(x)$.   As the reader should verify by applying the appropriate definitions, the conditional density functions have unit area:

$$\int_{a_2(y)}^{b_2(y)} \phi_1(x|y)\,dx = 1 \qquad \int_{a_1(x)}^{b_1(x)} \phi_2(y|x)\,dy = 1 \qquad (5\text{-}11)$$

The importance of the conditional distribution resides principally in the fact that, by means of it, certain multivariate problems can be reduced virtually to univariate problems and thus rendered easier to solve.

Let us now consider some specific examples.   In each case the joint distribution will be given, and we shall derive the marginal and conditional distributions.

EXAMPLE 5-3   (see Fig. 5-8).

$$f(x, y) = 24y(1 - x) \qquad 0 \leq x \leq 1 \qquad 0 \leq y \leq x$$

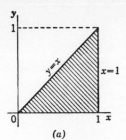 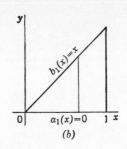 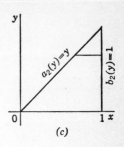

(a)　　　　　　(b)　　　　　　(c)

FIGURE 5-8
(a) Admissible region; (b) path of integration for marginal distribution of x;
(c) path of integration for marginal distribution of y.

Marginal density functions:

$$f_1(x) = \int_0^x f(x, y)\, dy = 24(1 - x) \int_0^x y\, dy = 12x^2(1 - x)$$

$$f_2(y) = \int_y^1 f(x, y)\, dx = 24y \int_y^1 (1 - x)\, dx = 12y(1 - y)^2$$

Marginal distribution of x:

$$f_1(x) = 12x^2(1 - x) \qquad (0, 1)$$

Marginal distribution of y:

$$f_2(y) = 12y(1 - y)^2 \qquad (0, 1)$$

Conditional density functions:

$$\phi_1(x\,|\,y) = \frac{24y(1 - x)}{12y(1 - y)^2} = \frac{2}{(1 - y)^2}\,(1 - x)$$

$$\phi_2(y\,|\,x) = \frac{24y(1 - x)}{12x^2(1 - x)} = \frac{2}{x^2}\,y$$

Conditional distribution of x:

$$\phi_1(x\,|\,y) = \frac{2}{(1 - y)^2}\,(1 - x) \qquad y \leq x \leq 1$$

Conditional distribution of y:

$$\phi_2(y\,|\,x) = \frac{2}{x^2}\,y \qquad 0 \leq y \leq x$$

Check:

$$\int_0^1 f_1(x)\, dx = 12 \int_0^1 x^2(1-x)\, dx = 12\beta(3,2) = \frac{12(2!)(1!)}{4!} = 1$$

$$\int_0^1 f_2(y)\, dy = 12 \int_0^1 y(1-y)^2\, dy = 12\beta(2,3) = 1$$

$$\int_y^1 \phi_1(x\,|\,y)\, dx = \frac{2}{(1-y)^2} \int_y^1 (1-x)\, dx = \frac{2}{(1-y)^2}\frac{(1-y)^2}{2} = 1$$

$$\int_0^x \phi_2(y\,|\,x)\, dy = \frac{2}{x^2} \int_0^x y\, dy = \frac{2}{x^2}\frac{x^2}{2} = 1 \qquad\qquad ////$$

This example brings out an important fact which readers sometimes overlook. Although the joint density function of independent variates factors into pure functions of each, factorization alone is not sufficient for independence; the factors must equal the respective marginal density functions. Moreover, whenever the conditional distribution of one variate depends either in form or in range of definition upon another variate, then the variates are dependent; however, when a conditional distribution does not involve another variate either explicitly in the equation of the density function or implicitly in the range of definition, the variates are independent. In the present example it is clear that $f(x, y) \neq f_1(x)f_2(y)$ despite the fact that $f(x, y)$ is a product of two pure functions, one in y and one in x. Here the conditional distributions depend both in form and in range upon the other variate, and neither agrees with the corresponding marginal distribution. Of course, x and y are independent if $F(x, y) = F_1(x) F_2(y)$.

EXAMPLE 5-4

$$f(x, y) = xe^{-x(1+y)} \qquad (0, \infty \text{ each})$$

Here the admissible region is the entire first quadrant, and so the limits of integration will be simply 0 to ∞ in each case. The total integral of the joint density function over the region for which it is defined equals unity, as it should, for

$$\int_0^\infty \int_0^\infty f(x, y)\, dx\, dy = \int_0^\infty e^{-x}\left(\int_0^\infty xe^{-xy}\, dy\right) dx = \int_0^\infty e^{-x}\, dx = 1$$

As already indicated by this integration, the marginal distribution of x is given by

$$f_1(x) = e^{-x} \qquad (0, \infty)$$

For the marginal density function of y, the integration with respect to x may be simplified by substituting $u = x(1 + y)$, $x = u/(1 + y)$, and since y is held fixed during the integration, $dx = du/(1 + y)$. Then

$$f_2(y) = \int_0^\infty f(x, y)\, dx = \frac{1}{(1 + y)^2} \int_0^\infty u e^{-u}\, du$$

$$= \frac{1}{(1 + y)^2} \Gamma(2) = \frac{1}{(1 + y)^2}$$

Thus the marginal distribution of y is given by

$$f_2(y) = \frac{1}{(1 + y)^2} \qquad (0, \infty)$$

The conditional distributions are

$$\phi_1(x \mid y) = \frac{f(x, y)}{f_2(y)} = (1 + y)^2 x e^{-x(1 + y)} \qquad (0, \infty)$$

and

$$\phi_2(y \mid x) = \frac{f(x, y)}{f_1(x)} = x e^{-xy} \qquad (0, \infty) \qquad\qquad ////$$

EXAMPLE 5-5 (see Fig. 5-9).

$$f(u, v) = \frac{1}{2u^2 v} \qquad 1 \le u < \infty\,; \frac{1}{u} \le v \le u$$

The marginal density function of u is given by

$$f_1(u) = \int_{1/u}^u f(u, v)\, dv = \frac{1}{2u^2} \Big[\ln v \Big]_{1/u}^u = \frac{\ln u}{u^2}$$

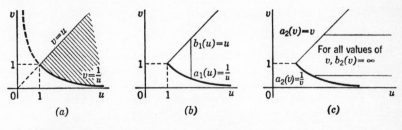

FIGURE 5-9
(a) Admissible region; (b) path of integration for marginal distribution of u; (c) paths of integration for marginal distribution of v.

and the marginal distribution of u is

$$f_1(u) = \frac{\ln u}{u^2} \qquad (1, \infty)$$

The marginal distribution of v, however, involves an extra step. For all values of v, the upper limit of integration on u is ∞, but the lower limit assumes two distinct functional forms, depending upon the value of v. When $v \le 1$ the lower limit lies on the hyperbola $u = 1/v$; but when $v \ge 1$ the lower limit lies on the straight line $u = v$; when $v = 1$ both functions yield the same value, $u = 1$. Consequently, the marginal density function will have two distinct equations, as follows:

Range 1 $(0 \le v \le 1)$:

$$f_2(v) = \int_{1/v}^{\infty} f(u, v)\, du = \frac{1}{2v}\left[-\frac{1}{u}\right]_{1/v}^{\infty} = \frac{1}{2}$$

Range 2 $(1 \le v < \infty)$:

$$f_2(v) = \int_{v}^{\infty} f(u, v)\, du = \frac{1}{2v}\left[-\frac{1}{u}\right]_{v}^{\infty} = \frac{1}{2v^2}$$

The marginal distribution of v thus requires two statements:

$$f_2(v) = \frac{1}{2} \qquad (0, 1) \qquad f_2(v) = \frac{1}{2v^2} \qquad (1, \infty)$$

Corresponding to these two equations, the conditional distribution of u given v assumes two separate forms:

If $0 \le v \le 1$ $\qquad \phi_1(u|v) = \dfrac{1}{u^2 v} \qquad \dfrac{1}{v} \le u < \infty$

and if $1 \le v < \infty$ $\qquad \phi_1(u|v) = \dfrac{v}{u^2} \qquad v \le u < \infty$

On the other hand, the conditional distribution of v given u requires only one statement:

$$\phi_2(v|u) = \frac{1}{2 \ln u}\frac{1}{v} \qquad \frac{1}{u} \le v \le u \qquad \text{////}$$

(Suggested Problems 5-10 to 5-20)

5-5 ILLUSTRATIONS OF THE USE OF GENERAL PRINCIPLES

The ideas developed in this chapter in conjunction with those from previous chapters now expand the range and type of problem that can be handled.   The following illustrative examples are presented to clarify these basic principles further and to indicate how these tools and concepts may be used in practical application.   It will be clear from the following that applications are very numerous so that those considered are merely indicative of a general class of problems.

EXAMPLE 5-6   The probability of success on a single trial is constant from trial to trial and equal to p.   After 10 trials it is known that the number of successes was greater than 4.   What is the probability of the various remaining possibilities?

The result must be a conditional-probability function which sums to unity over the remaining possible outcomes.   Therefore, the probability of exactly x successes is given by

$$P(x \mid x > 4) = \frac{C(10, x)p^x q^{10-x}}{1 - \sum_{y=0}^{4} C(10, y)p^y q^{10-y}} \qquad x = 5, 6, \ldots, 10 \qquad ////$$

EXAMPLE 5-7   The mail arrives at company A sometime between 8 and 9 A.M., and for practical purposes it is equally likely to be any time within that hour. The office boy must have the president's mail delivered to him by 9 A.M.   The boy apparently delivers the mail any time between its arrival at the company and 9 A.M., and again, the time of this delivery is equally likely within the limits of the possible time interval.   What is the probability that the mail will reach the president's desk between t and $t + dt$?

The probability that the mail will be delivered to the company between T and $T + dT$ is $dT/1$ (since the unit of time is an hour), and the conditional probability that it will be delivered to the president's desk between t and $t + dt$, knowing that it arrived at the company in the neighborhood of T, is $dt/(9 - T)$. By definition the joint probability that the mail will arrive at the company between T and $T + dT$ and be delivered to the president between t and $t + dt$ is

$$P(T, t)\, dT\, dt = \frac{dT\, dt}{9 - T} \qquad 8 < T < 9; T < t < 9$$

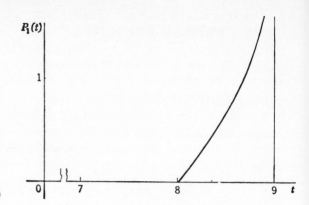

FIGURE 5-10

Having obtained the joint distribution, the two marginals can be evaluated by means of integration. However, the marginal or total probability of T is already known and the answer to our question involves the marginal distribution of t, which is

$$P_1(t)\,dt = dt \int_8^t \frac{dT}{9 - T} = -\ln(9 - t)\,dt \qquad 8 < t < 9$$

where again the unit of time is an hour (see Fig. 5-10). The area under the $P_1(t)$ curve is, of course, unity.   /////

EXAMPLE 5-8   The average number of cars per day that take a country-road connection between two main roads is 15. The only building located on this road is an antique shop, and the percentage of cars which go through this road and stop at the shop is .12. What is the chance that exactly x cars stop at the antique shop on a given day?

Since the average number of cars per day is given and no further information regarding their destination or reason for traveling the road is contained in the statement, the only reasonable assumption one can make is that the number of cars is Poisson-distributed on any given day. This would be equivalent to assuming that the cars go through this road individually and collectively at random. However, if it is known that n cars travel through this road on a given day with a constant probability of .12 of stopping, then the conditional probability of exactly x stops out of n cars must be binomially distributed. These two distributions are

$$P_1(n) = \frac{e^{-15}(15)^n}{n!} \qquad n = 0, 1, 2, \dots, \infty$$

and

$$P_2(x|n) = C(n, x)(.88)^{n-x}(.12)^x \qquad x = 0, 1, \dots, n$$

The joint probability of n cars and x stops will thus be the product of these two probabilities. The answer to our problem is, of course, the marginal distribution of x; this is obtained by summing over n. It is to be observed that n must be at least as large as x or otherwise it would be impossible to have available the x cars to stop at the antique shop. Therefore,

$$P_3(x) = \sum_{n=x}^{\infty} \frac{e^{-15}(15)^n}{n!} C(n, x)(.88)^{n-x}(.12)^x$$

By setting $n - x = y$ and observing that the resulting summation is merely the expansion for e^z, one obtains the result

$$P_3(x) = \frac{e^{-15}(.12)^x(15)^x}{x!} \sum_{y=0}^{\infty} \frac{(15)^y(.88)^y}{y!} = \frac{e^{-(15)(.12)}[(15)(.12)]^x}{x!}$$

The reader should note that this result is very general, and if the average number of cars were λ and the probability of a car stopping p, then the probability of exactly x stops on any one day would be given by a Poisson distribution with the parameter μ equal to λp. ////

EXAMPLE 5-9 For brand A the distribution of breaking strength (x) of synthetic fiber is given approximately by

$$f_1(x) = 3ax^2 e^{-ax^3} \qquad (0, \infty) \qquad a > 0$$

and for brand B the breaking strength (y) is given by

$$f_2(y) = 3by^2 e^{-by^3} \qquad (0, \infty) \qquad b > 0$$

If a piece of each brand of synthetic fiber is chosen at random and tested, what is the probability that the sample of brand A will withstand greater tension than the sample of brand B? If $a = b$, what is the probability that five such tests will all turn out to be favorable to brand A?

Although the actual breaking strength cannot be infinite, the cubic exponent in each density function makes the probability of very large values negligibly small, and for practical purposes the distributions may be accepted as described. The two distributions give the means of computing the probability of a sample value falling between x and $x + dx$ in the case of brand A, and between y and $y + dy$ for brand B. Since these events are independent, the joint probability is $f_1(x)f_2(y)\,dx\,dy$. The entire field of possible sample results (i.e., simultaneous values of x and y) is represented by the first quadrant of the xy plane.   However, all possible pairs of sample values (x, y) such that brand B breaks before brand A are represented by that portion of the first quadrant which lies below the line $y = x$ if we assume x is the abscissa and y the ordinate.   Therefore, the probability $P(E)$ of the event E that y be less than x is found by integrating the joint density function over this region; thus

$$P(E) = \int_0^\infty dx \int_0^x f_1(x)f_2(y)\, dy = \int_0^\infty f_1(x)(1 - e^{-bx^3})\, dx = 1 - \frac{a}{a + b}$$

In the case of $a = b$, we have $P(E) = \tfrac{1}{2}$ and, accordingly, if five tests were made, the probability that all will be favorable to brand A is $(\tfrac{1}{2})^5 = \tfrac{1}{32}$.   This latter result depends only on the fact that x and y have the same density function. Let us suppose that this is the case and that the common range of definition is (L, U).   Then the previous argument gives

$$P(E) = \int_L^U f_1(x)\, dx \int_L^x f_1(y)\, dy$$

If now we define w as

$$w = \int_L^x f_1(y)\, dy$$

then $dw = f_1(x)\, dx$.   Also, we have $w = 0$ when $x = L$ and $w = 1$ when $x = U$.

Consequently,
$$P(E) = \int_0^1 w\, dw = \tfrac{1}{2}$$

which shows that the result is independent of the form of the function.   ////

EXAMPLE 5-10   In a department store, customers arrive in the time interval from 9 A.M. to 5 P.M. according to some distribution, which may be interpreted as a density function of time at arrival.   Thus the total number of customers entering the store between τ and $\tau + d\tau$ is $Nf(\tau)\, d\tau$, where N is the total number of customers in any one day.   The probability that a person coming in at time τ will leave between t and $t + dt$ will be denoted by $g(t\,|\,\tau)\, dt$.   What is the distribution of the number of people in the store between the hours of 9 A.M. and 5 P.M.?

If we change the scale so that 0 corresponds to 9 A.M., and 1 to 5 P.M. and redefine our notation accordingly, we have the condition that the integral of $Nf(\tau)\, d\tau$ from 0 to 1 is equal to N.   We also have the condition that the integral of the conditional distribution $g(t\,|\,\tau)\, dt$ from τ to 1 must equal unity.

The average number of people actually in the store at any time T is given by $Nw(T)$, where $w(T)$ is the function to be determined.   Now

$$Nf(\tau)\,d\tau = \text{number who enter between } \tau \text{ and } \tau + d\tau$$

$$\int_\tau^T g(t\,|\,\tau)\,dt = \text{fraction of those people who come in at } \tau \text{ and leave}$$

$$\text{between } \tau \text{ and } T$$

$$1 - \int_\tau^T g(t\,|\,\tau)\,dt = \text{fraction who enter at } \tau \text{ and are still in store at}$$

$$\text{time } T$$

Then
$$Nw(T) = N \int_0^T f(\tau)\,d\tau \left[1 - \int_\tau^T g(t\,|\,\tau)\,dt\right]$$

which now permits $w(T)$ to be evaluated if $f(\tau)$ and $g(t\,|\,\tau)$ are known.   ////

EXAMPLE 5-11   Given a rectangular distribution $f(x) = 1$ $(0 < x < 1)$. Five observations are drawn at random from this distribution and then ordered by size so that s represents the smallest of the five.   What is the probability that s will lie between x_1 and $x_1 + dx_1$?   This type of problem, which comes under the general heading of ordered statistics, is solved by considering the general multinomial law introduced in Chap. 1.   It will be remembered that if the possible outcomes of a trial are $E_1, E_2, \ldots, E_r$ with corresponding probabilities $p_1, p_2, \ldots, p_r$ then the probability that in n trials E_1 occurs k_1 times, E_2 occurs k_2 times, etc., is

$$\frac{n!}{k_1!\,k_2!\cdots k_r!}\,p_1^{k_1}p_2^{k_2}\cdots p_r^{k_r} \qquad \text{where } k_1 + k_2 + \cdots + k_r = n$$

$$p_1 + p_2 + \cdots + p_r = 1$$

In this particular problem we desire the probability that, in picking x five times at random from the rectangular distribution, it falls 0 times in the interval $(0, x_1)$, once in $(x_1, x_1 + dx_1)$, and four times in $(x_1 + dx_1, 1)$.   Substituting the corresponding probabilities in the multinomial we have

$$\frac{5!}{0!\,1!\,4!}\left(\int_0^{x_1} dx\right)^0 (dx_1)^1 \left(\int_{x_1}^1 dx\right)^4 = 5(1 - x_1)^4\,dx_1 \qquad 0 \le x_1 < 1$$

which gives as the coefficient of dx_1 the probability density of the smallest observation.   It is to be noted that this distribution has a very high ordinate when $x_1 = 0$ and drops off very rapidly as x increases, reaching 0 when $x_1 \to 1$.

This then is the distribution obtained by sampling in groups of five and noting the smallest observation in each case. A large number of these smallest observations are now considered together, and they would generate the above density function. ////

EXAMPLE 5-12   Given the probability density function $ae^{-at}(0, \infty)$. Four observations are drawn at random from this population and ordered according to their magnitude. What is the joint distribution of the smallest (x_1) and the largest (x_4)?

In this particular case we have no observations between 0 and x_1, one observation between x_1 and $x_1 + dx_1$, two observations between $x_1 + dx_1$ and x_4, one between x_4 and $x_4 + dx_4$, and none between $x_4 + dx_4$ and infinity. Substituting the individual probabilities in the multinomial, we have the joint probability of x_1 and x_4 as follows:

$$f_1(x_1, x_4)\, dx_1\, dx_4 = \frac{4!}{1!2!1!}\, ae^{-ax_1}\, dx_1 \left(\int_{x_1}^{x_4} ae^{-ax}\, dx\right)^2 ae^{-ax_4}\, dx_4$$

$$= 12a^2 e^{-a(x_1 + x_4)}(e^{-ax_1} - e^{-ax_4})^2\, dx_1\, dx_4$$

$$0 < x_4 < \infty; 0 \le x_1 < x_4$$

where $f_1(x_1, x_4)$ represents the probability density of x_1 and x_4 simultaneously. ////

EXAMPLE 5-13   Five observations are drawn at random from the density function $f(x) = 6x(1 - x)$ $(0 < x < 1)$ and these observations are ordered from the smallest to the largest. What is the probability density function of the median (x_3)?

The solution follows a pattern of the previous two problems and is as follows:

$$f_1(x_3)\, dx_3 = \frac{5!}{1!2!2!}\left[6\int_0^{x_3} x(1 - x)\, dx\right]^2 6x_3(1 - x_3)\, dx_3\left[6\int_{x_3}^1 x(1 - x)\, dx\right]^2$$

$$= 30 \times 6^5 x_3(1 - x_3)\left(\frac{x_3^2}{2} - \frac{x_3^3}{3}\right)^2\left[\frac{1}{6} - \left(\frac{x_3^2}{2} - \frac{x_3^3}{3}\right)\right]^2 dx_3$$

////

EXAMPLE 5-14   In the inspection of light bulbs, vacuum tubes, condensers, and many other items, where a destructive test is necessary in order to determine their life or other characteristics, it is desirable often to simplify the testing techniques. For example, if one is testing electric light bulbs in order to determine the length of life, out of n bulbs tested, it is possible that some will continue to burn for a long while and that a test rack or shipment will be held up for this period. One technique to overcome this difficulty is to test a large number of items, but on the basis of the time to failure of the first few, determine the characteristics of the entire lot. Assuming a very simple life curve for electric light bulbs of the form $p(t) = (1/\tau)e^{-t/\tau}$, where τ is the mean life of a large lot, the probability that an individual bulb will burn out between t and $t + dt$ is $p(t)\, dt$. For a fixed value of τ, what is the probability of obtaining a pair of values for t_1 and t_2 which would be less probable than a given observed pair?

If n bulbs are placed on the test rack and the time to failure observed in all cases, we would produce a series of times $t_1, t_2, \ldots, t_n$ which we might consider in a sequence of order in which t_1 is the shortest and t_n the longest. The joint distribution of t_1 and t_2 may be obtained in the usual way as

$$p(t_1, t_2)\, dt_1\, dt_2 = \frac{n!}{1!\,1!\,(n-2)!} \left(\frac{1}{\tau}e^{-t_1/\tau}\, dt_1\right)\left(\frac{1}{\tau}e^{-t_2/\tau}\, dt_2\right)\left(\frac{1}{\tau}\int_{t_2}^{\infty} e^{-t/\tau}\, dt\right)^{n-2}$$

$$0 < t_1 < \infty$$
$$t_2 > t_1$$

$$= \frac{(n)(n-1)}{\tau^2} e^{-[t_1 + (n-1)t_2]/\tau}\, dt_1\, dt_2$$

It must be true, of course, that

$$\int_0^{\infty} \int_{t_1}^{\infty} p(t_1, t_2)\, dt_1\, dt_2 = 1$$

If the experiment is conducted as outlined, it means that the time of the failure of the first bulb is noted and we will call this value a, and the time to the second failure will be b. Thus, in each case the experiment would result in an actual pair of values (t_1, t_2) denoted as (a, b). Since the probability density function is constant, if the exponent of e is constant, all pairs of values which satisfy the condition

$$t_1 + (n-1)t_2 = a + (n-1)b = c$$

have the same probability of occurrence for a given value of τ as the observed point (a, b). In reference to Fig. 5-11, it is now possible to find the probability that (t_1, t_2) will fall in the shaded area, having observed the fixed point (a, b) as

$$p = \frac{n(n-1)}{\tau^2} \int_0^{c/n} e^{-t_1/\tau}\, dt_1 \int_{t_1}^{(c-t_1)/(n-1)} e^{-(n-1)t_2/\tau}\, dt_2$$

$$= 1 - \left(1 + \frac{c}{\tau}\right) e^{-c/\tau}$$

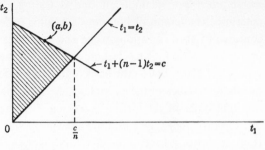

FIGURE 5-11

Since c is fixed by the observed value of the first and second bulb to burn out, and if we know the value of τ, then the area as found from the above expression will give the probability that we would observe by chance a sample which might be considered "worse" than that which was observed.   ////

EXAMPLE 5-15   The selling price of a specific article to the public will vary somewhere between \$.25 and \$.35 per item, depending upon the final cost of production as determined from the initial run.   It is assumed that any price within this range is equally likely to occur.   However, the volume of the sales as a probability density function is dependent upon this final retail figure and, expressed in millions of pieces, is of the form $f(v|c) = (c^2 v)e^{-cv}$ $(0 < v < \infty)$, where c is the cost per piece in dollars.   The effect of substituting values of c ranging from .25 to .35 in this equation is gradually to move the mode of the curve toward the y axis; thus, increasing the price reduces the volume of sales. What is the probability that the sales will exceed 9 million pieces?

Since the probability that the price will lie between c and $c + dc$ is $10\,dc$ and the conditional probability of $v|c$ is as indicated above, the joint probability may be directly computed as

$$f(v, c)\,dv\,dc = (10\,dc)(c^2 v e^{-cv}\,dv)$$

From this expression the marginal distribution of v can now be obtained by integrating over the variable c.   Thus

$$g(v)\,dv = 10v\,dv \int_{.25}^{.35} c^2 e^{-cv}\,dc = -\,10v\,dv \left[\frac{e^{-cv}}{v^3}(v^2 c^2 + 2cv + 2)\right]_{.25}^{.35}$$

The probability that the sales will exceed 9 million is thus given by the area underneath the marginal-distribution curve from 9 to infinity as

$$P(v > 9) = 10 \int_9^\infty \left\{ \frac{e^{-.25v}}{v^2} [(.25)^2 v^2 + 2(.25)v + 2] \right.$$

$$\left. - \frac{e^{-.35v}}{v^2} [(.35)^2 v^2 + 2(.35)v + 2] \right\} dv$$

The nonelementary parts of this integral offset each other in pairs (see $I(a)$ of Example 6-9), with the result that

$$P(v > 9) = {}^{10}\!/_9 (4.25e^{-2.25} - 5.15e^{-3.15})$$

$$= .2525$$

Thus the probability in question is about .25.

As an alternative method of solution, which is somewhat simpler, we may reverse the order of integration over the joint distribution. To this end let $a = .25$, $b = .35$, $M = 9$. Then

$$P(v > M) = \int_a^b 10c^2 \left(\int_M^\infty v e^{-cv} \, dv \right) dc$$

$$= 10 \int_a^b (Mc + 1)e^{-Mc} \, dc$$

$$= \frac{10}{M} [(Ma + 2)e^{-Ma} - (Mb + 2)e^{-Mb}]$$

Upon substitution of numerical values for a, b, M, this expression yields the same result as before.   ////

EXAMPLE 5-16   Particles are emanating from a radioactive source at an average rate of a impulses per unit of time.   Since these pulses occur at random, they can be described by a Poisson distribution which, for a given time period t, gives the following probability for exactly x emanations:

$$P(x|t) = \frac{(at)^x e^{-at}}{x!} \qquad x = 0, 1, 2, \ldots, \infty$$

Let us assume that time is an equally likely variable over any range and therefore its density function can be represented by $g(t) = 1/T$ $(0 < t < T)$.   Find the conditional distribution of t for a fixed value of x.

The joint probability that x emanations will be observed and the corresponding time t' falls between t and $t + dt$ is

$$f_1(x, t) \, dt = \frac{dt}{T} \frac{(at)^x e^{-at}}{x!}$$

Since this joint distribution is a function of x and t, it is now possible to integrate t over the appropriate limits and obtain the marginal distribution of x (the marginal distribution of t, of course, is known) as

$$f_2(x) = \frac{1}{T} \int_0^T \frac{(at)^x e^{-at} \, dt}{x!}$$

We know, however, that the required conditional probability $w(t|x) \, dt$ can be obtained by dividing the joint probability $f_1(x, t) \, dt$ by the marginal distribution $f_2(x)$. Thus

$$w(t|x) \, dt = \frac{(at)^x e^{-at} \, dt}{\int_0^T (at)^x e^{-at} \, dt}$$

It is now possible to let $T \to \infty$, thus permitting the denominator of the above expression to be evaluated as a gamma function. Thus the final probability that t' will lie between t and $t + dt$ at the same time that x (the number of emanations) assumes a given fixed value is given by

$$w(t|x) \, dt = \frac{a(at)^x e^{-at} \, dt}{x!} \qquad 0 < t < \infty$$

It is interesting to note that the form of this answer is the same as that of Example 4-8 and that if x is set equal to 0 we have the density function as an exponential ae^{-at}. Actually, of course, the expression found in this problem involves time and n emanations, whereas the previous example involved the time necessary for the emission of $n + 1$ particles. In the case of the exponential, the difference between these two concepts is merely a question of semantics, since the distribution of time for no emissions is the same as the time between individual emissions in this case. ////

EXAMPLE 5-17   The number of orders n per day for a certain chemical is Poisson-distributed (parameter μ), and the weight in pounds per order is exponentially distributed (parameter β). Assuming that the weights from order to order are independent, find the distribution of total weight of orders per day.

  Denote the weight of the ith order in a day by y_i and the total weight of n orders by w. Since $w \equiv 0$ if $n = 0$, and the latter event has a finite probability of occurrence ($e^{-\mu}$), the distribution of w is discontinuous at zero. Thus, for $w = 0$ we have

$$P(w = 0) = P(n = 0) = e^{-\mu}$$

However, a density function exists for $w > 0$.

Consider a fixed number of orders $n > 0$. Then $w = y_1 + y_2 + \cdots + y_n$ and, by Theorem 4-1, the conditional distribution of w becomes

$$\phi(w|n) = \frac{1}{\beta^n \Gamma(n)} w^{n-1} e^{-w/\beta} \qquad (0, \infty)$$

Since n is an integer, we may replace $\Gamma(n)$ by $(n - 1)!$ and write the joint distribution of n and w for $n > 0$ as follows:

$$f(n, w) = e^{-\mu} \frac{\mu^n}{n!} \frac{1}{\beta^n(n - 1)!} w^{n-1} e^{-w/\beta}$$

$$= \frac{\mu}{\beta} e^{-\mu} e^{-w/\beta} \frac{(\mu w/\beta)^{n-1}}{n!(n - 1)!} \qquad n = 1, 2, \ldots, \infty; 0 < w < \infty$$

Therefore, the marginal distribution of w in the range $w > 0$ becomes

$$h(w) = \sum_{n=1}^{\infty} f(n, w) = \frac{\mu}{\beta} e^{-\mu} e^{-w/\beta} \sum_{k=0}^{\infty} \frac{(\mu w/\beta)^k}{k!(k + 1)!} \qquad 0 < w < \infty$$

where $k = n - 1$.

Readers familiar with advanced calculus may be interested in the fact that this series can be identified with the modified Bessel function of the first kind of order unity. The latter function [symbol $I_1(u)$] may be expressed as follows:

$$I_1(u) = \frac{u}{2} \sum_{k=0}^{\infty} \frac{(u/2)^{2k}}{k!(k + 1)!}$$

Substituting $\mu w/\beta = (u/2)^2$ we thus obtain

$$h(w) = e^{-\mu} \frac{\sqrt{\mu}}{\sqrt{\beta}} w^{-1/2} e^{-w/\beta} I_1\left(2\sqrt{\frac{\mu w}{\beta}}\right) \qquad 0 < w < \infty$$

and this density function together with the equation $P(w = 0) = e^{-\mu}$ defines the distribution of w.

In the more general case in which the weight per order is a gamma variate (parameters α, β) we would still have $P(w = 0) = e^{-\mu}$, and for $n > 0$ the conditional distribution of w would be given by

$$\phi(w|n) = \frac{1}{\beta^{n\alpha+n} \Gamma(n\alpha + n)} w^{n\alpha+n-1} e^{-w/\beta} \qquad 0 < w < \infty$$

and the marginal density function would be

$$h(w) = e^{-\mu} e^{-w/\beta} \sum_{n=1}^{\infty} \frac{\mu^n w^{n\alpha+n-1}}{n! \beta^{n\alpha+n} \Gamma(n\alpha + n)} \qquad 0 < w < \infty$$

In this case, however, the series expansion for the marginal density does not reduce to a Bessel function. ////

EXAMPLE 5-18   Ten observations are drawn at random from an arbitrary density function $f(x)$ which may be defined over a finite or infinite range of the variate x. Three more observations are drawn at random from $f(x)$. What is the probability that all 3 of the latter observations are larger than any of the previous 10? The probability that the largest of the original 10 observations falls at x_{10} is $g(x_{10})\,dx_{10}$, where

$$g(x_{10})\,dx_{10} = \frac{10!}{9!}\left[\int_{-\infty}^{x_{10}} f(x)\,dx\right]^9 f(x_{10})\,dx_{10}$$

If the largest of the group of 10 is at x_{10}, the conditional probability that the three observations drawn subsequently will be greater than this given value is

$$\left[\int_{x_{10}}^{\infty} f(x)\,dx\right]^3$$

and the corresponding joint probability of the two events is then

$$g(x_{10})\,dx_{10}\left[\int_{x_{10}}^{\infty} f(x)\,dx\right]^3$$

Since x_{10} can take on any value in the range of definition of the variate—under our assumption $f(x)$ is continuous—the total probability of the event as stated will be

$$P(E) = \int_{-\infty}^{\infty} g(x_{10})\,dx_{10}\left[\int_{x_{10}}^{\infty} f(x)\,dx\right]^3$$

$$= 10\int_{-\infty}^{\infty} \left[\int_{x_{10}}^{\infty} f(x)\,dx\right]^3 \left[\int_{-\infty}^{x_{10}} f(x)\,dx\right]^9 f(x_{10})\,dx_{10}$$

Let $u = \int_{-\infty}^{x_{10}} f(x)\,dx$ and then $du = f(x_{10})\,dx_{10}$; thus

$$P(E) = 10\int_{0}^{1} (1-u)^3 u^9\,du = 10\beta(4, 10) = \tfrac{1}{286}$$

Thus the probability of the event is independent of the form of $f(x)$. Applications of this form of analysis to the evaluation of observational data are obvious and form a class of techniques which are called *nonparametric*.   ////

EXAMPLE 5-19   Let $f(x)$ be the density function of an arbitrary probability distribution. Find the probability that $n + 2$ independent observations can be drawn from the distribution in the following way: If x_1 and x_n are the minimum and maximum of the first n observations, the $(n + 1)$st and $(n + 2)$nd observations must lie outside the interval $x_1 \le x \le x_n$.

Following the argument in Example 5-18, it follows that

$$P = (n)(n-1) \int_{-\infty}^{\infty} f(x_1)\, dx_1 \int_{x_1}^{\infty} f(x_n) \left[\int_{x_1}^{x_n} f(x)\, dx \right]^{n-2} \left[1 - \int_{x_1}^{x_n} f(x)\, dx \right]^2 dx_n$$

Let

$$u = \int_{-\infty}^{x_1} f(x)\, dx \text{ and } v = \int_{-\infty}^{x_n} f(x)\, dx \ (u \le v \le 1).$$

Then

$$v - u = \int_{x_1}^{x_n} f(x)\, dx, \ du = f(x_1)\, dx_1,$$

and $dv = f(x_n)\, dx_n$. Since, when $x_n = x_1$, $u = v$, we have

$$n(n-1) \int_0^1 du \int_u^1 (v-u)^{n-2} [1 - (v-u)]^2 \, dv$$

and

$$n(n-1) \int_0^1 du \left[\frac{(v-u)^{n-1}}{n-1} - 2\frac{(v-u)^n}{n} + \frac{(v-u)^{n+1}}{n+1} \right]_u^1 = \frac{6}{(n+1)(n+2)}$$

It is interesting to note that, when $n = 23$, the probability is equal to .01.   ////

EXAMPLE 5-20   A message is to be sent simultaneously over three communication channels connecting the points A and B. The probability density function for the time of transit for the message is $f_1(t), f_2(t),$ and $f_3(t)$ $(0 \le t < \infty)$ for the three channels separately. What is the probability density for the shortest time of arrival for this message at B? Let u be the shortest time or first arrival and $f(u)$, the density of u; then

$$f(u)\, du = f_1(u)\, du \int_u^{\infty} f_2(t)\, dt \int_u^{\infty} f_3(t)\, dt + f_2(u)\, du \int_u^{\infty} f_1(t)\, dt \int_u^{\infty} f_3(t)\, dt$$

$$+ f_3(u)\, du \int_u^{\infty} f_1(t)\, dt \int_u^{\infty} f_2(t)\, dt$$

This equation merely states that, if u is the shortest time, the message may arrive at that time u on any one of the three channels and at a later time on the remaining two.

The total area under the curve should equal unity and so for verification

$$\int_0^{\infty} f(u)\, du = \int_0^{\infty} \left[f_1(u) \int_u^{\infty} f_2(t)\, dt \int_u^{\infty} f_3(t)\, dt + f_2(u) \int_u^{\infty} f_1(t)\, dt \int_u^{\infty} f_3(t)\, dt \right.$$

$$\left. + f_3(u) \int_u^{\infty} f_1(t)\, dt \int_u^{\infty} f_2(t)\, dt \right] du$$

If the variables are changed so that

$$x = \int_u^\infty f_1(t)\, dt \qquad y = \int_u^\infty f_2(t)\, dt \qquad \text{and} \qquad z = \int_u^\infty f_3(t)\, dt$$

then

$$dx = -f_1(u)\, du \qquad dy = -f_2(u)\, du \qquad \text{and} \qquad dz = -f_3(u)\, du$$

so that, recognizing the exact differential $d(xyz)$,

$$\int_0^\infty f(u)\, du = -\int_{1,1,1}^{0,0,0} (yz\, dx + xz\, dy + xy\, dz) = -\left[xyz\right]_{1,1,1}^{0,0,0} = 1$$

A specific example of this might be the application to a two-channel case with the following time density functions:

$$f_1(t) = 1 \qquad 0 \le t \le 1 \qquad f_2(t) = \tfrac{1}{2} \qquad 0 \le t \le 2$$

Then

$$f(u)\, du = du \int_u^2 \frac{1}{2}\, dt + \frac{du}{2} \int_u^1 dt = \frac{du}{2}(3 - 2u) \qquad 0 \le u \le 1 \qquad \text{////}$$

(*Suggested Problems 5-21 to 5-31*)

5-6 POINT ESTIMATION OF A PARAMETER

In Sec. 4-12, we considered the idea of confidence limits as an interval estimation of a parameter. In this section we shall consider the best estimate of a single value for the parameter as a function of observational data. Fisher's method of maximum likelihood often yields a "best estimate" in the sense that it satisfies certain criteria which are intuitively desirable. The reader interested in the mathematical and intuitive foundations of maximum likelihood estimates should consult a test on statistical methods. The technique consists of computing the compound-probability function (density function if a continuous variable) of a specific observed sample and then maximizing this function, called the *likelihood function*, with respect to the parameter. Assuming that a high turning point exists in the meaningful range of θ, this is accomplished in the usual way by computing the derivative and setting it equal to zero to find the stationary point. Thus, if we have the probability density $f(x, \theta)$ and observations $x_1, x_2, \ldots, x_n$ the joint-probability density, assuming independence of the x's, would be

$$L = f(x_1, \theta)f(x_2, \theta) \cdots f(x_n, \theta) = \prod_{i=1}^n f(x_i, \theta) \qquad (5\text{-}12)$$

Since the logarithm of L is a maximum for the same value of θ as L itself, it is often convenient to take the natural logarithm of both sides before differentiation. Accordingly, we may put

$$\ln L = \ln f(x_1, \theta) + \ln f(x_2, \theta) + \cdots + \ln f(x_n, \theta) \tag{5-13}$$

and
$$\frac{\partial(\ln L)}{\partial \theta} = \frac{1}{f(x_1, \theta)} \frac{\partial f(x_1, \theta)}{\partial \theta} + \cdots + \frac{1}{f(x_n, \theta)} \frac{\partial f(x_n, \theta)}{\partial \theta} = 0 \tag{5-14}$$

This yields a single equation for the solution of θ in terms of the x's. If more than one parameter is to be estimated, one can differentiate with respect to each parameter partially, creating simultaneous equations in the unknown parameters when each partial derivative is set equal to zero.

EXAMPLE 5-21 Observations $x_1, x_2, \ldots, x_n$ are drawn at random from a population which has the functional form $f(x) = ke^{-kx}$ $(0 < x < \infty)$. What is the maximum likelihood estimate of k? Computing the likelihood function, we have

$$L = k^n e^{-k(x_1 + x_2 + \cdots + x_n)}$$

or
$$\ln L = n \ln k - k(x_1 + x_2 + \cdots + x_n)$$

The $\ln L$ is now differentiated with respect to k and a stationary point evaluated:

$$\frac{\partial}{\partial k}(\ln L) = \frac{n}{k} - (x_1 + x_2 + \cdots + x_n) = 0$$

and so
$$k = \frac{1}{(x_1 + x_2 + \cdots + x_n)/n} = \frac{1}{\bar{x}}$$

Thus the maximum likelihood estimator for k is $1/\bar{x}$. ////

EXAMPLE 5-22 A set of observations $x_1, x_2, \ldots, x_n$ are drawn at random from a normal population with unknown mean μ but a known standard deviation σ. What combination of the x's is the maximum likelihood estimator for μ? Here, employing the more convenient notation $\exp[-h(x)]$ to represent $e^{-h(x)}$ we have

$$L = \frac{1}{(\sqrt{2\pi})^n \sigma^n} \exp\left[-\sum_{i=1}^{n} \frac{(x_i - \mu)^2}{2\sigma^2}\right]$$

$$\ln L = -\frac{n}{2} \ln 2\pi - n \ln \sigma - \frac{1}{2\sigma^2} \sum (x_i - \mu)^2$$

$$\frac{\partial(\ln L)}{\partial \mu} = \frac{2}{2\sigma^2} \sum_{i=1}^{n} (x_i - \mu) = 0$$

Since this equation is satisfied by setting $\sum x_i/n = \mu$, the sample mean is the maximum likelihood estimator of μ for a normal population. ////

EXAMPLE 5-23   If in Example 5-22 the parameter μ is known but σ is unknown, what is the maximum likelihood estimator of σ?   Here,

$$L = (2\pi\sigma^2)^{-n/2} e^{-\Sigma(x_i - \mu)^2/2\sigma^2}$$

or

$$\ln L = -n \ln \sigma - \left(\frac{n}{2}\right)\ln 2\pi - \frac{\Sigma(x_i - \mu)^2}{2\sigma^2}$$

$$\frac{\partial(\ln L)}{\partial \sigma} = -\frac{n}{\sigma} + \frac{\Sigma(x_i - \mu)^2}{\sigma^3} = 0$$

Thus the maximum-likelihood solution is

$$\sigma^2 = \frac{\Sigma(x_i - \mu)^2}{n} \qquad\qquad ////$$

EXAMPLE 5-24   Consider the Poisson distribution with the unknown parameter λ as $p(x) = e^{-\lambda}\lambda^x/x!$.   Observations $x_1, x_2, \ldots, x_n$ are available from experimentation where, of course, each x_i represents the number of successes observed during a single experiment.   What is the maximum likelihood estimate of λ?   Here

$$L = \frac{e^{-n\lambda}\lambda^{(x_1 + x_2 + \cdots + x_n)}}{x_1! x_2! \cdots x_n!}$$

or

$$\ln L = -n\lambda + (x_1 + x_2 + \cdots + x_n)\ln \lambda - \ln \prod_{i=1}^{n} x_i!$$

$$\frac{\partial(\ln L)}{\partial \lambda} = -n + \frac{x_1 + x_2 + \cdots + x_n}{\lambda} = 0$$

and therefore

$$\lambda = \frac{x_1 + x_2 + \cdots + x_n}{n} = \bar{x} \qquad\qquad ////$$

(*Suggested Problems 5-32 and 5-33*)

5-7   BAYESIAN APPROACH TO ESTIMATION

In Sec. 4-12 we discussed confidence limits as a means of evaluating the range within which we might expect a particular parameter to lie, given a certain specified probability for the occurrence of the event.   The assumption is that the parameter has a unique value and that the purpose of a particular experiment is to estimate this single value.   Another approach, which had many adherents in the early part of the century, involved the concept that the parameter had a probability-distribution function in its own right. This amounted to asserting that the degree of ignorance of the value of such a parameter can be represented by this distribution function. Because of the concepts involved in modern physics, particularly those connected with indeterminism, there has been a return to this basic idea, with freedom and discernment having also been built into the model in order to incorporate the technical acumen of the investigator into the analysis. Bayes' formula, as discussed in Chap. 1, is still the basic concept used, but the a priori probabilities in the formula are now estimated by the investigator himself. Bayes' formula can easily be adapted to take care of the cases where continuous variables are involved also.

This entire approach to estimation then starts with an a priori distribution of the parameter which is estimated and then this a priori is altered by the results of an experiment to produce an a posteriori distribution of the parameter. One ends up then with a different distribution function for the degree of ignorance of the parameter, and this new distribution now can be considered a priori for later experiments. Actually the distribution function for the final degree of ignorance after a large amount of experimentation would be the same irrespective of the initial assumption of the a priori distribution and so it is not critical in that sense. However, the better the original a priori estimate, the less experimentation is necessary; this makes for efficiency. If certain values of the parameter are excluded because of physical limitations of the problem and thus completely eliminated from the prior distribution function, they will never occur in any a posteriori estimates, but this fact is not true of the technique discussed in Sec. 4-12. There are therefore three phases to this approach: (1) the a priori estimate of the distribution function for the degree of ignorance; (2) experimentation and the use of these experimental results to calculate the probability that possible values of the parameter could have produced these results; and (3) calculation of the new distribution function for the parameter a posteriori. These three phases will now be discussed more fully.

1 The investigator is usually well acquainted with the field in which he is conducting experiments and therefore, of his own knowledge, has ideas about the possible parameter values and the likelihood of their occurrence. He also is well aware of the literature and what results and ideas other investigators have had about the particular parameter he proposes to measure experimentally. His total judgment about the situation then determines the prior. For example, he may believe that he has no knowledge at all except that the parameter must have a lower and upper value determined from physical considerations. Therefore, he might assume an equally likely density function over the range of possibilities. Generally, the prior is bounded in its domain because of problem considerations such as the gravitational pull of the moon, which certainly must be within certain values for the moon to maintain its observable orbit. The percentage of the population voting for a particular candidate has to be somewhere between 0 and 100 percent, etc. The prior may be peaked in the vicinity of what is considered its most likely value and then fall off on each side in a continuous manner down to the limiting boundaries of the parameter. It may not be symmetric, as the investigator may also believe that there should be some bias built into the model. In any event, a distribution function is hypothesized for the parameter and, except for minor details, all those closely connected with the field would probably construct the prior very similarly in domain and shape. There are some advantages to specifying a definite functional form for the prior and adjusting the range and shape through the change in the parameters in this functional form. We shall discuss this advantage in phase 3, but actually, with computers available in most cases, any distribution can be easily handled, including one that is merely sketched in by hand. Computationally, then, it is often easier to assume a discrete set of possible values for the parameter although they must of course constitute a very large set of points over the domain if they are to simulate a continuous parameter.

2 Experimentation is now conducted, and it becomes necessary to know the probability of obtaining the observed results for all possible values of the parameter. This means the conditional probability of the observed results (or at least a measure of this conditional probability) given that the actual value of the parameter is, say, μ. This is a very straightforward process if only one observation is taken on the population. For example, in a discrete case where the conditional distribution is Poisson and we observe four successes, the conditional probability of exactly four successes given that the parameter value is μ, is therefore

$$\frac{e^{-\mu}\mu^4}{4!}$$

If a continuous distribution for the errors of experimentation are hypothesized, such as a normal distribution with a known σ, we could use the likelihood value as described in Sec. 5-6. For example, if $\sigma = 5$ and the observed value was 10.2, the likelihood value for this observation with a given parameter value μ would be

$$\frac{1}{\sqrt{2\pi(25)}} \exp - \frac{(10.2 - \mu)^2}{2(25)}$$

As a matter of fact, we could use other measures of this conditional probability, depending upon our evaluation of the importance of certain observed results on the prior. It would be possible to calculate the probability of observing a deviation as large as 10.2 or larger with a given μ and use this value as a measure of the conditional probability such as

$$\frac{2}{\sqrt{2\pi}\,5}\int_{10.2}^{\infty} \exp -\frac{(x-\mu)^2}{2(25)}\,dx \qquad \text{if} \quad \mu < 10.5$$

or

$$\frac{1}{\sqrt{2\pi}\,5}\int_{-\infty}^{10.5} \exp -\frac{(x-\mu)^2}{2(25)}\,dx \qquad \text{if} \quad \mu > 10.5$$

These values may be obtained from the normal tables for all μ in which one is interested. Obviously, to assume μ as discrete is an advantage in this case.

When more than one observation is available as a measurement of the population parameter, the likelihood function evaluated for a specific parametric value can be used as a measure of the conditional probability for both the discrete and continuous case, as was done in Sec. 5-6.

The question always arises whether some function of these same n observations would not be as or more effective in determining the parameter than using all n observations and thus perhaps reducing the number of terms in the likelihood function to be used for calculations. For example, if n observations $(x_1, x_2, \ldots, x_n)$ are drawn from a normal distribution with mean μ and standard deviation σ, the likelihood function would be

$$\frac{1}{\sqrt{(2\pi)^n \sigma^n}} \exp -\sum_{i=1}^{n} \frac{(x_i-\mu)^2}{2\sigma^2}$$

Let us assume that we are attempting to estimate μ and that σ is known. If we observe that $\sum_{i=1}^{N}(x_i-\mu)^2 = \sum_{i=1}^{n}[(x_i-\bar{x})+(\bar{x}-\mu)]^2$, where $\bar{x} = x_1 + x_2 + \cdots x_n$, the likelihood function could be written as

$$\left[\frac{\sqrt{n}}{\sqrt{2\pi}\,\sigma} \exp -\frac{(\bar{x}-\mu)^2}{2(\sigma/\sqrt{n})^2}\right]\left[\frac{\sqrt{n}}{(\sqrt{2\pi})^{n-1}\sigma^{n-1}} \exp -\sum_{i=1}^{n}\frac{(x_i-\bar{x})^2}{2\sigma^2}\right]$$

The left-hand factor contains only $\bar{x}$ and μ; the right-hand factor is a function of the observations only with no μ present. Thus $\bar{x}$ is what is called a sufficient statistic to estimate μ (a sufficient estimator), and no additional information is obtained by using the observations individually. In this case, therefore, it would be logical to use the density function for the mean as the likelihood of the occurrence of the n observations; this would simplify the problem from n dimensions to 1. This reduction of the dimensions necessary to compute the likelihood function is dependent upon the form of the distribution. It is obviously very difficult to obtain such a function in a complicated situation and, in fact, such a function may not exist.

3   Let us first consider a situation where the parameter θ can take on only a set of discrete values $\theta_1, \theta_2, \ldots, \theta_n$. Let $f_1(\theta_i)$ be the estimate of the observer, a priori, of the probability that θ takes on the value θ_i, where $\sum_{i=1}^{n} f_1(\theta_i) = 1$. Let n random observations be made from the population and the likelihood function $\bar{L}(x_1, x_2, \ldots, x_n | \theta_i)$ be computed. Then, if $g_1(\theta_i | x_1, x_2, \ldots, x_n)$ is the a posteriori probability that the value of the parameter is θ_i, given that the n observations were $x_1, x_2, \ldots, x_n$, we have by Bayes' formula

$$g_1(\theta_i | x_1, x_2, \ldots, x_n) = \frac{f_1(\theta_i)\bar{L}(x_1, x_2, \ldots, x_n | \theta_i)}{\sum_{j=1}^{n} f_1(\theta_j)\bar{L}(x_1, x_2, \ldots, x_n | \theta_j)} \qquad (5\text{-}15)$$

which permits the estimation of all other a posteriori values of the probability for each possible value of θ_i. If at a later date a new set of observations $y_1, y_2, \ldots, y_n$ are taken on the population, the $g_1(\theta_i | x_1 \cdots x_n)$ may now be considered as a new prior $f_2(\theta_i)$ and the process repeated in order to obtain a better estimate of the probabilities for each θ_i.

In the continuous case, the a priori probability that θ lies between θ and $\theta + d\theta$ is $f_1(\theta)\, d\theta$ $(a < \theta < b)$ and Bayes' formula gives for $g_1(\theta | x_1, x_2, \ldots, x_n)\, d\theta$ the a posteriori value of the probability,

$$g_1(\theta | x_1, x_2, \ldots, x_n)\, d\theta = \frac{f_1(\theta)\, d\theta \bar{L}(x_1, x_2, \ldots, x_n | \theta)}{\int_a^b \bar{L}(x_1, x_2, \ldots, x_n | \theta) f_1(\theta)\, d\theta} \qquad (5\text{-}16)$$

where the domain of possible values of θ is still between a and b. Considerable research has gone into the problem of how to choose $f_1(\theta)$ and the functions generating the likelihood functions so that the denominator of Eq. (5-16) is summable or integrable, depending on which of the two above forms our interest is centered on. Also a very useful property would be that the actual functional form of $g(\theta)$ would be the same as that for $f_1(\theta)$, with the only difference being parametric changes in this function. Some success has been achieved in this direction; the reader is referred to a text on this subject.

EXAMPLE 5-25   An investigator is aware, from physical considerations, that the phenomenon he is about to observe is a Poisson process with an occurrence rate of either 1 per unit time or 3 per unit time.   From his knowledge of the physical situation, his estimate of the probabilities of these two possibilities is $\frac{1}{4}$ and $\frac{3}{4}$, respectively.   He conducts three experiments and observes the occurrence for a unit of time in each case and records the events 6, 2, and 4. What is the a posteriori distribution of the possible parametric values?   The likelihood of the occurrence of these observations, if the parameter is 1 or 3, respectively, would be

$$P(6, 2, 4 \mid 1) = \frac{e^{-1}}{6!} \frac{e^{-1}}{2!} \frac{e^{-1}}{4!}$$

$$P(6, 2, 4 \mid 3) = \frac{e^{-3}3^6}{6!} \frac{e^{-3}3^2}{2!} \frac{e^{-3}3^4}{4!}$$

and thus the a posteriori probability of the value 1 for the parameter, by Eq. (5-15), would be

$$P(1 \mid 6, 2, 4) = \frac{\frac{1}{4}(e^{-3}/6!2!4!)}{\frac{1}{4}(e^{-3}/6!2!4!) + \frac{3}{4}(e^{-9}3^{12}/6!2!4!)} = \frac{1}{1 + 3^{13} \cdot e^{-6}}$$

$$\approx \frac{1}{1 + 3952} \approx \frac{1}{3953}$$

and thus

$$P(3 \mid 6, 2, 4) \approx \frac{3952}{3953}$$

The average of the three observations would be 4; using the technique of confidence limits and the normal approximation with $\sigma = \sqrt{\mu}$, the limits for a Poisson process with an average of 4 and a probability of .99 would be -1.2 to 9.2, which tends to indicate the same result but in a much less convincing manner.   ////

EXAMPLE 5-26   A scientist knows very little about a parameter except that it is greater than zero and less than 4.   He therefore assumes it is equally likely over this region and thus has a density $f_1(\mu) = \frac{1}{4}$ $(0 < \mu < 4)$.   Unfortunately, the measuring device at his disposal is *biased* (inasmuch as it never yields a value x less than the true value x_0 it is measuring).   Thus the value indicated on the meter is always greater than x_0 and has a density $f(x \mid x_0) = 2e^{-2(x-x_0)}$ $(x \geq x_0)$.   Two observations are made by the scientist on his population, and

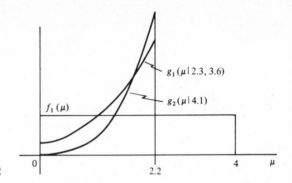

FIGURE 5-12

they turn out to be 2.2 and 3.6, respectively.   What is the change caused by these results in the density function representing his ignorance of the parameter? He makes a third observation that happens to be 4.1.   How does this observed value further alter the distribution?

The a posteriori value of the parameter μ that lies between μ and $\mu + d\mu$, when the observations 2.2 and 3.6 are given, would be [by Eq. (5-16)]

$$g_1(\mu|2.2, 3.6)\,d\mu = \frac{2e^{-2(2.2-\mu)}2e^{-2(3.6-\mu)}\frac{1}{4}\,d\mu}{4\int_0^{2.2} e^{-2(5.8-2\mu)}\frac{1}{4}\,d\mu} \qquad 0 \leq \mu \leq 2.2$$

$$= \frac{4e^{4\mu}}{e^{8.8} - 1}\,d\mu \qquad 0 \leq \mu \leq 2.2$$

since the parameter must be less than the smallest observed value, because of the manner in which the equipment operates.   The distribution of $g_1(\mu|2.2, 3.6)$ is shown in Fig. 5-12.

If we now consider $g_1(\mu) = f_2(\mu)$ as the prior and consider the fact that 4.1 is now observed, we have

$$g_2(\mu|4.1)\,d\mu = \frac{2e^{-2(4.1-\mu)}4e^{4\mu}/(e^{8.8} - 1)}{\int_0^{2.2} 2e^{-2(4.1-\mu)}4e^{4\mu}/(e^{8.8} - 1)\,d\mu}$$

$$= \frac{e^{6\mu}\,d\mu}{\int_0^{2.2} e^{6\mu}\,d\mu} = \frac{6e^{6\mu}\,d\mu}{e^{13.2} - 1} \qquad 0 \leq \mu \leq 2.2$$

which is shown also in Fig. 5-12.

This would seem to indicate that the probability of the true value of the parameter being somewhere between 2 and 2.2 is

$$P(2 < \mu < 2.2) = \int_2^{2.2} \frac{6e^{6\mu}}{e^{13.2} - 1}\,d\mu = \frac{e^{13.2} - e^{12.0}}{e^{13.2} - 1}$$

$$\approx 1 - e^{-1.2} = 0.70$$

This is as far as one can go in pinpointing the parameter at this stage of the investigation. More experimentation would be enlightening. ////

(Suggested Problems 5-34 to 5-36)

5-8 SEARCH THEORY AND DISTRIBUTION OF EFFORT

A subject which has contemporary interest and is also probabilistic in formulation is known as the *theory of search* or, analogously, *best distribution of effort*. This field of study, which has been exploited in many different directions, in both material content and mathematical depth, is beyond the scope of this text. However, it seems in order to consider here the simplified one-dimensional case since the solution has interesting probabilistic consequences and at the same time the mathematical approach to the solution is quite different. Although the formulation here is in terms of searching for an object, the reader can intuitively see that the ideas can equally well apply to problems such as how to find customers, how to distribute research and development money, how one should divide an advertising budget between daily papers, magazines, and direct mail, etc.

An object is known to be somewhere along the x axis. The probability that it actually lies between x and $x + dx$ is given, a priori, by the density function $p(x)$. We have a total amount of effort E at our disposal to locate this object, where E itself may be expressed as plane hours, dollars, or any other meaningful unit. How shall we expend this effort along the x axis in order to maximize the probability of finding the object (using some reasonable hypothesis of the effectiveness of the expenditure of effort at a point on the x axis)?

A reasonable hypothesis, then, for our original problem would be that the probability of success at a point (i.e., at least one sighting) is an exponential function of the density of effort expended at that point. (See Examples 2-11 and 2-12.)

Since our original a priori density function for the position of the object was $p(x)$, the probability that the object will be observed at least once as a function of any chosen $\phi(x)$ is given by

$$P[\phi(x)] = \int_{-\infty}^{\infty} p(x)(1 - e^{-\phi(x)}) \, dx \qquad (5\text{-}17)$$

where $\int_{-\infty}^{\infty} \phi(x) \, dx = E(\phi(x) \geq 0)$ and $\int_{-\infty}^{\infty} p(x) \, dx = 1$.

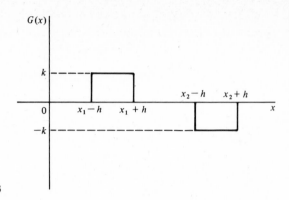

FIGURE 5-13

The problem then reduces itself to one of choosing $\phi(x)$ to maximize P subject to the necessary restrictions. This procedure is equivalent to maximizing the percentage of objects sighted, assuming there were a large number of such objects all distributed over x with density $p(x)$.

In order to solve this problem we need to define a function $G(x)$ (actually not only a function of x but also of four parameters x_1, x_2, h, and k) as in Fig. 5-13, which function is zero except in the neighborhood of x_1 and x_2. From the definition of $G(x)$, as shown in Fig. 5-13, it is apparent that $\int_{-\infty}^{\infty} [\phi(x) + tG(x)]\,dx = E$, where $0 \le t \le 1$. This function $G(x)$ obviously transfers a certain amount of search effort from x_2 to x_1. This discussion is a modification and less rigorous treatment of the subject than that given by B. O. Koopman[1] related to problems in The Theory of Search.

On the assumption that $\phi(x)$ is the "best" distribution of effort in the sense that it will maximize the probability of success, P, then

$$P[\phi(x)] - P[\phi(x) + tG(x)] \ge 0$$

or

$$\int_{-\infty}^{\infty} [p(x)(1 - e^{-\phi(x)}) - p(x)(1 - e^{-[\phi(x) + tG(x)]})]\,dx \ge 0$$

or

$$\int_{-\infty}^{\infty} p(x)e^{-\phi(x)}(e^{-tG(x)} - 1)\,dx \ge 0$$

Expanding $e^{-tG(x)}$ in a power series of t and assuming t is very small but positive, so that we may neglect all terms in the expansion but the first one, we have

$$\int_{-\infty}^{\infty} p(x)e^{-\phi(x)}[-tG(x)]\,dx \ge 0 \qquad t \text{ very small but positive} \qquad (5\text{-}18)$$

[1] B. O. Koopman: The Theory of Search, III. The Optimum Distribution of Searching Effort, *Operations Research*, vol. 5, no. 5, October, 1957.

Assuming x_2 is any point where some search is to be conducted [that is, $\phi(x) > k > 0$, since otherwise transfer of a positive amount of effort would be impossible] and simplifying the integral by remembering $G(x)$ is zero except in the neighborhood of x_1 and x_2, we may write Eq. (5-18) as

$$\int_{x_1-h}^{x_1+h} p(x)e^{-\phi(x)}(-tk)\,dx = \int_{x_2-h}^{x_2+h} p(x)e^{-\phi(x)}(tk)\,dx \geq 0 \qquad (5\text{-}19)$$

Now, utilizing the mean-value theorem for integrals, we have

$$-tk[2hp(\xi)e^{-\phi(\xi)} - 2hp(\eta)e^{-\phi(\eta)}] \geq 0 \qquad (5\text{-}20)$$

where

$$x_1 - h < \xi < x_1 + h \qquad \text{and} \qquad x_2 - h < \eta < x_2 + h$$

As $h \to 0$, $\xi \to x_1$, $\eta \to x_2$, and k is constant and positive by definition but can be assumed as small as we wish, and so

$$-t[p(x_1)e^{-\phi(x_1)} - p(x_2)e^{-\phi(x_2)}] \geq 0 \qquad (5\text{-}21)$$

The expression in Eq. (5-21) is true only if the quantity inside the brackets is negative or

$$p(x_1)e^{-\phi(x_1)} - p(x_2)e^{-\phi(x_2)} \leq 0$$

and thus

$$p(x_1)e^{-\phi(x_1)} \leq p(x_2)e^{-\phi(x_2)} \qquad (5\text{-}22)$$

If, however, we had decided to let x_1 be a point where effort was to be expended [that is, $\phi(x_1) > k > 0$], then we would have interchanged x_1 and x_2 in our argument and thus shown that

$$p(x_2)e^{-\phi(x_2)} \leq p(x_1)e^{-\phi(x_1)}$$

which is inconsistent unless the equality is true.   Apparently, then, all the points for which $\phi(x) > k$ (k small as we wish) satisfy the equality

$$p(x)e^{-\phi(x)} = C \qquad \text{or} \qquad p(x) = Ce^{\phi(x)} \qquad C \text{ a constant}$$

since equality for all x is tantamount to the expression itself being equal to a constant.

When $\phi(x) = 0$, $p(x) = C$ and it can be proved rigorously that search should be conducted at all points for which $p(x) \geq C$ and no search conducted where $p(x) < C$.

The constant C is determined from the condition that the search is limited or $\int \phi(x)\,dx = E$.   Since $\phi(x) = \ln[p(x)] - \ln C$, $\int \{\ln[p(x)] - \ln C\}\,dx = E$, where again the integration is carried over those values of x for which search is

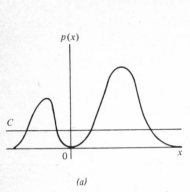

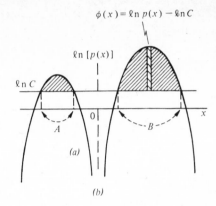

FIGURE 5-14

to be conducted.   Geometrically the answer to our problem is evident.   In Fig. 5-14a we have the original a priori function $p(x)$ and in Fig. 5-14b we have the graph of the $\ln[p(x)]$.

If we are going to search only for values of $p(x) \geq C$, obviously the values are above the line $\ln C$ in Fig. 5-14b and then there are two regions, A and B, for this particular $p(x)$ which are to be included in the search effort.   Since $\phi(x) = \ln[p(x)] - \ln C$, the density of search is equal to the length of the element as shown in Fig. 5-14b and the total shaded area represents the value of $E = \int \{\ln[p(x)] - \ln C\} \, dx$, where the values of x included in the integration are those included in the regions A and B as illustrated for this $p(x)$.   It can be proved rigorously (but clearly from the geometry) that there always exists a finite domain of x for a finite E (otherwise $\int \{\ln[p(x)] - \ln C\} \, dx$ would not exist) and that one can always adjust C so that the area between $\ln[p(x)]$ and $\ln C$ is E.   Furthermore, the increase in this area can be shown to be monotonic as the $\ln C$ line is moved downward in Fig. 5-14b.   Thus the final result is unique (i.e., only one C for a given E).

The shaded area in Fig. 5-15a represents the probability p that the search was successful after an expenditure of effort E and thus the unshaded area under $p(x)$ or $1 - p$ represents the probability of failure.   If the curve bounding the unshaded area $p'(x)$ (solid outline in Fig. 5-15b) is now divided by $1 - p$ in order to bring it back to unit area as in Fig. 5-15b (dotted outline), we now have the conditional a priori probability of the position of the object, given that it was not found in the original search.   If additional effort E' becomes available at a later date, we can use this curve $p'(x)/(1 - p)$ in exactly the same manner as $p(x)$ was handled originally in determining how to distribute this new resource E' in the best possible manner.

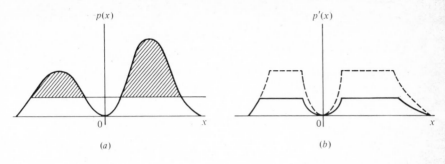

FIGURE 5-15

Here it is important to note, however, that the same overall result would be attained if we had $E + E'$ at the start and allotted it properly. This is true only for the exponential effectiveness of search $(1 - e^{-\phi(x)})$ and reminds us of some of the unique properties of the exponential density function associated with the Poisson hypothesis.

EXAMPLE 5-27 The a priori position of an object situated somewhere along the x axis is given by the probability density $p(x) = \frac{1}{2}e^{-|x|}(-\infty < x < \infty)$. What is the probability of finding the object if four units of search are available for the effort? The graph of $p(x)$ is shown in Fig. 5-16a and the $\ln[p(x)]$ in 5-16b. The shaded area in Fig. 5-16b must equal 4 so that the intersection of $\ln[p(x)]$ with the $\ln \lambda$ must be at the points $(-2, -2 - \ln 2)$ and $(2, -2 - \ln 2)$. Thus $\ln \lambda = -2 - \ln 2$ or $\lambda = \frac{1}{2}e^{-2}$, and the search should take place between $x - 2$ and $x = 2$ only with the density of search, $\phi(x)$, in that domain as

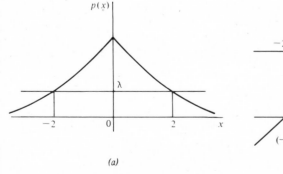

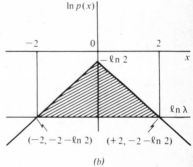

(a)

(b)

FIGURE 5-16

$\phi(x) = 2 - |x|$. The probability of finding the object with four units of search available would then be

$$P = 2 \int_0^2 \left(\frac{e^{-x}}{2} - \frac{e^{-2}}{2} \right) dx = 1 - 3e^{-2} \approx .61$$

////

(Suggested Problems 5-37 to 5-40)

PROBLEMS

In Probs. 5-1 to 5-3 use the type of reasoning employed in establishing Theorems 4-1 and 4-2 rather than attempt an analytic proof.

5-1 The variate x has two admissible values 0, 1 and $f(0) = q, f(1) = p$. Given that $x_1, x_2, \ldots, x_n$ are random observations of x (hence independently distributed as x in repeated trials) and $y = x_1 + x_2 + \cdots + x_n$, show that the distribution of y in repeated independent trials is binomial with parameters n, p.

5-2 The variate u has the binomial distribution with parameters n_1, p and the independent variate v has the binomial distribution with parameters n_2, p (that is, p is the same in both). If $z = u + v$, show that the distribution of z is binomial with parameters n, p, where $n = n_1 + n_2$.

5-3 The distribution of the variate x is geometric with parameter p. Given that $x_1, x_2, \ldots, x_n$ are random observations of x and $y = x_1 + x_2 + \cdots + x_n$, show that y has the Pascal distribution with the same parameter p.

5-4 The discrete variate x has the distribution $f(x) = \frac{1}{4}$ ($x = 2, 4, 6, 8$). If two random observations are taken and $y = x_1 + x_2$, find the distribution of y.

5-5 A home-to-home salesman finds that he has a probability of .6 of finding a woman at home during the day and a probability of .7 of selling her something if she is at home. If he visits 30 houses a day, what is the marginal distribution of the number of sales that he makes?   *Ans.* $f(x) = C(30, x)(.42)^x(.58)^{30-x}$.

5-6 Independent variates x_1, x_2 have the Poisson distribution with respective parameters μ_1, μ_2. Show that the variate $y = x_1 + x_2$ has the Poisson distribution with parameters $\mu = \mu_1 + \mu_2$. Extend by induction to any number of independent Poisson variates.

  (Hint: Set up the joint distribution of x_1, x_2; replace x_2 by $y - x_1$; multiply and divide by $y!$; then sum over x_1 from $x_1 = 0$ to $x_1 = y$, using the binomial theorem.)

5-7 Using the same approach as in Prob. 5-6, show analytically that the sum of two independent binomial variates with respective parameters (n_1, p) and (n_2, p) is binomially distributed with parameters (n, p), where $n = n_1 + n_2$.

5-8   The variate x has the Poisson distribution with parameter μ and the conditional distribution of y, given x is binomial with parameters (x, p), that is, $\phi(y|x) = C(x, y)p^y q^{x-y}$ $(y = 0, 1, 2, \ldots, x)$. Show that the marginal distribution of y is Poisson with parameter μp.

(*Hint:* Set up the joint distribution of x, y. Introduce a new variable $u = x - y$ and sum over u from 0 to ∞ by recognizing the series expansion.)

5-9   Independent variates x_1, x_2, x_3, x_4 have the Poisson distribution with respective parameters $\mu_1 = .21$, $\mu_2 = 1.08$, $\mu_3 = .61$, $\mu_4 = .10$. Using the results of Prob. 5-6, determine the probability that $y = 3$, where $y = x_1 + x_2 + x_3 + x_4$.

Ans. $\frac{4}{3}e^2$.

5-10   The variates x and y are jointly distributed as follows:

$$f(x, y) = x \sin y \qquad 0 \le y \le \pi, 0 \le x \le 1$$

Find the marginal distribution of each variate and show that they are independent.   *Ans.* $f_1(x) = 2x \ (0, 1)$, $f_2(y) = \frac{1}{2} \sin y \ (0, \pi)$.

5-11   The variates x and y are jointly distributed as follows:

$$f(x, y) = Kx^2 y(2 - y) \qquad (0, 2) \text{ each}$$

Find K and the marginal distribution of each.   *Ans.* $K = \frac{9}{32}$.

5-12   Find the marginal and the conditional distributions of x and y which are jointly distributed as follows:

$$f(x, y) = (6/5)x^2(4xy + 1) \qquad (0, 1) \text{ each}$$

5-13   The joint distribution of x and y is

$$f(x, y) = \frac{1}{8}(x^2 - y^2)e^{-x} \qquad 0 \le x < \infty; \ -x \le y \le x$$

Find the marginal distribution of each variate and the conditional distribution of y for a fixed value of x.   *Ans.* $\phi(y|x) = 3(x^2 - y^2)/4x^3 \quad (-x \le y \le x)$

5-14   A random variable x has a probability density function $f(x) = 3x^2 (0 \le x \le 1)$. An observation of x is made which we will call x_1. A second observation x_2 is now made from the range $x_1 \le x_2 \le 1$. Thus x_2 is drawn from a conditional distribution depending upon the original occurrence of x_1. (*a*) What is the conditional density function $\phi(x_2|x_1)$? (*b*) What is the marginal distribution of x_2? (*c*) What is the probability that x_2 is less than or equal to $\frac{1}{2}$?

Ans. $3x_2^2/(1 - x_1^3)$; $-3x_2^2 \ln (1 - x_2^3)$; .008.

5-15   Find the marginal distribution of x if x and y are jointly distributed as follows: $f(x, y) = (2/\pi) (1 - x^2 - y^2) (0 \le x^2 + y^2 \le 1)$. Also show that the integral of $f(x, y)$ over its region of definition is unity.

Ans. $8(1 - x^2)^{3/2}/3\pi \quad (-1, 1)$.

5-16   The variates x and y are jointly distributed as follows (where $\alpha > 0$):

$$f(x, y) = \frac{\sqrt{3}}{6\pi\alpha^2} e^{-(x^2 - xy + y^2)/3\alpha^2} \qquad (-\infty, \infty) \text{ each}$$

Find the marginal distributions of x and y and the conditional distribution of each variate for a given value of the other. Integrate by completing the square in the exponent.

5-17 Prove that a necessary and sufficient condition for the independence of two variates x, y is the twofold requirement (a) that their joint density function $f(x, y)$ factor into pure functions of the separate variates [as $f(x, y) = kg(x)h(y)$, $k = $ constant] and (b) that the admissible region be rectangular (as $A \leq x \leq B$; $a \leq y \leq b$, where A, B, a, b, are constants).

5-18 Prove that a necessary and sufficient condition for the independence of the polar variates r, θ is the twofold requirement that (a) their joint-distribution function be factorable into pure functions of the separate variates and (b) that the admissible region be of the form $A \leq \theta \leq B$, $a \leq r \leq b$, where a, b, A, B are constants.

5-19 The marginal distribution of x is normal with parameters $\mu = 0$, $\sigma = 1$; the conditional distribution of y, given x, is normal with parameters $\mu = kx$, $\sigma = \sqrt{1 - k^2}$; and the conditional distribution of z, given any pair of values of x and y, is normal with parameters $\mu = ky$, $\sigma = \sqrt{1 - k^2}$. (a) State the joint distribution of x, y, z. (b) Find the marginal distributions of y and z. (c) Find the conditional distribution of x, given y. (d) State the joint distribution of y and z. (e) Find the conditional distribution of z, given x (but not given y). Integrations can be carried out by completing the square in the exponent. Make use of formal similarities.

5-20 In the joint distribution of two independent unit normal variates x, y show that the probability that a random point will fall within the square enclosed by the lines $x = -a$, $x = a$, $y = -a$, $y = a$ is less than the probability that it will fall within the circle of equal area defined by $x^2 + y^2 = 4a^2/\pi$. With this fact in mind, prove the inequality

$$\frac{1}{\sqrt{2\pi}} \int_{-a}^{a} e^{-x^2/2} \, dx \leq \sqrt{1 - e^{-2a^2/\pi}}$$

(Hint: Express the square of the required integral as the product of two integrals of identical form but with one in x and the other in y. Rewrite this as a double integral, and then, changing to polar coordinates, integrate over the circle $r = 2a/\sqrt{\pi}$.)

5-21 The joint distribution of x and y is given by $f(x, y) = 1/x^2y^2$ [$(1, \infty)$ each]. A point x', y' is chosen at random in the admissible region and the product $u = x'y'$ and the quotient $v = x'/y'$ of its coordinates are computed. By integrating $f(x, y)$ over the appropriate portions of the admissible region, find (a) $P(u \leq 4)$, (b) $P(v \leq 2)$, (c) the probability of the compound event $u \leq 4$, $v \leq 2$. (d) Do any points of the admissible region satisfy the simultaneous inequalities $u < 2$, $v > 2$?

5-22 The waiting time for the first occurrence of an event governed by a Poisson distribution (with $\mu = pt$) is given by $f(t) = pe^{-pt}$. Suppose that the orders for a certain article of merchandise come from two main sources, the demands per unit time being independent Poisson variates with respective parameters μ_1, μ_2. Show that the elapsed time between orders is exponentially distributed with $\beta = 1/(\mu_1 + \mu_2)$.

5-23 Given that the waiting time t for some chance event is exponentially distributed, and letting t_0 be an arbitrary fixed value of t, find the conditional distribution of waiting time under the restriction that $t > t_0$. Denoting a random value of t by t' deduce that $P(t' > t_0 + x \,|\, t' > t_0) = P(t > x)$. Express this proposition in words.

5-24 The proposition proved in Prob. 5-23 for the exponential distribution may be stated analytically as follows:

$$f(x + y) \bigg/ \int_x^\infty f(z)\,dz = f(y) \qquad y \geq 0$$

Prove, conversely, that if this identity holds for all $y \geq 0$, and the variate is nonnegative, the distribution is exponential. This fact that the conditional probability of exceeding a given arbitrary value by a stated margin (or more) depends only on the margin stipulated and not on the arbitrary starting point is thus a definitive property of the exponential distribution.

(*Hint:* Simplify and differentiate with respect to x. To prove uniqueness, it is then sufficient to show that the differential equation obtained by setting $y = 0$ is satisfied only by an exponential function.)

5-25 In the theory of failure of mechanical systems, business enterprises, etc., the notation $Z(t) = f(t)/[1 - F(t)]$ is used to denote the conditional density function of failure between time t and $t + dt$, given nonfailure prior to time t. For certain retail businesses, Lomax[1] found that this function is well represented by an expression of the form $Z(t) = b/(t + a)$ $(a, b > 0)$. Find $f(t)$, the probability density function of failure time. *Ans.* $(b/a)[a/(t + a)]^{b+1}$ $(0, \infty)$,

5-26 From an arbitrary continuous distribution, 100 observations are made. What is the probability that an additional 4 will all be greater than the smallest one of the hundred? *Ans.* $100/104$.

5-27 Given a variate having a continuous-probability density function for some domain where $x = 0$. What is the probability that if five observations are drawn consecutively the last two will be less than all of the first three? *Ans.* $1/10$.

5-28 Given $f(x) = 2x$ $(0 < x < 1)$. Five observations are drawn at random from this distribution and ordered from the smallest to the largest. Find the distribution of (*a*) the second smallest, (*b*) the median, (*c*) the largest.

5-29 Four observations are drawn at random from the distribution $f(x) = \alpha e^{-\alpha x}$ $(0, \infty)$. These are ordered from the smallest to the largest. (*a*) What is the joint distribution of the second and fourth? (*b*) What is the distribution of the largest?

5-30 Given the density function $f(x) = x$ $(0 \leq x \leq 1)$; $f(x) = 2 - x$ $(1 \leq x \leq 2)$. Three observations are drawn at random from this distribution and arranged in order of magnitude. What is the distribution of the median?

5-31 Prove that the probability that the $(n + 1)$st observation is larger than the greatest of the first n is $1/(n + 1)$ for any continuous distribution.

[1] *J. Am. Stat. Assoc.*, December, 1954.

5-32 Given the density function

$$f(x) = \frac{1}{\beta^{\alpha+1}\Gamma(\alpha+1)}(x-A)^{\alpha}e^{-(x-A)/\beta} \qquad x > 0$$

with α and A also known ($\alpha > 0$). Find the maximum likelihood estimate of β from independent observations $x_1, \ldots, x_n$.   *Ans.* $\sum_i (x_i - A)/n(\alpha + 1)$.

5-33 Given the probability distribution $f(x) = p(1 - p)^{x-1}$ ($x = 1, 2, 3, \ldots$). Find the maximum likelihood estimate of the probability p from the n independent observations $x_1, x_2, \ldots, x_n$.   *Ans.* $1/\bar{x}$.

5-34 The probability of obtaining a head on an individual toss of a perhaps biased coin can be any one of three values .4, .5, or .6. An a priori estimate for these three possibilities are .2, .6, and .2, respectively. The coin is now tossed and a tail is obtained on the first two tosses. After that, the next $n - 2$ tosses produce heads only. Determine the a posteriori values for the parameter as a function of $n \geq 2$. Sketch for $n = 2, 3,$ and 4.

5-35 For the same coin as in Prob. 5-34, an a priori density function for the parameter θ (probability of a head) is assumed of the form

$$f(\theta) = 140(1 - \theta)^3\theta^3 \qquad 0 \leq \theta \leq 1$$

Five observations are now made by tossing the coin five times, and two heads and three tails are observed. What is the a posteriori density function for θ? If a second experiment is conducted and out of 10 trials 2 heads and 8 tails are observed, what is the new distribution for our ignorance of the parameter θ?   *Ans.* (*a*) $5544\theta^5(1 - \theta)^6$.

5-36 Our state of ignorance regarding a certain parameter λ makes a gamma distribution for the density a reasonable prior as

$$\frac{\lambda^{k-1}e^{-\lambda/B}}{B^k(k-1)!} \qquad k \text{ an integer}$$

An experiment indicates a value of $\lambda = 10$. What is the a posteriori density function for λ? Comment on how the mathematical form of the density function might change when new data are incorporated into the estimate.

5-37 The a priori position of an object is given by the density function

$$f(x) = \begin{cases} \frac{1}{4} & 0 < x < 1 \\ \frac{3}{4} & 1 < x < 2 \end{cases} \qquad f(x) = 0 \text{ elsewhere}$$

What is the a posteriori probability that the object is between $x = .5$ and $x = 1.5$, given that three units of search were expended and failed to find the object?   *Ans.* $\frac{1}{2}$.

5-38 The a priori position of an object is given by the density function

$$f(x) = \sqrt{2/\pi}\, e^{-x^2/2} \qquad 0 \leq x < \infty$$

How much search effort is necessary in order that the probability of finding the object will be .90?

5-39 If the a priori position of the object in Prob. 5-38 was $f(x) = e^{-x}$ $(0 \leq x < \infty)$, how different would the answer be?

5-40 The a priori probability for the position of a given object is given by the density function

$$f(x) = \begin{cases} \frac{1}{8} & 0 < x < 1 \\ \frac{5}{8} & 1 < x < 2 \\ \frac{2}{8} & 2 < x < 3 \\ 0 & \text{elsewhere} \end{cases}$$

After 1.5 units of search were expended, the object was not located. What is the probability that the actual position of the object is now between $x = .5$ and $x = 1.5$? (Assume the exponential saturation law and that the effort of 1.5 units was distributed most advantageously.)   *Ans.*   .385.

6

DERIVED DISTRIBUTIONS

6-1 INTRODUCTION

The province of any mathematical theory is the extension of knowledge by deduction from accepted premises. By this means, science is able to make full use of established principles and avoid degeneration into sheer empiricism. An important deductive problem in the field of random variables is the derivation of the distribution of some function (or the joint distribution of several functions) of initial variates, for which the joint distribution is already known.

In this chapter, we shall study this problem beginning with a function of a single variate and progressing to multivariate functions. In the univariate case, suppose that y is a known function of another variate x. Given the distribution of x, the problem is to determine the distribution of y. A general solution of this problem can be obtained in terms of density functions, provided that y is a monotonic function of x and its derivative with respect to x exists.

6-2   MONOTONIC FUNCTIONS

If y increases whenever x increases, it is said to be a *monotonically increasing function;* if y decreases whenever x increases, it is said to be a *monotonically decreasing function.*   For example, e^x is a monotonically increasing function, whereas e^{-x} is a monotonically decreasing function.   The function x^2 is not monotonic if x can have negative as well as positive values.   However, if $x \geq 0$, x^2 is monotonically increasing, whereas if $x \leq 0$, x^2 is monotonically decreasing.

The nature of a monotonic function necessitates a unique correspondence between the dependent and independent variable.   Consequently, for mathematical purposes, the latter can be regarded in turn as a function of the former.   Moreover, if the initial function can be differentiated with respect to x, the inverse function can be differentiated with respect to y.   The distribution function $F(x)$, although not strictly monotonic in the above sense, is nevertheless monotonic in a broader sense in that it never decreases.

6-3   THE DENSITY FUNCTION OF A MONOTONICALLY INCREASING FUNCTION

Assume that x has a density function $f(x)$ and let $F(x)$ and $G(y)$ denote, respectively, the distribution functions of x and y.   Suppose that y is a differentiable, monotonically increasing function of x and that $y = b$ when $x = a$.   Then since $y \leq b$ when and only when $x \leq a$, we have at once $P(y \leq b) = P(x \leq a)$, whence $G(b) = F(a)$.   As stated in the previous paragraph, we may regard x as a function of y, say $\phi(y)$, so that $a = \phi(b)$.   Differentiating both sides of the equation $G(b) = F(a)$ with respect to b yields:

$$\frac{dG(b)}{db} = \frac{dF(a)}{db} = \frac{dF(a)}{da}\frac{da}{db} = \frac{dF(a)}{da}\frac{d\phi(b)}{db} \qquad (6\text{-}1)$$

By definition, however, the derivative of the distribution function is the density function.   Hence

$$g(b) = f(a)\frac{d\phi(b)}{db} = f[\phi(b)]\frac{d\phi(b)}{db} \qquad (6\text{-}2)$$

Or, replacing b by y,

$$g(y) = f[\phi(y)]\frac{d\phi(y)}{dy} \qquad (6\text{-}3)$$

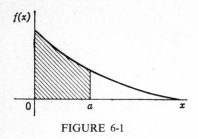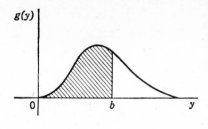

FIGURE 6-1

EXAMPLE 6-1 Suppose that x has the probability density function $f(x) = e^{-x}$ $(0, \infty)$ and we wish to derive the probability density function of y where $y = \sqrt[3]{x}$. Here $x = y^3$ and $dx/dy = 3y^2$. Thus the distribution of y is given by

$$g(y) = 3y^2 e^{-y^3} \qquad (0, \infty)$$

The two density functions are shown in Fig. 6-1. To carry the example further, consider the probability that x lies between 0 and ln 2. This quantity is

$$P(x \leq \ln 2) \equiv F(\ln 2) = \int_0^{\ln 2} e^{-x}\, dx = \tfrac{1}{2}$$

At the same time that $x \leq \ln 2$ we have $y \leq \sqrt[3]{\ln 2}$, and

$$P\left(y \leq \sqrt[3]{\ln 2}\right) \equiv G\left(\sqrt[3]{\ln 2}\right) = \int_0^{\sqrt[3]{\ln 2}} 3y^2 e^{-y^3}\, dy = \tfrac{1}{2}$$

Thus, the two probabilities are equal and, in general, the two shaded areas of Fig. 6-1 are equivalent. ////

EXAMPLE 6-2 Rounding errors in a computed quantity are often uniformly distributed over a range extending from $-.5$ to $.5$ unit in the last decimal place carried. Thus, if x represents the error in such a computed quantity, we may say $f(x) = 1$ $(-.5, .5)$ where it is understood that the scale of x is chosen so that one unit of x represents one unit in the last decimal place carried. Suppose that the quantity under consideration is the speed S of a moving particle and the corresponding computed value is C. Then, neglecting all sources of error other than that due to the computing process itself, we may put

$$x = S - C \quad \text{and} \quad f(x) = 1 \quad (-.5, .5)$$

Assume that on the scale of x both the actual and the computed values S, C are considerably greater than unity. Now let us determine the behavior of the

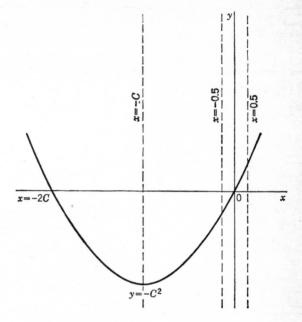

FIGURE 6-2

errors in the computed value of *kinetic energy*. Here the true value is $(m/2)S^2$, whereas the computed value is $(m/2)C^2$, so that the error is

$$\frac{m}{2}(S^2 - C^2)$$

For the present let us put $y = S^2 - C^2$ and derive the distribution of y; afterward we can easily find the distribution of the error itself, which is $(m/2)y$. Rewriting S as $C + x$ we obtain

$$y = (C + x)^2 - C^2 = 2Cx + x^2$$

This equation (Fig. 6-2) represents a parabola with vertex (turning point) at $x = -C$, $y = -C^2$, and since $C > 1$ it follows that y is a monotonically increasing function of x over the permissible range $-.5 \le x \le .5$. The corresponding limits on y are $-C + .25 \le y \le C + .25$. From the equation of the parabola we have at once

$$(C + x)^2 = C^2 + y$$

and on the *right* side of the vertex,

$$x = -C + \sqrt{C^2 + y}$$

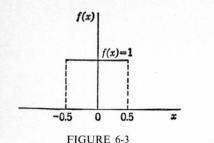
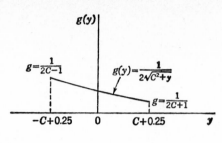

FIGURE 6-3

Since $f(x)$, being equal to unity, is independent of x, we have $f[\phi(y)] = 1$ and the only variable part of the density function of y is contributed by the derivative

$$\frac{dx}{dy} = \frac{1}{2\sqrt{C^2 + y}}$$

Hence (Fig. 6-3),
$$g(y) = \frac{1}{2\sqrt{C^2 + y}} \qquad (-C + .25, C + .25)$$

Sometimes readers are confused because there is no variable term to manipulate in $f(x)$. Therefore we shall present an alternative derivation of $g(y)$ based on the distribution function $G(y)$. Let $y = b$ when $x = a$. Then $y \le b$ when, and only when, $x \le a$; and so

$$G(b) = F(a) = \int_{-.5}^{a} dx = a + .5$$

But the relation between x and y yields $a = -C + \sqrt{C^2 + b}$. Hence,

$$G(b) = -C + \sqrt{C^2 + b} + .5$$

Accordingly,
$$g(b) = \frac{dG(b)}{db} = \frac{1}{2\sqrt{C^2 + b}}$$

and in general,
$$g(y) = \frac{1}{2\sqrt{C^2 + y}} \qquad (-C + .25, C + .25)$$

Finally, denoting the actual error $(m/2)y$ by z, we have $y = (2/m)z$ and the density function of z, say $h(z)$, is

$$h(z) = \frac{1}{\sqrt{2m}} \frac{1}{\sqrt{k + z}} \qquad \left(-\frac{mC}{2} + \frac{m}{8}, \frac{mC}{2} + \frac{m}{8}\right)$$

where $k = mC^2/2$.

////

6-4 THE DENSITY FUNCTION OF A
MONOTONICALLY DECREASING FUNCTION

If y is a monotonically decreasing function, and $y = b$ when $x = a$, then

$$P(y \le b) = P(x \ge a)$$

Thus, according to the previous assumptions and notation,

$$G(b) = 1 - F(a) \qquad g(b) = -f[\phi(b)] \frac{d\phi(b)}{db} \qquad (6\text{-}4)$$

But if y is a monotonically decreasing function of x, $\phi(y)$ is likewise a monotonically decreasing function of y. Consequently, $d\phi(b)/db$ is always negative, and $-d\phi(b)/db$, being always positive, can be regarded as the absolute value of the derivative. Therefore, the equation for $g(b)$ can be rewritten as $g(b) = f[\phi(b)]\,|d\phi(b)/db|$ and, in general,

$$g(y) = f[\phi(y)] \left| \frac{d\phi(y)}{dy} \right| \qquad (6\text{-}5)$$

Note that Eq. (6-5) applies equally well to either case.

EXAMPLE 6-3 Consider the function $y = 1/x$ where x is distributed as follows:

$$f(x) = \frac{6x}{(1 + x)^4} \qquad (0, \infty)$$

Here $x = 1/y$, $|dx/dy| = 1/y^2$, and

$$g(y) = \frac{6/y}{(1 + 1/y)^4} \frac{1}{y^2} = \frac{6y}{(1 + y)^4}$$

Since $y \to 0$ as $x \to \infty$, and $y \to \infty$ as $x \to 0$, the range of y is $(0, \infty)$; we do not say $(\infty, 0)$ because we are interested in the distribution of y, and we always represent the range of any variate as the interval from its lowest value to its highest value. In this example we have arrived at the rather unusual result that the reciprocal of a variate has the same form of distribution (Fig. 6-4) as the variate itself:

$$g(y) = \frac{6y}{(1 + y)^4} \qquad (0, \infty) \qquad ////$$

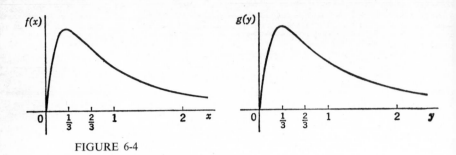

FIGURE 6-4

EXAMPLE 6-4   Let x be normally distributed with parameters μ_1, σ_1 and let y be a linear function of x, say $y = a + bx$.   If the coefficient b were zero, the linear function would degenerate to a constant, and we shall exclude this trivial case.   For nonzero values of b, the function is monotonically increasing if b is positive and monotonically decreasing if b is negative.   In either case, we have

$$x = \frac{y - a}{b}, \qquad \left|\frac{dx}{dy}\right| = \frac{1}{|b|}$$

In the density function of x,

$$f(x) = \frac{1}{\sigma_1 \sqrt{2\pi}} e^{-(x - \mu_1)^2 / 2\sigma_1^2} \qquad (-\infty, \infty)$$

we must express x in terms of y and then multiply by $1/|b|$.   Thus we put

$$x - \mu_1 = \frac{y - a - b\mu_1}{b}, \qquad \frac{(x - \mu_1)^2}{2\sigma_1^2} = \frac{(y - a - b\mu_1)^2}{2b^2\sigma_1^2} = \frac{(y - \mu)^2}{2\sigma^2}$$

where $\mu = a + b\mu_1$ and $\sigma = |b|\sigma_1$.   Accordingly,

$$g(y) = \frac{1}{|b|} \frac{1}{\sigma_1 \sqrt{2\pi}} e^{-(y - \mu)^2 / 2\sigma^2} = \frac{1}{\sigma \sqrt{2\pi}} e^{-(y - \mu)^2 / 2\sigma^2} \qquad (-\infty, \infty)$$

The range of y is $(-\infty, \infty)$ because a constant times infinity yields infinity. We have therefore proved that any nontrivial linear function of a normal variate is normally distributed.   ////

This result is important enough to warrant a formal statement as a theorem.

Theorem 6-1   If x is normally distributed with parameters μ_1, σ_1, and y is any linear function $y = a + bx$ ($b \neq 0$), then y is normally distributed with corresponding parameters $\mu = a + b\mu_1$, $\sigma = |b|\sigma_1$.   ////

6-5   THE PROBABILITY TRANSFORMATION

A variable change of fundamental importance in distribution theory is known as the *probability transformation* and is defined by the equation $y = F(x)$, where $F(x)$ is the distribution function of x. This transformation exists for all distributions, and in the class for which probability density functions exist, it has the interesting property of being rectangularly distributed. It is obvious from the definition that the range of y is 0 to 1, and in case $f(x)$ exists we have

$$\frac{dy}{dx} = f(x) \qquad \frac{dx}{dy} = \frac{1}{f(x)}$$

whence
$$g(y) = \frac{f(x)}{f(x)} = 1 \qquad 0 \le y \le 1 \qquad (6\text{-}6)$$

The probability transformation forms a theoretical link between any pair of distributions having probability densities. Theoretically, any distribution for which the probability density exists can be transformed into any other by applying the probability transformation to each and thus bringing the variates into one-to-one correspondence. The value of this lies in the fact that an initial variate having a distribution which is not suitable for certain types of standard statistical analyses can be replaced by another which is derived from the first by a definite transformation and which has a distribution that satisfies the conditions under which the proposed techniques apply. Even when the distribution functions cannot be expressed as explicit closed functions of the variate values, they can be obtained in tabular form by numerical integration. The transformation is then arrived at in the form of a table exhibiting the corresponding values of the two variates for selected values of the equated distribution functions.

EXAMPLE 6-5   Find a function of x having the distribution $g(y) = 3(1 - \sqrt{y})$ $(0, 1)$, given that $f(x) = 6x(1 - x)$, $(0, 1)$. The distribution functions are

$$F(x) = \int_0^x f(x)\,dx = 3x^2 - 2x^3$$

$$G(y) = \int_0^y g(y)\,dy = 3y - 2y^{3/2}$$

Setting $G(y) = F(x)$ we get

$$3y - 2y^{3/2} = 3x^2 - 2x^3$$

Since this must be satisfied identically, the solution is unique and is found by inspection to be $y = x^2$. ////

(Suggested Problems 6-1 to 6-4)

6-6 DENSITY FUNCTIONS OF THE SQUARE AND THE ABSOLUTE VALUE

If y is not a monotonic function of x, the foregoing formulas do not hold. We shall consider two cases of interest, which also indicate the general procedure in all cases where the variable change is not monotonic. First let $y = x^2$. Then $x = \pm\sqrt{y}$ and

$$G(y) = \int_{-\sqrt{y}}^{\sqrt{y}} f(x)\,dx$$

whence, differentiating the definite integral,

Density function of $y = x^2$:

$$g(y) = \frac{f(\sqrt{y}) + f(-\sqrt{y})}{2\sqrt{y}} \qquad (6\text{-}7a)$$

Second, let $y = |x|$. Then,

$$G(y) = \int_{-|x|}^{|x|} f(x)\,dx = \int_{-y}^{y} f(x)\,dx$$

Density function of $y = |x|$:

$$g(y) = f(y) + f(-y) \qquad (6\text{-}7b)$$

EXAMPLE 6-6   Let $y = x^2$ and

$$f(x) = \frac{1}{\sqrt{2\pi}}\,e^{-x^2/2} \qquad (-\infty, \infty)$$

Then $f(\sqrt{y}) = (1/\sqrt{2\pi})e^{-y/2} = f(-\sqrt{y})$ and so

$$g(y) = \frac{1}{\sqrt{2\pi}}\,y^{-1/2}e^{-y/2} \qquad (0, \infty)$$

Again, let $z = |x|$ with x the same as before. Then $f(z) = 1/\sqrt{2\pi}\,e^{-z^2/2} = f(-z)$ and

$$g(z) = \frac{2}{\sqrt{2\pi}}\,e^{-z^2/2} \qquad (0, \infty) \qquad ////$$

EXAMPLE 6-7   The distribution of x is rectangular over the range $(-\tfrac{1}{2}, \tfrac{3}{2})$, that is,

$$f(x) = \tfrac{1}{2} \qquad (-\tfrac{1}{2}, \tfrac{3}{2})$$

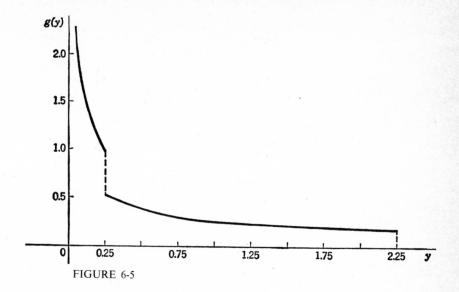

FIGURE 6-5

and $y = x^2$. The total range of y is from 0 to $\frac{9}{4}$, but the fact that negative values of x cut off at $-\frac{1}{2}$ means that $f(-\sqrt{y}) \equiv 0$ as soon as y exceeds $\frac{1}{4}$. Therefore we must distinguish two ranges:

Range 1 ($0 \leq y \leq \frac{1}{4}$):

$$g(y) = \frac{\frac{1}{2} + \frac{1}{2}}{2\sqrt{y}} = \frac{1}{2}y^{-1/2}$$

Range 2 ($\frac{1}{4} < y \leq \frac{9}{4}$):

$$g(y) = \frac{0 + \frac{1}{2}}{2\sqrt{y}} = \frac{1}{4}y^{-1/2}$$

Therefore the distribution of y (Fig. 6-5) has a discontinuity at $y = \frac{1}{4}$. The complete distribution of y is stated by writing these two equations in tandem:

$$g(y) = \frac{1}{2}y^{-1/2} \qquad 0 \leq y \leq \frac{1}{4} \qquad g(y) = \frac{1}{4}y^{-1/2} \qquad \frac{1}{4} < y \leq \frac{9}{4} \qquad ////$$

(*Suggested Problems 6-5 to 6-7*)

6-7 THE DISTRIBUTION OF A FUNCTION OF TWO OR MORE INDEPENDENT VARIATES

Most derived distributions involve more than a simple variable change. Typically, we are interested in variables which cannot be measured directly but can be constructed mathematically as functions (such as sums, products, quotients) of variables which are directly measurable. This introduces the consideration of a multidimensional space, but for the time being we shall confine our analysis to two dimensions. The general principles, however, apply to any number of dimensions.

Consider two independent variates x, y with marginal density functions $f_1(x)$ and $f_2(y)$, respectively. By hypothesis, their joint density function $f(x, y)$ is equal to the product of their marginal density functions, that is, $f(x, y) = f_1(x)f_2(y)$. Let u be a monotonic function of x and y in the sense that u behaves as a monotonic function of x when y is held fixed and as a monotonic function of y when x is held fixed. Some examples are $u = x + y$, $u = xy$, $u = x/y$, $u = ye^{-x}$. To obtain the distribution of u, we imagine that one of the initial variates x, y actually is held fixed at some arbitrary value, and the other varies over its possible range, according to its own marginal distribution. For definiteness, assume that x is fixed and y is free to vary. The probability distribution of u which then results is, by definition, the conditional distribution of u given x. Moreover, as we have said, u becomes a monotonic function of the single variable y, although the functional equation relating u and y will contain parameters which depend on x. Therefore, the conditional density function $\phi(u|x)$ can be derived from $f_2(y)$ by the univariate method previously presented. Having done this, we may then construct the joint density function of x and u—say $g(x, u)$—by using the formula

$$g(x, u) = f_1(x)\phi(u|x)$$

Finally, the marginal density of u is obtained by integrating this joint density function with respect to x. The best way to clarify the method is through the consideration of specific examples.

EXAMPLE 6-8. Let us suppose that we have a random variable x which is rectangularly distributed in the region from $x = 0$ to $x = 1$ and thus $f(x) = 1$ $(0, 1)$. Let us draw two observations at random from this infinite population and call them x_1 and x_2. The problem is to find the distribution of the sum of x_1 and x_2, say $u = x_1 + x_2$. What is meant by this is that two observations are drawn at random and added together and the sum computed. This process is repeated over and over again, each time obtaining a new value for the sum. We now ask what the probability distribution of all these values of the sum would be if we could carry out this experiment. Since the population is infinite, its probability distribution is not disturbed by sampling. Hence the sample values behave independently, and the sampling process is equivalent to drawing the first observation from one rectangular distribution and the second from another, both of which run from 0 to 1.

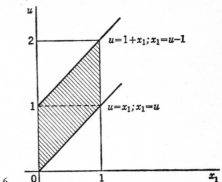

FIGURE 6-6

As the first step in deriving the distribution of u, we regard x_1 as fixed and change the variable in the distribution of x_2, which is given by $f_2(x_2) = 1\,(0, 1)$. The derivative of x_2 with respect to u holding x_1 fixed will, of course, be a *partial* derivative. Thus we have

$$x_2 = u - x_1 \qquad \frac{\partial x_2}{\partial u} = 1$$

whence $\quad \phi(u \mid x_1) = [f_2(u - x_1)]\dfrac{\partial x_2}{\partial u} = (1)(1) = 1 \qquad x_1 \le u \le 1 + x_1$

The range of u is obtained by noting that the equation $u = x_1 + x_2$ yields $u = x_1$ when $x_2 = 0$ and $u = x_1 + 1$ when $x_2 = 1$.

The next step is to obtain the joint distribution of x_1 and u. This is given by

$$g(x_1, u) = f_1(x_1)\phi(u \mid x_1) = 1 \qquad 0 \le x_1 \le 1; x_1 \le u \le 1 + x_1$$

The bivariate region of definition is obtained merely by supplementing the original range of x_1 with a statement of the conditional limits of u. From the diagram of this region (shaded portion of Fig. 6-6) we see that the integration on x_1 must be performed in two steps, which implies that the marginal density of u requires two functions for its definition.

In the range $0 \le u \le 1$ the values of x_1 run from 0 at the left-hand boundary to u at the right-hand boundary, and the corresponding limits of integration are from 0 to u. In the range $1 \le u \le 2$ the left-hand boundary is the line $x_1 = u - 1$ and the right-hand boundary is the vertical line $x_1 = 1$; thus in this

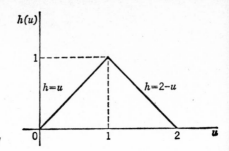

FIGURE 6-7

range, the limits are from $u - 1$ to 1. Hence we obtain the marginal distribution of u as follows:

Range 1 ($0 \le u \le 1$):

$$h(u) = \int_0^u dx_1 = u$$

Range 2 ($1 \le u \le 2$):

$$h(u) = \int_{u-1}^1 dx_1 = 2 - u$$

whereupon,

$$h(u) = u \qquad 0 \le u \le 1 \qquad h(u) = 2 - u \qquad 1 \le u \le 2$$

This triangular distribution is shown in Fig. 6-7.   ////

EXAMPLE 6-9 Two independent random variables x, y are distributed as follows:

$$f_1(x) = \tfrac{1}{2} \quad (1, 3) \qquad \text{and} \qquad f_2(y) = e^{-(y-2)} \quad (2, \infty)$$

Find the distribution of $z = x/y$ (see Fig. 6-8).

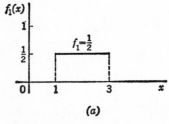

(a)

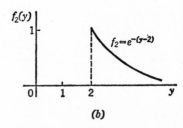

(b)

FIGURE 6-8

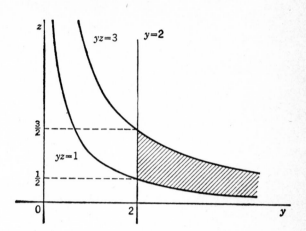

FIGURE 6-9

Since $f_1(x)$ is simpler than $f_2(y)$ it will be easier to determine $\phi(z|y)$ than the alternative. Accordingly, holding y fixed we find the conditional distribution of $z = x/y$ regarded as a function of x alone. Here $x = yz$, $\partial x/\partial z = y$, and

$$[f_1(yz)]y = \tfrac{1}{2}y$$

Now when $x = 1$, $z = 1/y$, and when $x = 3$, $z = 3/y$. Therefore, the conditional distribution of z given y is

$$\phi(z|y) = \tfrac{1}{2}y \qquad \left(\frac{1}{y}, \frac{3}{y}\right)$$

and the joint distribution of y and z is

$$g(y, z) = \tfrac{1}{2}ye^{-(y-2)} \qquad \left(2 \le y < \infty; \frac{1}{y} \le z \le \frac{3}{y}\right)$$

The admissible region (shaded in Fig. 6-9) consists of that portion of the first quadrant bounded on the left by the vertical line $y = 2$, above by the hyperbola $z = 3/y$, below by $z = 1/y$, and open on the right.

The least possible value of z is 0, which is the limit approached as $y \to \infty$. The greatest possible value of z is $\tfrac{3}{2}$, which corresponds to the point of intersection of the higher boundary curve $z = 3/y$ and the line $y = 2$. Once again, the integration must be done in two steps. The lower hyperbola cuts the left boundary at $z = \tfrac{1}{2}$. Below this level, y ranges between the two hyperbolas, the limits being $y = 1/z$ to $y = 3/z$. Above the level $z = \tfrac{1}{2}$, the left-hand boundary is the line $y = 2$, and the limits are $y = 2$ to $y = 3/z$. Thus the distribution of z must be expressed by two different functions, each of which applies to a stated range.

Range 1 $(0 \leq z \leq \frac{1}{2})$:

$$h(z) = \int_{1/z}^{3/z} g(y, z) \, dy = \frac{-e^2}{2} \left[e^{-y}(y+1) \right]_{1/z}^{3/z}$$

$$= \frac{e^2}{2z} [e^{-1/z}(1+z) - e^{-3/z}(3+z)]$$

Range 2 $(\frac{1}{2} \leq z \leq \frac{3}{2})$:

$$h(z) = \int_{2}^{3/z} g(y, z) \, dy = \frac{-e^2}{2} \left[e^{-y}(y+1) \right]_{2}^{3/z}$$

$$= \frac{3}{2} - \frac{e^2}{2z} e^{-3/z}(3+z)$$

We note that when $z = \frac{1}{2}$ both functions yield the same value

$$h\left(\frac{1}{2}\right) = \frac{3 - 7e^{-4}}{2}$$

As a check let us verify that the density function has unit area. To this end consider, first, the following indefinite integral:

$$I(a) = \int \left(1 + \frac{a}{z}\right) e^{-a/z} \, dz$$

If we set $u = -1/z$, this integral becomes

$$I(a) = \int (1 - au)e^{au} \frac{du}{u^2} = \int \frac{e^{au}}{u^2} \, du - a \int \frac{e^{au}}{u} \, du$$

On the other hand,

$$a \int \frac{e^{au}}{u} \, du = \frac{e^{au}}{u} + \int \frac{e^{au}}{u^2} \, du$$

From this, we easily get

$$I(a) = \frac{-e^{au}}{u} = ze^{-a/z}$$

Now, integrating $h(z)$,

$$\int_{0}^{3/2} h(z) \, dz = \frac{e^2}{2} \left[I(1) - I(3) \right]_{0}^{1/2} + \left[\frac{3z}{2} - \frac{e^2}{2} I(3) \right]_{1/2}^{3/2} = 1 \qquad ////$$

EXAMPLE 6-10 Find the distribution of the sum $y = x_1 + x_2$ of two independent gamma variates x_1, x_2, the respective parameters of which are (α_1, β) and (α_2, β); that is, the exponential parameter β is common to both distributions. The distributions of x_1 and x_2 are

$$f_1(x_1) = k_1 x_1^{\alpha_1} e^{-x_1/\beta} \qquad (0, \infty)$$

$$f_2(x_2) = k_2 x_2^{\alpha_2} e^{-x_2/\beta} \qquad (0, \infty)$$

where
$$k_1 = \frac{1}{\beta^{\alpha_1 + 1}\Gamma(\alpha_1 + 1)} \qquad k_2 = \frac{1}{\beta^{\alpha_2 + 1}\Gamma(\alpha_2 + 1)}$$

and no restrictions are imposed on the parameters α_1, α_2 beyond the fundamental requirement for the existence of the distributions, namely, $\alpha_1 > -1$, $\alpha_2 > -1$.

Holding x_1 fixed in the equation $y = x_1 + x_2$ we put

$$x_2 = y - x_1 \qquad \frac{\partial x_2}{\partial y} = 1$$

and note that y ranges from x_1 to ∞, respectively, as x_2 goes from 0 to ∞. Thus the conditional distribution of y given x_1 is

$$\phi(y \mid x_1) = [f_2(y - x_1)]\frac{\partial x_2}{\partial y} = k_2(y - x_1)^{\alpha_2} e^{-(y - x_1)/\beta} \qquad x_1 \leq y < \infty$$

and as the joint distribution of x_1 and y we obtain

$$g(x_1, y) = k_1 k_2 x_1^{\alpha_1}(y - x_1)^{\alpha_2} e^{-x_1/\beta} e^{-(y - x_1)/\beta}$$

$$= k_1 k_2 e^{-y/\beta} x_1^{\alpha_1}(y - x_1)^{\alpha_2} \qquad 0 \leq x_1 < \infty; x_1 \leq y < \infty$$

The admissible region, shown in Fig. 6-10, consists of the second octant of the plane, i.e., that portion of the first quadrant included between the y axis and the line $y = x_1$. As indicated by the figure, the extreme limits of y are 0 to ∞. The limits of integration on x_1 are from 0 to y, and the integration can be performed in one step as a beta function, with the substitution $x_1 = ty$.

$$h(y) = k_1 k_2 e^{-y/\beta} \int_0^y x_1^{\alpha_1}(y - x_1)^{\alpha_2} \, dx_1 = k y^{\alpha_1 + \alpha_2 + 1} e^{-y/\beta} \qquad (0, \infty)$$

where
$$k = \frac{1}{\beta^{\alpha_1 + \alpha_2 + 2}\Gamma(\alpha_1 + \alpha_2 + 2)}$$

Therefore, the sum of two independent gamma variates with common parameter β is another gamma variate with the same β and with $\alpha = \alpha_1 + \alpha_2 + 1$.

////

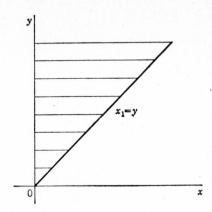

FIGURE 6-10

We may generalize the result of this example. Suppose we had the sum of three independent gamma variates, $z = x_1 + x_2 + x_3$ with common β. We could rewrite this identically as $z = y + x_3$ and, by applying the proposition we just proved, deduce that the distribution of z is gamma with parameters

$$\alpha = \alpha_1 + \alpha_2 + \alpha_3 + 2$$

and β. By induction, we obtain the following theorem, a special case of which was proved in Chap. 4:

Theorem 6-2 If $x_1, x_2, \ldots, x_k$ are independent gamma variates with α parameters $\alpha_1, \alpha_2, \ldots, \alpha_k$ and a common value of β, then their sum $x = x_1 + x_2 + \cdots + x_k$ is a gamma variate with the same β and with $\alpha = \alpha_1 + \alpha_2 + \cdots + \alpha_k + k - 1$.   ////

EXAMPLE 6-11 The independent variates x, y are normally distributed with parameters (μ_1, σ_1) and (μ_2, σ_2), respectively:

$$f_1(x) = \frac{1}{\sigma_1 \sqrt{2\pi}} e^{-(x-\mu_1)^2/2\sigma_1^2} \qquad (-\infty, \infty)$$

$$f_2(y) = \frac{1}{\sigma_2 \sqrt{2\pi}} e^{-(y-\mu_2)^2/2\sigma_2^2} \qquad (-\infty, \infty)$$

Find the distribution of the sum $z = x + y$.

Holding x fixed, we obtain the conditional distribution of z given x by changing variables in $f_2(y)$. Here $y = z - x$, $\partial y/\partial z = 1$, and so

$$\phi(z \mid x) = \frac{1}{\sigma_2 \sqrt{2\pi}} e^{-(z-x-\mu_2)^2/2\sigma_2^2} \qquad (-\infty, \infty)$$

The range of z is $(-\infty, \infty)$ because y has this range, and for any fixed value of x the quantity $x + y$ will approach infinity in either direction with y. Now let us put

$$K = \frac{1}{2\pi\sigma_1\sigma_2} \qquad Q = \frac{(x - \mu_1)^2}{2\sigma_1^2} + \frac{(z - x - \mu_2)^2}{2\sigma_2^2}$$

Then the joint distribution of x, z becomes

$$g(x, z) = Ke^{-Q} \qquad (-\infty, \infty \text{ each})$$

and the marginal density function of z will be given by

$$h(z) = K \int_{-\infty}^{\infty} e^{-Q} \, dx$$

To perform the integration, we first complete the square on x obtaining, after simplification,

$$Q \equiv \frac{\sigma_1^2 + \sigma_2^2}{2\sigma_1^2\sigma_2^2} \left(x - \frac{\sigma_2^2\mu_1 + \sigma_1^2 z - \sigma_1^2\mu_2}{\sigma_1^2 + \sigma_2^2} \right)^2 + \frac{(z - \mu_1 - \mu_2)^2}{2(\sigma_1^2 + \sigma_2^2)}$$

When we integrate on x, we hold z fixed. Hence for integration purposes, we may change variables, setting

$$v = x - \frac{\sigma_2^2\mu_1 + \sigma_1^2 z - \sigma_1^2\mu_2}{\sigma_1^2 + \sigma_2^2} \qquad dx = dv$$

Also, for convenience let us put

$$\mu = \mu_1 + \mu_2 \qquad \sigma^2 = \sigma_1^2 + \sigma_2^2$$

Then

$$h(z) = Ke^{-(z-\mu)^2/2\sigma^2} \int_{-\infty}^{\infty} e^{-\sigma^2 v^2/2\sigma_1^2\sigma_2^2} \, dv$$

By use of the gamma function, we find that the latter integral reduces to $(\sigma_1\sigma_2\sqrt{2\pi})/\sigma$. Hence, multiplying by K we obtain

$$h(z) = \frac{1}{\sigma\sqrt{2\pi}} e^{-(z-\mu)^2/2\sigma^2} \qquad (-\infty, \infty)$$

showing that the sum of any two independent normally distributed variates is again normally distributed. ////

By induction, this proposition can be extended to any number of independent normal variates. This fact, together with Theorem 6-1, yields the following theorem:

Theorem 6-3   If $x_1, x_2, \ldots, x_n$ are independent, normally distributed variates with respective parameters $(\mu_1, \sigma_1), (\mu_2, \sigma_2), \ldots, (\mu_n, \sigma_n)$, then the *weighted* sum $x = a_0 + a_1 x_1 + a_2 x_2 + \cdots + a_n x_n$ (where at least one of the coefficients $a_1, a_2, \ldots, a_n$ is not zero) is normally distributed with parameters (μ, σ), where

$$\mu = a_0 + a_1 \mu_1 + a_2 \mu_2 + \cdots + a_n \mu_n$$

and
$$\sigma = \sqrt{a_1{}^2 \sigma_1{}^2 + a_2{}^2 \sigma_2{}^2 + \cdots + a_n{}^2 \sigma_n{}^2}$$ ////

A very important special case of Theorem 6-3 arises when a series of independent random observations $x_1, x_2, \ldots, x_n$ of the same normal variate x are taken and averaged to obtain the sample mean $\bar{x} = (x_1 + x_2 + \cdots + x_n)/n$. If the parameters of x are (μ, σ), then the observations can be regarded as particular values of independent normal variates all having the same parameters (μ, σ). The coefficients $a_1, a_2, \ldots, a_n$ are all equal to $1/n$ and $a_0 = 0$. Thus the formula $a_0 + \sum_1^n a_i \mu_i$ reduces to μ and the formula $\sum_1^n a_i{}^2 \sigma_i{}^2$ reduces to σ^2/n. Consequently $\bar{x}$, in repeated trials, will be normally distributed with parameters $(\mu, \sigma/\sqrt{n})$. Although a corollary of the latter theorem, this result is important enough to be classed as a theorem itself.

Theorem 6-4   In repeated independent trials, the sample mean $\bar{x}$ of n independent random observations of the same normal variate x with parameters (μ, σ) is normally distributed with corresponding parameters $(\mu, \sigma/\sqrt{n})$.   ////

(Suggested Problems 6-8 to 6-12)

6-8   SOME DERIVED DISTRIBUTIONS USED IN STATISTICS

Three basic distributions are employed over and over again in statistical applications of mathematical probability. These are known as the chi-square distribution, the t distribution, and the F distribution. They are used to answer such questions as whether an assumed type of probability distribution fits the observational facts, whether two or more variates are similarly distributed, or, more specifically, whether two or more variates have a common average value. At present, we shall not attempt to show the connection between the formal definitions and the practical applications—this subject is developed later—but we shall derive the distributions of the variates from the definitions we are about to state. All three distributions are offshoots from the normal.

6-9 THE CHI-SQUARE DISTRIBUTION

Chi-square (χ^2) is defined as the sum of the squares of n independent unit normal variates. Let x be a unit normal variate. Then putting $y = x^2$, the distribution of y, already found in Example 6-6, is

$$f(y) = \frac{1}{\sqrt{2\pi}} \, y^{-1/2} e^{-y/2} \qquad (0, \infty) \qquad (6\text{-}8)$$

This, then, is the distribution of chi-square when $n = 1$. Noticing that this is merely a gamma distribution with $\alpha = -\frac{1}{2}$, $\beta = 2$ we may apply Theorem 6-2 and state the general distribution of chi-square at once,

Chi-square distribution with n degrees of freedom:

$$f(\chi^2) = \frac{1}{2^{n/2} \Gamma(n/2)} (\chi^2)^{n/2-1} e^{-\chi^2/2} \qquad (0, \infty) \qquad (6\text{-}9)$$

We mention, in passing, that for reasons founded on the effective number of independent variates on which a derived function is based, the parameter n is called the *degrees of freedom* of chi-square. Thus we say that the equation

$$\chi^2 = x_1{}^2 + x_2{}^2 + \cdots + x_n{}^2$$

where $x_1, x_2, \ldots, x_n$ are independent unit normal variates, or the equivalent equation

$$\chi^2 = y_1 + y_2 + \cdots + y_n$$

where the y's are independently distributed according to Eq. (6-8), defines chi-square with n degrees of freedom, and its distribution is given by Eq. (6-9). An obvious deduction from the definition is that the sum of two independent chi-squares with, say, n_1 and n_2 degrees of freedom, respectively, is in turn distributed as chi-square with $n_1 + n_2$ degrees of freedom and, of course, this proposition (which is merely a special case of Theorem 6-2) can be extended to any number of independent chi-squares. The term "degrees of freedom" is used in several connections, but in the probability sense it nearly always means the maximum number of variates in a given system which are mutually independent under stated conditions.

The t distribution and the F distribution are developed through auxiliary functions based on chi-square, and we shall call these functions the *mean square* and the *root mean square*. While chi-square is the *sum* of the squares of n independent unit normal variates, the *average* of the squares (usually called the mean square), which we shall denote by w, is equal to the sum of the squares divided by n. The square root of the average of the squares (usually called the root mean square), which we shall denote by r for the time being, is the positive square root of w.   For simplicity, we shall denote both density functions by the letter f, even though the expressions are different.   By simple changes of variable in Eq. (6-9) with $w = \chi^2/n$ and $r = \sqrt{\chi^2/n}$, we obtain

Mean-square distribution with n degrees of freedom:

$$f(w) = \frac{(n/2)^{n/2}}{\Gamma(n/2)} w^{n/2-1} e^{-nw/2} \qquad (0, \infty) \qquad (6\text{-}10)$$

Root-mean-square distribution with n degrees of freedom:

$$f(r) = \frac{2(n/2)^{n/2}}{\Gamma(n/2)} r^{n-1} e^{-nr^2/2} \qquad (0, \infty) \qquad (6\text{-}11)$$

6-10   THE t DISTRIBUTION

The variate t may be defined as the quotient of two independent variates x and r, where x is unit normal and r is the root mean square of n other independent unit normal variates; that is $t = x/r$.   To derive the distribution of t let us hold r fixed and change variables in the distribution of x by setting $x = rt$.   Making this substitution in

$$f_1(x) = \frac{1}{\sqrt{2\pi}} e^{-x^2/2} \qquad (-\infty, \infty)$$

we obtain
$$\phi(t|r) = \frac{r}{\sqrt{2\pi}} e^{-r^2t^2/2} \qquad (-\infty, \infty)$$

and since the marginal distribution of r is given by Eq. (6-11), the joint distribution of r and t is

$$g(r, t) = \frac{2(n/2)^{n/2}}{(\sqrt{2\pi})\Gamma(n/2)} r^n e^{-r^2(n+t^2)/2} \qquad 0 \le r < \infty, \ -\infty < t < \infty$$

Setting $u = (r\sqrt{n+t^2})/\sqrt{2}$, we may express the marginal distribution of t as follows:

$$h(t) = \int_0^\infty g(r, t)\, dr = \frac{n^{n/2}}{\Gamma(n/2)\sqrt{\pi}} \frac{1}{(n+t^2)^{(n+1)/2}} \int_0^\infty 2u^n e^{-u^2}\, du$$

$$= \frac{n^{n/2}\Gamma[(n+1)/2]}{\Gamma(n/2)\sqrt{\pi}} \frac{1}{(n+t^2)^{(n+1)/2}} \qquad (-\infty, \infty)$$

The distribution of t is usually rewritten by expressing $n + t^2$ as $n(1 + t^2/n)$. The result is

t Distribution with n degrees of freedom:

$$h(t) = \frac{\Gamma[(n+1)/2]}{\Gamma(n/2)\sqrt{n\pi}} \frac{1}{(1 + t^2/n)^{(n+1)/2}} \qquad (-\infty, \infty) \qquad (6\text{-}12)$$

Equation (6-12) gives the distribution of t with n degrees of freedom.

6-11   THE F DISTRIBUTION

The variate F may be defined as the quotient of the respective mean squares of a and b independent unit normal variates. Thus $f_1(w_1)$ is given by Eq. (6-10) with $n = a$, and $f_2(w_2)$ is given by setting $n = b$. Since $F = w_1/w_2$, we have $w_1 = w_2 F$, and holding w_2 fixed, we obtain the conditional distribution of F from $f_1(w_1)$.

$$\phi(F \mid w_2) = \frac{(a/2)^{a/2}}{\Gamma(a/2)} w_2^{a/2} F^{a/2-1} e^{-aw_2 F} \qquad (0, \infty)$$

The joint distribution is

$$g(w_2, F) = \frac{(a/2)^{a/2}(b/2)^{b/2}}{\Gamma(a/2)\Gamma(b/2)} F^{a/2-1} w_2^{a/2+b/2-1} e^{-w_2(aF+b)/2} \qquad (0, \infty \text{ each})$$

By setting $u = w_2(aF + b)/2$ and $c = a + b$, we may express the marginal distribution of F as follows,

F Distribution with a and b degrees of freedom:

$$h(F) = \frac{a^{a/2} b^{b/2}}{\Gamma(a/2)\Gamma(b/2)} \frac{F^{a/2-1}}{(aF+b)^{c/2}} \int_0^\infty u^{c/2-1} e^{-u}\, du \qquad (0, \infty)$$

whence

$$h(F) = \frac{a^{a/2} b^{b/2}}{\beta(a/2, b/2)} \frac{F^{a/2-1}}{(aF+b)^{c/2}} \qquad (0, \infty) \qquad (6\text{-}13)$$

This can be simplified a little by dividing numerator and denominator by $a^{c/2}$. The result is

Alternative form of F distribution with a and b degrees of freedom:

$$h(F) = \frac{(b/a)^{b/2}}{\beta(a/2, b/2)} \frac{F^{a/2-1}}{(F + b/a)^{c/2}} \qquad (0, \infty) \qquad (6\text{-}14)$$

The variate F, as here defined, is also known as the *variance ratio* with a and b degrees of freedom.

(*Suggested Problems 6-13 to 6-16*)

6-12   THE DISTRIBUTION OF A FUNCTION OF NONINDEPENDENT VARIATES

The general principles discussed in Sec. 6-7 for deriving the distribution of a function of independent variates can be extended to nonindependent variates by substituting conditional distributions for marginal distributions.

Consider a pair of nonindependent variates x, y and a monotonic function of them u. Denote the joint density function of x and y by $f(x, y)$, the marginal density function of x by $f_1(x)$, and the conditional density function of y given x by $\phi(y|x)$. Then

$$f(x, y) = f_1(x)\phi(y|x)$$

Holding x fixed, the conditional density function of u given x, say $\theta(u|x)$, is derived from $\phi(y|x)$ by the regular univariate method. We then have

$$g(x, u) = f_1(x)\theta(u|x)$$

and the remaining steps are the same as before.

This, at any rate, is the logic of the method. In practice it is possible to eliminate some steps, but this is best explained with reference to a specific example.

EXAMPLE 6-12   Find the distribution of $u = y/\sqrt{x}$ given that x and y are jointly distributed as follows:

$$f(x, y) = 120y(x - y)(1 - x) \qquad 0 \le x \le 1; 0 \le y \le x$$

The marginal distribution of x is

$$f_1(x) = \int_0^x f(x, y)\, dy = 20x^3(1 - x) \qquad (0, 1)$$

and so the conditional distribution of y given x is

$$\phi(y|x) = \frac{6y}{x^3}(x - y) \qquad 0 \le y \le x$$

As y goes from 0 to x, the function $u = y/\sqrt{x}$ goes from 0 to $\sqrt{x}$, and to change variables we put

$$y = u\sqrt{x} \qquad \frac{\partial y}{\partial u} = \sqrt{x}$$

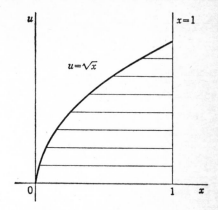

FIGURE 6-11

Accordingly,

$$\theta(u|x) = [\phi(u\sqrt{x}|x)]\sqrt{x} = \frac{6u}{x^{3/2}}(\sqrt{x} - u) \qquad 0 \le u \le \sqrt{x}$$

$$g(x, u) = 120x^{3/2}(1 - x)u(\sqrt{x} - u) \qquad 0 \le x \le 1; 0 \le u \le \sqrt{x}$$

The admissible region is shown in Fig. 6-11.

Integrating the joint density function $g(x, u)$ from $x = u^2$ to 1, we find the marginal distribution of u to be as follows:

$$h(u) = \tfrac{2}{7} u(35 - 48u + 28u^6 - 15u^8) \qquad (0, 1)$$

The fact that a density function cannot be negative implies that this polynomial cannot have any roots inside the range (0, 1) for which it represents the distribution of u. By algebraic factorization we obtain

$$h(u) = \tfrac{2}{7}u(1 - u)^3(15u^5 + 45u^4 + 62u^3 + 66u^2 + 57u + 35) \qquad (0, 1)$$

From this form, it is evident that there are, in fact, no roots inside the range (0, 1).   ////

With regard to the possibility of saving some steps, we notice that the actual operations are as follows:

1. $\phi(y|x) = f(x, y)/f_1(x)$.
2. $\theta(u|x) = [\phi(y^*|x)]|\partial y/\partial u|$, where y^* denotes the functional expression of y in terms of u and x. In the foregoing example this would mean writing $u\sqrt{x}$ in place of y.
3. $g(x, u) = f_1(x)\theta(u|x)$. At this step, $f_1(x)$ cancels out and we are left with

$$g(x, u) = [f(x, y^*)]\left|\frac{\partial y}{\partial u}\right|$$

In other words, we may write the joint distribution of x and u directly without bothering with the intermediate marginal and conditional distributions. We merely substitute for y in $f(x, y)$ and multiply the result by the absolute value of the partial derivative of y with respect to u. The limits on y are replaced by the corresponding limits on u, which are found directly from the equation by which u is defined. Thus, in the foregoing example we could have written

$$f(x, u\sqrt{x}) = 120(u\sqrt{x})(x - u\sqrt{x})(1 - x)$$
$$= 120x - (1\,x)u(\sqrt{x} - u)$$

and when this is multiplied by the partial derivative $\sqrt{x}$ we get the same expression for $g(x, u)$ as before. This technique is especially convenient when the function considered involves more than two variables. A three-variable illustration is given in the following example.

EXAMPLE 6-13   Given the trivariate distribution

$$f(x_1, x_2, x_3) = (x_2 - x_3)e^{-x_1}   0 \le x_1 < \infty; 0 \le x_2 \le x_1; 0 \le x_3 \le x_2$$

find the distribution of the function $y = x_1 - x_2 + x_3$.

Replacing x_3, we substitute $x_3 = y + x_2 - x_1$ and then multiply by $\partial x_3 / \partial y = 1$. As x_3 goes from 0 to x_2, the function $y = x_1 - x_2 + x_3$ goes from $x_1 - x_2$ to x_1. Hence the joint distribution of x_1, x_2, y becomes

$$g(x_1, x_2, y) = [x_2 - (y + x_2 - x_1)]e^{-x_1}(1)$$
$$= (x_1 - y)e^{-x_1}   0 \le x_1 < \infty; 0 \le x_2 \le x_1; x_1 - x_2 \le y \le x_1$$

An equivalent set of limits, defining exactly the same admissible region, is $0 \le y < \infty$; $y \le x_1 < \infty$; $x_1 - y \le x_2 \le x_1$. Accordingly, the marginal distribution of y is given by

$$h(y) = \int_y^\infty (x_1 - y)e^{-x_1}\,dx_1 \int_{x_1-y}^{x_1} dx_2 = ye^{-y}   (0, \infty)   ////$$

EXAMPLE 6-14   Under the topic of ordered statistics in Chap. 5 the joint distribution of the smallest and largest observations in a random sample was derived in Example 5-12 for the particular case of a sample of four observations from an exponential distribution. The same line of argument is readily generalized to any sample size n and any density function $f(x)$ $(A \le x \le B)$.

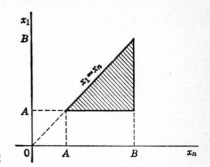

FIGURE 6-12

Denoting the (algebraically) smallest and largest observations by x_1 and x_n, respectively, and noticing that the permutation factor $n!/1!(n-2)!1!$ reduces to $n(n-1)$, we obtain (Fig. 6-12)

$$g(x_1, x_n) = n(n-1)f(x_1)\left[\int_{x_1}^{x_n} f(x)\, dx\right]^{n-2} f(x_n)$$

$$A \leq x_n \leq B; A \leq x_1 \leq x_n \qquad ////$$

Now suppose we wished to know the distribution of the range R (i.e., the difference between the largest and smallest observations) in repeated random samples of n observations. Since x_1 and x_n are not independent, the range $R = x_n - x_1$ provides a simple but important illustration of a function of non-independent variates. Replacing x_n by $x_1 + R$ we obtain (Fig. 6-13)

$$h(x_1, R) = n(n-1)f(x_1)\left[\int_{x_1}^{x_1+R} f(x)\, dx\right]^{n-2} f(x_1 + R)$$

$$A \leq x_1 \leq B; 0 \leq R \leq B - x_1$$

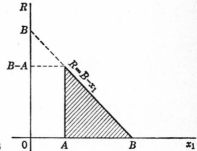

FIGURE 6-13

Whence, the marginal distribution of R is given by

$$p(R) = n(n-1) \int_A^{B-R} f(x_1) f(x_1 + R) \left[\int_{x_1}^{x_1+R} f(x)\, dx \right]^{n-2} dx_1$$

$$0 \le R \le B - A$$

As a specific illustration, let us take the exponential distribution $f(x) = (1/\beta)e^{-x/\beta}$ $(0, \infty)$. Here the integration can be performed explicitly and yields

$$p(R) = \frac{n-1}{\beta} e^{-R/\beta} (1 - e^{-R/\beta})^{n-2} \qquad 0 \le R < \infty$$

(*Suggested Problems 6-17 to 6-23*)

6-13 JOINT DISTRIBUTIONS OF MULTIVARIATE FUNCTIONS

When two or more quantities are computed from the same data, or, equivalently, when two or more functions are determined from a common set of initial variates, their joint distribution in repeated independent trials may become an important issue in the problem considered. For instance, the question of independence can be settled only by referring to the joint distribution. The methods thus far expounded yield the distribution of a single function of initial variates but do not apply to the joint distribution of two or more functions. The latter comes under the heading of variate transformations, and we shall borrow what we need from the standard material of advanced calculus.

Suppose the initial variates (whether independent or not) are x_1, x_2, x_3 and their joint density function $f(x_1, x_2, x_3)$ is defined over some region of space R_x. Let y_1, y_2, y_3 be single-valued functions of x_1, x_2, x_3 such that the x's can be recovered uniquely if the y's are given. A necessary property of $g(y_1, y_2, y_3)$, the joint density function of the y's, is that its total integral over its own region of definition R_y is unity. Now the theory of transforming integrals has already been developed, and we have only to adapt it to our own purposes. If we begin with the definite integral

$$\iiint_{R_x} f(x_1, x_2, x_3)\, dx_1\, dx_2\, dx_3 = 1 \qquad (6\text{-}15)$$

and decide to change to new variables y_1, y_2, y_3, then the integrand proper $f(x_1, x_2, x_3)$ is transformed by direct substitution, but the differential element

$dx_1 \, dx_2 \, dx_3$ has to be replaced by the correct volume element in the new coordinate system. In advanced calculus, it is proved that the new volume element becomes

$$\left[\text{Absolute value of } \frac{\partial(x_1, x_2, x_3)}{\partial(y_1, y_2, y_3)}\right] dy_1 \, dy_2 \, dy_3$$

where the expression $\partial(x_1, x_2, x_3)/\partial(y_1, y_2, y_3)$ is called the *Jacobian* of the x system with respect to the y system and is defined as the determinant of first partial derivatives. That is,

$$\frac{\partial(x_1, x_2, x_3)}{\partial(y_1, y_2, y_3)} \equiv \begin{vmatrix} \dfrac{\partial x_1}{\partial y_1} & \dfrac{\partial x_1}{\partial y_2} & \dfrac{\partial x_1}{\partial y_3} \\[2mm] \dfrac{\partial x_2}{\partial y_1} & \dfrac{\partial x_2}{\partial y_2} & \dfrac{\partial x_2}{\partial y_3} \\[2mm] \dfrac{\partial x_3}{\partial y_1} & \dfrac{\partial x_3}{\partial y_2} & \dfrac{\partial x_3}{\partial y_3} \end{vmatrix} \qquad (6\text{-}16)$$

Following accepted engineering usage, we shall represent the Jacobian of the initial variables with respect to the new by the symbol J. Thus $dx_1 \, dx_2 \, dx_3$ is replaced by $|J| \, dy_1 \, dy_2 \, dy_3$ (note *absolute* value) and this absolute value of the Jacobian becomes the analog of the absolute value of the derivative employed in changing a single variable. Consequently, in order to state that

$$\iiint_{R_x} f(x_1, x_2, x_3) \, dx_1 \, dx_2 \, dx_3 = \iiint_{R_y} g(y_1, y_2, y_3) \, dy_1 \, dy_2 \, dy_3 = 1 \qquad (6\text{-}17)$$

we must incorporate the term $|J|$ into the definition of the new density function. Just as dx/dy is the reciprocal of dy/dx, it can be demonstrated also that the Jacobian of the x's with respect to the y's is the reciprocal of the Jacobian of the y's with respect to the x's. Before illustrating the derivation of joint distributions, we shall give an example of the Jacobian itself.

EXAMPLE 6-15   In changing from rectangular to polar coordinates in two dimensions we have $x = r \cos \theta$, $y = r \sin \theta$ and

$$\frac{\partial(x, y)}{\partial(r, \theta)} = \begin{vmatrix} \dfrac{\partial x}{\partial r} & \dfrac{\partial x}{\partial \theta} \\[2mm] \dfrac{\partial y}{\partial r} & \dfrac{\partial y}{\partial \theta} \end{vmatrix} = \begin{vmatrix} \cos \theta & -r \sin \theta \\[1mm] \sin \theta & r \cos \theta \end{vmatrix}$$

$$= r \cos^2 \theta + r \sin^2 \theta = r$$

Accordingly, as the reader already knows, $dx\,dy$ transforms into $r\,dr\,d\theta$. Again, to go from polar to rectangular coordinates, we put $r = \sqrt{x^2 + y^2}$, $\theta = \tan^{-1}(y/x)$ whence

$$\frac{\partial(r, \theta)}{\partial(x, y)} = \begin{vmatrix} \dfrac{x}{\sqrt{x^2 + y^2}} & \dfrac{y}{\sqrt{x^2 + y^2}} \\[2ex] \dfrac{-y}{x^2 + y^2} & \dfrac{x}{x^2 + y^2} \end{vmatrix}$$

$$= \frac{x^2 + y^2}{(x^2 + y^2)^{3/2}} = \frac{1}{r}$$

which is the reciprocal of first Jacobian. Thus from $dx\,dy$ we go to $r\,dr\,d\theta$ and from $r\,dr\,d\theta$ we go to $r(1/r)\,dx\,dy = dx\,dy$.   ////

We are now prepared to derive some joint distributions by variate transformations.

EXAMPLE 6-16   Given two random observations x_1, x_2 of the variate x, where $f(x) = 1/x^2$ $(1, \infty)$, find the joint and marginal distributions of u and v where $u = x_1 x_2$ and $v = x_1/x_2$.

Since the initial variates are independent, their joint distribution is given by

$$f(x_1, x_2) = f_1(x_1)f_2(x_2) = \frac{1}{x_1{}^2 x_2{}^2} \qquad (1, \infty \text{ each})$$

Noticing that the denominator of this joint density function is equal to u^2, we may write

$$f(x_1, x_2) = \frac{1}{u^2}$$

Next, we must find the Jacobian. In this case, the labor of differentiation can be shortened by deriving the Jacobian of the new variates with respect to the initial ones and then taking the reciprocal. Hence we put

$$\frac{\partial(u, v)}{\partial(x_1, x_2)} = \begin{vmatrix} x_2 & x_1 \\[1ex] \dfrac{1}{x_2} & \dfrac{-x_1}{x_2{}^2} \end{vmatrix} = -2\frac{x_1}{x_2} = -2v$$

whereupon

$$J = -\frac{1}{2v} \qquad |J| = \frac{1}{2v}$$

and

$$g(u, v) = \frac{1}{2vu^2}$$

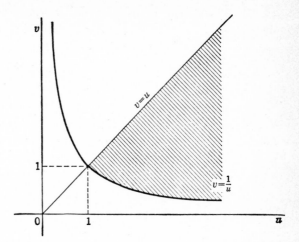

FIGURE 6-14

This gives us the equation for the joint density function, but we have yet to establish the admissible region. The general procedure is to express the initial variates in terms of the new ones and then determine the loci which the boundary curves of the initial region trace out in the new system of coordinates.

From the product of u and v, we get

$$uv = x_1{}^2 \qquad x_1 = \sqrt{uv}$$

and from the quotient we get

$$\frac{u}{v} = x_2{}^2 \qquad x_2 = \frac{\sqrt{u}}{\sqrt{v}}$$

Now the original boundaries were $x_1 = 1$, $x_2 = 1$. Accordingly, the new ones are

$$\sqrt{uv} = 1 \qquad \text{or} \qquad v = \frac{1}{u}$$

$$\frac{\sqrt{u}}{\sqrt{v}} = 1 \qquad \text{or} \qquad v = u$$

These curves, together with the fact that u cannot be less than unity (since this is the product of the minimum values of x_1 and x_2), describe the admissible region, indicated by shading in Fig. 6-14.

The complete statement of the joint distribution, then, is

$$g(u, v) = \frac{1}{2vu^2} \qquad 1 \le u < \infty ; \frac{1}{u} \le v \le u$$

The marginal distribution of u is

$$g_1(u) = \frac{1}{2u^2} \int_{1/u}^{u} \frac{dv}{v} = \frac{\ln u}{u^2} \qquad (1, \infty)$$

However, the marginal distribution of v has two parts.   Thus

Range 1 $(0 \le v \le 1)$:

$$g_2(v) = \frac{1}{2v} \int_{1/v}^{\infty} \frac{du}{u^2} = \frac{1}{2}$$

Range 2 $(1 \le v < \infty)$:

$$g_2(v) = \frac{1}{2v} \int_{v}^{\infty} \frac{du}{u^2} = \frac{1}{2v^2} \qquad ////$$

EXAMPLE 6-17   From a unit normal population, we draw two random observations x_1, x_2 and compute their sum $u = x_1 + x_2$ and their difference $v = x_1 - x_2$.   Find the joint distribution of u and v which would be obtained if this experiment could be repeated for an indefinitely large number of independent trials.   The initial variates are jointly distributed as follows:

$$f(x_1, x_2) = f_1(x_1)f_2(x_2) = \frac{1}{2\pi} e^{-(x_1^2 + x_2^2)/2} \qquad (-\infty, \infty \text{ each})$$

By definition, $x_1 + x_2 = u$; $x_1 - x_2 = v$.   Therefore, by addition, $x_1 = (u + v)/2$ and by subtraction, $x_2 = (u - v)/2$.   The Jacobian is

$$J = \begin{vmatrix} \frac{1}{2} & \frac{1}{2} \\ \frac{1}{2} & -\frac{1}{2} \end{vmatrix} = -\frac{1}{2}$$

and $|J| = \frac{1}{2}$.   Also

$$x_1^2 + x_2^2 = \frac{u^2 + 2uv + v^2}{4} + \frac{u^2 - 2uv + v^2}{4}$$

$$= \frac{u^2 + v^2}{2}$$

whence the joint distribution of u and v is

$$g(u, v) = \frac{1}{4\pi} e^{-(u^2 + v^2)/4} \qquad (-\infty, \infty \text{ each})$$

Both variates have a total range of $(-\infty, \infty)$ because the original variates have this range, and the new ones can go to infinity in either direction with them.

The fact that u and v have this range independently of each other may be seen by fixing the value of one of them, say u, at an arbitrary value c. Then

$$x_1 + x_2 = c \qquad x_2 = c - x_1 \qquad v = x_1 - x_2 = 2x_1 - c$$

and the latter function will vary from minus to plus infinity with x_1 irrespective of c. Similarly, fixing v at an arbitrary value k, we have $u = 2x_2 + k$, which will vary from minus to plus infinity with x_2 irrespective of k.

The marginal density functions can be written at once by applying Theorem 6-3 or, alternatively, by integrating the joint density function. Denoting the marginal densities by $g_1(u)$ and $g_2(v)$, we find

$$g_1(u) = \frac{1}{2\sqrt{\pi}}\, e^{-u^2/4} \qquad (-\infty, \infty) \qquad\qquad g_2(v) = \frac{1}{2\sqrt{\pi}}\, e^{-v^2/4} \qquad (-\infty, \infty)$$

Therefore $g(u, v) = g_1(u)g_2(v)$ and the variates are independent, even though constructed from the same data. ////

EXAMPLE 6-18 Given x_1, x_2 as in Example 6-17, find the joint and marginal distributions of the sample average $\bar{x} = (x_1 + x_2)/2$ and the sum of squares of deviations from the sample average

$$S = (x_1 - \bar{x})^2 + (x_2 - \bar{x})^2$$

Also find the distribution of the variate $z = \bar{x}\sqrt{2}/\sqrt{S}$.

Here the original variates cannot be recovered uniquely from $\bar{x}$ and S, because the latter is quadratic. We therefore proceed indirectly. To simplify the expression for S, rewrite $x_1 - \bar{x}$ as follows:

$$x_1 - \bar{x} = x_1 - \frac{x_1 + x_2}{2} = \frac{x_1 - x_2}{2}$$

Similarly,

$$x_2 - \bar{x} = \frac{x_2 - x_1}{2} = -\frac{x_1 - x_2}{2}$$

Whence, by squaring and adding, we obtain

$$S = \frac{(x_1 - x_2)^2}{2}$$

Referring now to the constructed variates u, v of the previous example, we recognize that $\bar{x} = u/2$, $S = v^2/2$. But since u and v are independent, it follows that

$\bar{x}$ and S, being pure functions of them, are also independent. Hence their marginal distributions, found by changing variables in $g_1(u)$ and $g_2(v)$, are

$$h_1(\bar{x}) = \frac{1}{\sqrt{\pi}} e^{-\bar{x}^2} \qquad (-\infty, \infty)$$

and (from the intermediate distribution of v^2)

$$h_2(S) = \frac{1}{\sqrt{2\pi}} S^{-1/2} e^{-S/2} \qquad (0, \infty)$$

Their joint distribution, of course, is

$$h(\bar{x}, S) = \frac{1}{\pi\sqrt{2}} e^{-\bar{x}^2} S^{-1/2} e^{-S/2} \qquad -\infty < \bar{x} < \infty; 0 \le S < \infty$$

By inspection we see that the variate $\bar{x}$ is normally distributed with $\mu = 0$, $\sigma^2 = \frac{1}{2}$ (which checks with Theorem 6-4), and S is independent of $\bar{x}$ and is distributed as chi-square with *one* degree of freedom. Thus a derived variate can be distributed as chi-square even though it is not originally defined as the sum of squares of independent unit normal variates. In this case, of course, S is distributed as chi-square with 1 degree of freedom precisely because $v/\sqrt{2}$ is unit normal. And although we started with two unit normal variates, x_1 and x_2, we have only 1 degree of freedom for S. The other degree of freedom was absorbed by $\bar{x}$.

Turning now to z, its numerator $\bar{x}\sqrt{2}$ is readily seen to be a unit normal variate. The denominator, $\sqrt{S}$, is the root-mean-square variate with 1 degree of freedom. Therefore, the variate $z = \bar{x}\sqrt{2}/\sqrt{S}$ is distributed as t with 1 degree of freedom:

$$f(z) = \frac{1}{\pi} \frac{1}{1 + z^2} \qquad (-\infty, \infty) \qquad ////$$

EXAMPLE 6-19 From a general normal population (μ, σ arbitrary), three observations x_1, x_2, x_3 are taken at random, and from them the quantities $u = x_1 - x_3$, $v = x_2 - x_3$ are computed. If an indefinite number of independent trials of this experiment could be made, find the joint distribution of u and v. Also find the distribution of the quotient $z = u/v$.

The distribution of single random observations from this population is

$$f(x) = \frac{1}{\sigma\sqrt{2\pi}} e^{-(x-\mu)^2/2\sigma^2} \qquad (-\infty, \infty)$$

Hence, the joint distribution of three random observations is

$$f(x_1, x_2, x_3) = \frac{1}{(\sigma\sqrt{2\pi})^3} e^{-Q} \qquad (-\infty, \infty \text{ all variates})$$

where

$$Q = \frac{(x_1 - \mu)^2 + (x_2 - \mu)^2 + (x_3 - \mu)^2}{2\sigma^2}$$

Although the problem prescribes two variates u, v we must introduce a third, say w, in order to make a complete change of variables. Otherwise, the Jacobian is undefined. This third variable can be any function which suits our convenience, for instance, one of the initial variables. The only restriction is that the original variables must be recoverable from the new. In this problem, a simple and convenient choice is

$$w = (x_1 - \mu) + (x_2 - \mu) + (x_3 - \mu)$$

Then, since we may write

$$u = x_1 - x_3 \equiv (x_1 - \mu) - (x_3 - \mu)$$

$$v = x_2 - x_3 \equiv (x_2 - \mu) - (x_3 - \mu)$$

the following results are readily obtained by substitution:

$$x_1 - \mu = \frac{2u - v + w}{3} \qquad x_2 - \mu = \frac{2v - u + w}{3} \qquad x_3 - \mu = \frac{w - u - v}{3}$$

The Jacobian (which we may simplify by adding the first two rows to the third) is

$$J = \begin{vmatrix} \tfrac{2}{3} & -\tfrac{1}{3} & \tfrac{1}{3} \\ -\tfrac{1}{3} & \tfrac{2}{3} & \tfrac{1}{3} \\ -\tfrac{1}{3} & -\tfrac{1}{3} & \tfrac{1}{3} \end{vmatrix} = \begin{vmatrix} \tfrac{2}{3} & -\tfrac{1}{3} & \tfrac{1}{3} \\ -\tfrac{1}{3} & \tfrac{2}{3} & \tfrac{1}{3} \\ 0 & 0 & 1 \end{vmatrix} = \tfrac{1}{3}$$

and the quadratic exponent of e becomes

$$Q = \frac{u^2 - uv + v^2}{3\sigma^2} + \frac{w^2}{6\sigma^2} = Q' + \frac{w^2}{6\sigma^2}$$

where Q' represents that part of Q which depends on u, v. By an argument similar to that used in Example 6-17, we conclude that all three variates range from $-\infty$ to ∞ without restriction. Therefore, their joint distribution is

$$g(u, v, w) = \frac{1}{3(\sigma\sqrt{2\pi})^3} e^{-Q'} e^{-w^2/6\sigma^2} \qquad (-\infty, \infty \text{ all})$$

and the joint distribution of u and v is

$$h(u, v) = \int_{-\infty}^{\infty} g(u, v, w)\, dw = \frac{\sqrt{3}}{6\pi\sigma^2} e^{-Q'} \qquad (-\infty, \infty \text{ each})$$

This time, we find the rather common result that two variates u, v derived from the same data are not independent but dependent. Nevertheless, the third variate w, as we have defined it, does turn out to be independent of u and v.

Now to obtain the distribution of $z = u/v$ we could follow the method used in previous sections, but to illustrate variate transformations let us choose a second variable $y = v$. Then $u = vz = yz$ and

$$J = \begin{vmatrix} z & y \\ 1 & 0 \end{vmatrix} = -y \qquad |J| = |y|$$

The quantity Q' in $h(u, v)$ becomes

$$Q' = \frac{y^2 z^2 - y^2 z + y^2}{3\sigma^2} = \frac{y^2(z^2 - z + 1)}{3\sigma^2}$$

and the joint distribution of y and z is

$$f(y, z) = \frac{\sqrt{3}}{6\pi\sigma^2} |y| e^{-y^2(z^2 - z + 1)/3\sigma^2} \qquad (-\infty, \infty \text{ each})$$

Since the integrand is symmetrical about zero (a result made possible by the absolute value of the Jacobian), we obtain, for the marginal distribution of z,

$$g(z) = \frac{\sqrt{3}}{6\pi\sigma^2} \int_0^\infty 2y e^{-y^2(z^2 - z + 1)/3\sigma^2} \, dy$$

$$= \frac{\sqrt{3}}{2\pi} \frac{1}{z^2 - z + 1} \qquad (-\infty, \infty)$$

Therefore the distribution of z is independent of the values of μ and σ. Because $h(u, v)$ is a symmetrical function of u and v, we would obtain the same distribution for the alternative quotient v/u—a fact which is also verified by changing from z to its reciprocal in $g(z)$.

By completing the square in the density function of z, we obtain the following equivalent expression

$$g(z) = \frac{\sqrt{3}}{2\pi} \frac{1}{(z - \frac{1}{2})^2 + \frac{3}{4}} \qquad (-\infty, \infty)$$

which shows that the modal value of z is $\frac{1}{2}$. At first sight this fact is puzzling because it is difficult to see intuitively why the value $\frac{1}{2}$ has any special significance for the variate considered. One way of understanding how this result comes about is to define an analogous quotient

$$q = \frac{x_1 - x_2}{x_3 - x_2}$$

and note the identity

$$z \equiv \frac{x_1 - x_2}{x_2 - x_3} \equiv 1 - \frac{(x_1 - x_2)}{x_3 - x_2} \equiv 1 - q$$

From the formal interchangeability of the variates x_1, x_2, x_3 it is obvious that q has exactly the same distribution as z. Hence, both must have the same modal value m. On the other hand, if the mode of q is m, that of z must be $1 - m$, since $z = 1 - q$. Therefore

$$m = 1 - m \qquad \text{and} \qquad m = \tfrac{1}{2} \qquad ////$$

EXAMPLE 6-20   The variates x_1, x_2 are said to be "normally correlated" when their joint density function is of the form

$$f(x_1, x_2) = ke^{-Q/2} \qquad (-\infty, \infty \text{ each variate}) \qquad (6\text{-}18)$$

where $k = \dfrac{1}{2\pi\sigma_1\sigma_2\sqrt{1 - \rho^2}}$

$$Q = \frac{1}{1 - \rho^2}\left[\frac{(x_1 - \mu_1)^2}{\sigma_1^2} - \frac{2\rho(x_1 - \mu_1)(x_2 - \mu_2)}{\sigma_1\sigma_2} + \frac{(x_2 - \mu_2)^2}{\sigma_2^2}\right]$$

It can be shown that the marginal distribution of x_1 is normal with parameters μ_1, σ_1 and that of x_2 is normal with parameters μ_2, σ_2. The parameter ρ is called the *coefficient of linear correlation*. In the present context, this quantity must be less than unity in absolute value, but in connection with the general theory of correlation, the upper limit unity is theoretically possible. The locus of points having a fixed probability density can be obtained by setting Q equal to a constant. By analytical geometry, it can be shown that this locus is an ellipse. Let us determine the probability P that a random point x_1', x_2' will fall outside any given contour ellipse corresponding to a fixed value of Q, say $Q = C_1$. Although this can be done by geometric considerations, we shall use the method of variate transformations. In either approach, the trick is to notice that the probability in question can be obtained in terms of the distribution function of Q.

Since Q is a function of the variates x_1, x_2 it is a random variable itself. If a random point falls outside a given contour ellipse $Q = C_1$, there is another contour ellipse $Q = C_2$, with larger semiaxes, which passes through it. Hence the probability that x_1', x_2' fall outside of $Q = C_1$ is exactly equal to the probability that $Q > C_1$.

To obtain the distribution of Q let us first make the transformation

$$y_1 = \frac{x_1 - \mu_1}{\sigma_1} \qquad y_2 = \frac{x_2 - \mu_2}{\sigma_2}$$

The Jacobian of x_1, x_2 with respect to y_1, y_2 is $\sigma_1\sigma_2$ and the quadratic expression becomes

$$Q = \frac{1}{1 - \rho^2}(y_1^2 - 2\rho y_1 y_2 + y_2^2)$$

Thus the joint distribution of y_1, y_2 is

$$g(y_1, y_2) = \frac{1}{2\pi\sqrt{1 - \rho^2}}\, e^{-Q/2} \qquad (-\infty, \infty \text{ each})$$

Next, complete the square in Q by putting

$$z_1 = \frac{y_1 - \rho y_2}{\sqrt{1 - \rho^2}} \qquad z_2 = y_2$$

This gives

$$Q = z_1^2 + z_2^2$$

and since the Jacobian of the y's with respect to the z's is equal to $\sqrt{1 - \rho^2}$, the joint distribution of the z's reduces to

$$h(z_1, z_2) = \frac{1}{2\pi}\, e^{-(z_1^2 + z_2^2)/2} \qquad (-\infty, \infty \text{ each})$$

Therefore, z_1 and z_2 are independent unit normal variates, and Q, being equal to their sum of squares, is distributed as chi-square with 2 degrees of freedom. Consequently, denoting the density function and distribution function of Q by $\lambda(Q)$ and $\Lambda(Q)$, respectively, we have

$$\lambda(Q) = \tfrac{1}{2}e^{-Q/2} \qquad 0 \le Q < \infty \qquad (6\text{-}19)$$

$$\Lambda(Q) = 1 - e^{-Q/2} \qquad 0 \le Q < \infty \qquad (6\text{-}20)$$

whereupon $$P = 1 - \Lambda(C_1) = e^{-C_1/2} \qquad (6\text{-}21)$$

This answers the original question. As far as the marginal distributions of x_1 and x_2 are concerned, the fact that the distribution of $y_2 \equiv z_2$ is unit normal implies immediately that x_2 is normal with parameters μ_2, σ_2. By interchanging subscripts, we also infer that x_1 is normal with parameters μ_1, σ_1. ////

EXAMPLE 6-21 As an exercise in establishing a multivariate admissible region, let us derive the joint distribution of y_1, y_2, y_3, where

$$y_1 = x_1 + x_2 \qquad y_2 = x_2 + x_3 \qquad y_3 = x_1 + x_3$$

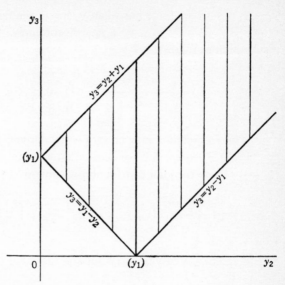

FIGURE 6-15
Cross section made by plane $y_1 = $ constant.

given that the x's are random observations of the variate x, the distribution of which is $f(x) = e^{-x} (0, \infty)$.   Here the problem is almost wholly one of determining the region of definition.

Solving for x_1, x_2, x_3 in terms of y_1, y_2, y_3, we readily obtain

$$x_1 = \frac{y_1 - y_2 + y_3}{2} \qquad x_2 = \frac{y_1 + y_2 - y_3}{2} \qquad x_3 = \frac{y_2 + y_3 - y_1}{2}$$

and the Jacobian of the x's with respect to the y's is $\frac{1}{2}$.   Since the joint density function of the x's is

$$f(x_1, x_2, x_3) = e^{-(x_1 + x_2 + x_3)} \qquad (0, \infty \text{ each})$$

and $x_1 + x_2 + x_3 = (y_1 + y_2 + y_3)/2$, the joint density function of the y's is

$$g(y_1, y_2, y_3) = \tfrac{1}{2}e^{-(y_1 + y_2 + y_3)/2}$$

The boundary of the admissible region in the x system is given by $x_1 = 0$, $x_2 = 0$, $x_3 = 0$.   Hence, substituting for the x's, the boundary of the y system will be given by the planes

$$y_1 - y_2 + y_3 = 0 \qquad y_1 + y_2 - y_3 = 0 \qquad y_2 + y_3 - y_1 = 0$$

These planes intersect at the origin and cut out an infinite inclined wedge having semi-infinite rectangular cross sections (Fig. 6-15) when cut by any plane $y_1 = $ constant.   Therefore, when y_2 lies in the range $0 \leq y_2 \leq y_1$, the range of y_3 is

$y_1 - y_2 \leq y_3 \leq y_1 + y_2$, but when y_2 lies in the range $y_1 \leq y_2 < \infty$, the range of y_3 is $y_2 - y_1 \leq y_3 \leq y_2 + y_1$. Accordingly, we may express the joint distribution as follows:

$$g(y_1, y_2, y_3) = \frac{e^{-(y_1 + y_2 + y_3)/2}}{2} \quad \begin{array}{l} 0 \leq y_1 < \infty \\ (1) \quad 0 \leq y_2 \leq y_1 ; \; y_1 - y_2 \leq y_3 \leq y_1 + y_2 \\ (2) \quad y_1 \leq y_2 < \infty ; \; y_2 - y_1 \leq y_3 \leq y_2 + y_1 \end{array}$$

This formulation may be verified by deriving the marginal distribution of y_1. From Theorem 6-2 we know that the correct result is $g_1(y_1) = y_1 e^{-y_1}$ $(0, \infty)$. From the joint distribution we obtain

$$g_1(y_1) = \int_0^{y_1} \left[\int_{y_1 - y_2}^{y_1 + y_2} g(y_1, y_2, y_3) \, dy_3 \right] dy_2 + \int_{y_1}^{\infty} \left[\int_{y_2 - y_1}^{y_2 + y_1} g(y_1, y_2, y_3) \, dy_3 \right] dy_2$$
$$(0, \infty)$$

The integration is easy to carry out and the result is

$$g_1(y_1) = y_1 e^{-y_1} \quad (0, \infty)$$

which checks. The reader is invited to test his own understanding by deriving the marginal distribution of y_2 or y_3 from $g(y_1, y_2, y_3)$.   ////

(*Suggested Problems 6–24 to 6–29*)

6-14 SIGNIFICANCE TESTS

A significance test is a mathematical procedure applied to empirical data for deciding, on grounds of probability, whether or not a suitably formulated hypothesis is tenable. An observational result is said to be *significant*, and the hypothesis in question is *rejected*, when the obtained result belongs to an objectively specified unfavorable class having a fixed, small probability of occurrence in random samples from the hypothesized population. Two key words here are *random* and *class*. Probability is trivial or meaningless unless there is some element of randomness involved; hence, within the framework of observational conditions (either natural or controlled), there must be freedom for the "chips to fall as they may." This means that the sample must be random as far as the variable (usually called the test variate) used to test the hypothesis is concerned, although the raw data need not be wholly random. The notion of a class is needed, because in almost any population containing a large number

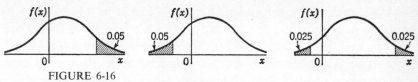

FIGURE 6-16
Possible critical regions of probability .05 for use in significance tests.

of distinct values or types, the probability of drawing any *prescribed* value or type by random selection is extremely small. For example, when 13 playing cards are dealt from a standard deck without jokers, it is certain that the outcome will be a legitimate hand of bridge. Now, the probability of obtaining any preassigned hand as a random draw from the deck is less than 1 in 600 billion (the number of distinct hands being about 6.35×10^{11}), and if the hand were specified in advance, the event of drawing it would be overwhelmingly significant of skill. Unforeseen in its particular composition, however, the same hand would signify nothing more or less than the fact that some combination or other is bound to occur, and that the number of distinct combinations is extremely large. A continuous variate presents this situation in its logical ultimate. Significance, therefore, involves not only the principle of small probability but also the issue of logical interpretation with reference to the essential import of the hypothesis. Such interpretation is made by dividing all possible values of the test variate (each one being a random draw from the hypothesized population) into two classes: the *acceptance region* and the *critical region*, also called the *rejection region*. The boundaries of the classes are set in such a way that the total probability (unity) is appropriately divided between them, say .95, .05, or .99, .01. Subject to this arbitrary division, the acceptance region is then defined as the class of all results that are construed as favorable to the hypothesis, whereas the critical region or rejection region is the complementary class of all results that are construed as unfavorable. The probability assigned to the critical region, commonly either .05 or .01, is called the *significance level*. Although other values than these two are sometimes used, depending upon the problem itself, the significance level should always be chosen before the experiment is conducted. The critical region ordinarily includes the more extreme values of a variate. Depending on what is logically relevant to the hypothesis, the critical region might include only the highest bracket of values, or only the lowest bracket; or it might be divided in a suitable manner (usually in equal amounts as to probability) between the highest and lowest brackets. Typical situations are illustrated graphically in Fig. 6-16 wherein, for the sake of simplicity, the probability function $f(x)$ has been represented as a smooth curve.

Although the typical working hypothesis cannot be subjected to a direct probabilistic examination, the same end can be achieved indirectly. The equivalent of establishing a given hypothesis is to discredit its logical alternative. In place of a working hypothesis that a certain tendency exists, we may examine the contrary proposition that the alleged tendency is absent. From the latter viewpoint, all possible empirical results are regarded as occurring at random within a definite framework. This assumption lends itself to precise expression in terms of a probability distribution, and because of its negative character, it is commonly called the *null hypothesis*. It is essential, however, that the variable actually used in the significance test be such that its probability distribution would be known if the null hypothesis were true.

Thus in the modus operandi of hypothesis testing, as applied to the analysis of observational data, there usually exists a four-step procedure which should be followed.

1   A hypothesis is made by the observer to describe a mathematical model that would generate a set of observations similar to those observed if the hypothesis were true. This hypothesis is usually in the form of a prescribed distribution function of a random variable.

2   A test variate is now chosen which is some computable function of the experimental observations. The choice of a test variate depends upon what population characteristics of the experimental data the observer wishes to emphasize as a reason for his acceptance or rejection of the chosen hypothesis. This freedom of choice offers considerable latitude to the observer; thus different individuals may use different test variates in any given situation. The reader should observe that a hypothesis may be accepted if one test variate is used as a criterion and rejected on another test variate although the experimental data are, of course, identical in each case. The test variate must be chosen in advance of the experiment if the laws of probability are to be operative and not after the investigator has examined his data.

For example, the hypothesis to be tested might be that the population giving rise to the data has a density function $f(x) = 1$ $(0 \leq x \leq 1)$, and the actual data would consist of five observations from this same population. A large number of test variates are now possible. A few of them are the average of all five observations, the value of the largest one of the five, the range (largest minus smallest), the sum of squares of all five, the sum of squares of the next to largest and next to smallest, etc.

3   A critical region for the test variate must be chosen in advance of the examination of the recorded observations. Here again, the scientific acumen of the observer plays an important part in this choice since consider-able latitude is available in assigning the area and positions of this critical region. Considerable weight will be given to the seriousness (cost or other implications) of arriving at the wrong conclusion by either accepting the hypothesis when it is not true or rejecting the hypothesis when in reality it is true.

4 The experiment is now conducted and the test variate computed for the observed sample; the value either lies in the critical region or it does not. If it does, the hypothesis is rejected; otherwise the hypothesis is accepted.

Since the design and analysis of experimental or collected data are beyond the scope of this book, we shall merely illustrate the relationship of hypothesis testing to probability theory by considering problems where the test variate and the critical region are given as part of the statement of the problem.

EXAMPLE 6-22 A particular measuring device has a distribution of error measurements about a true value which is normally distributed with a standard deviation of 5. Is the average of 25 observations whose value was 10 signifi-cantly different from a theoretical value of 13 at the 1 percent level if a two-sided, critical region of .005 on each tail is used? Since the average of the 25 observations is also normally distributed with $\sigma = 5/\sqrt{25} = 1$, the unit normal standard deviate is

$$z = \frac{10 - 13}{1} = -3$$

From the table of normal deviates the critical region is outside ± 2.6 if the re-quired area is .01. Since -3 is within the critical region, we reject the hypo-thesis and assert that the data differ significantly from the theoretical value of 13 at the 1 percent level.   ////

EXAMPLE 6-23 The hypothesis to be tested is that a certain population has the form $f(x) = 1$ $(0 \leq x \leq 1)$, and the chosen test variate is the next to smallest observation of five observed at random from this population. Using a 10 per-cent significance level on the left tail and the fact that the five observations were .14, .96, .08, .77, and .63, is the hypothesis rejected by these experimental results?

The probability that the next to smallest will lie between y and $y + dy$ is

$$g(y)\,dy = \frac{5!}{3!}\left(\int_0^y dx\right)\left(\int_y^1 dx\right)^3 dy = 20y(1-y)^3\,dy$$

If the area to the left of some point y_0 is to be .10, then y_0 is determined from the equation

$$20\int_0^{y_0} y(1-y)^3\,dy = .10$$

or

$$10y_0^2[1 - 2y_0 + 1.5y_0^2 - .4y_0^3] = .10$$

which gives $y_0 \approx .11$.

Since .14 lies to the right of .11, it does not lie in the critical region and so the hypothesis is accepted, not rejected.   ////

EXAMPLE 6-24   A paint manufacturer has favorable laboratory reports on an ingredient purported to improve the wearing quality of shellac.   Desiring a limited field trial, he arranges with the owners of six buildings to allow experimental tests on their property.   Each building has a much-used wooden staircase with three flights of 12 steps each.   Half of the 36 steps are to be coated with regular shellac and the other half with fortified shellac.   Possible extraneous factors are neutralized and independence of the observations ensured by a technique of randomization.

The odd-numbered steps are treated with one finish or another, depending (let us say) upon the toss of a balanced coin, and the next even-numbered step is given the alternative treatment.   In this way, both treatments are used the same number of times, and both have equal chances of exposure to any outside influences that might be operative.   As a measure of wear after 3 months of use, the damaged area for each treatment is recorded as a percentage of the total area covered, and separate figures are reported for each of the six buildings. It was proved in Example 5-9, although it is intuitively obvious, that if any two continuous variates y_1, y_2 have the same distribution function, then the probability is $\frac{1}{2}$ that a random observation of y_1 will exceed a random observation of y_2.   Under the null hypothesis, the two wear percentages in the same building are random observations from the same population.   Thus a possible test variate x would be the number of positive differences.   Besides the hypothesis that positive and negative differences are equally likely, there is the tacit assumption that the six differences are mutually independent.   Using this hypothesis

Table 6-1 PERCENTAGE DAMAGE AFTER
THREE MONTHS

Building	Regular Shellac	Fortified Shellac	Difference reg.—fort.
1	5	2	3
2	11	7	4
3	4	5	−1
4	14	8	6
5	9	4	5
6	12	11	1

and a critical region of .05 situated on the right-hand tail of the distribution, would one accept or reject this hypothesis on the basis of the experimental results shown in Table 6-1?

Here the distribution of the test variate, under the null hypothesis, is

$$f(x) = \frac{C(6, x)}{2^6} = \frac{C(6, x)}{64} \qquad x = 0, 1, \ldots, 6$$

For the reader's convenience, this distribution is exhibited in Table 6-2. This test is known as the *sign test*.

Since we have chosen a one-sided critical region and .05 as the level of significance, the boundary of this region will be determined by starting with the largest value of x and working downward until the total probability is as close as possible to .05 without exceeding it. Here we find that the only admissible value is $x = 6$, since the next lower value 5 would bring the total probability up to .11. The results shown in Table 6-1 yield five positive differences and one negative difference; hence $x = 5$, and this value does not fall in the critical region which we preassigned before the experiment was conducted. Therefore we accept the null hypothesis and conclude that this test does not indicate a significant difference between the wearing qualities of fortified and regular shellac. In other words, the possibility of obtaining five or more positive differences by

Table 6-2 DISTRIBUTION OF THE NUMBER OF POSITIVE DIFFERENCES UNDER THE NULL HYPOTHESIS

x	0	1	2	3	4	5	6	Sum
$f(x)$	.016	.094	.234	.312	.234	.094	.016	1.000

chance alone is such that one would not admit that the result was due to the difference in the shellacs. From the practical point of view this means that if a difference does exist more observations or else a more sensitive test would be necessary in order to demonstrate it.   ////

EXAMPLE 6-25 Sometimes it is possible to construct from the sample itself a synthetic population to use as the basis of a test of significance. This is illustrated below using the observations in Table 6-1. The sign test takes no account of the magnitude of the differences. If we have any way of knowing the probability distribution of the differences under the assumption of no improvement, we could apply a more sensitive test by incorporating this information. In default of such knowledge, we can construct a population of differences based upon the observations themselves. The idea here is that the two quantities in each pair might just as well have been interchanged. Thus the synthetic population consists of six possible pairs of differences: $(3, -3)$; $(4, -4)$; $(1, -1)$; $(6, -6)$; $(5, -5)$; $(1, -1)$. And by hypothesis, we are drawing a random sample of six numbers—one from each pair. This implies that the population of distinct samples consists of 64 possibilities, ranging from one extreme with all signs negative to the other, with all signs positive. These are shown in Table 6-3. This time, as our test variate we choose the sum of the six differences. For our immediate purpose, it is not necessary to construct the entire distribution of x, but for the sake of clarity we have done so, with the results shown in Table 6-4. The nearest we can come to the significance level .05 is to set the boundary of the critical region at $x = 18$, which yields $P(x \geq 18) = \frac{3}{64} \approx .047$. With the test variate selected, the actual sample value is $x = 18$, which falls within the critical region. Thus we reject the null hypothesis at the .047 level of significance and infer that the added ingredient *does* improve the wearing quality of shellac. This procedure is sometimes referred to as the *magnitude test*.

Most of the computational labor of this analysis can be avoided by constructing the critical region directly. The largest sample sum is obtained when all signs are positive, which can happen in just one way. The next largest sum is obtained when the smallest difference is negative. The smallest difference is unity, and there are two instances of this magnitude. Therefore, there are two possible samples having a sum equal to 18 and one sample having a sum of 20, making a total of three samples with a sum of 18 or greater. With only 64 distinct possibilities, we cannot enlarge the critical region by even one more element without exceeding .05. Hence, the critical region is $x = 18, 20$ and the actual significance level is $\frac{3}{64}$.   ////

Table 6-3 SYNTHETIC POPULATION

Serial number	Sample	Sum	Serial number	Sample	Sum
1	$-3, -4, -1, -6, -5, -1$	-20	33	$+3, -4, -1, -6, -5, -1$	-14
2	$-3, -4, -1, -6, -5, +1$	-18	34	$+3, -4, -1, -6, -5, +1$	-12
3	$-3, -4, -1, -6, +5, -1$	-10	35	$+3, -4, -1, -6, +5, -1$	-4
4	$-3, -4, -1, -6, +5, +1$	-8	36	$+3, -4, -1, -6, +5, +1$	-2
5	$-3, -4, -1, +6, -5, -1$	-8	37	$+3, -4, -1, +6, -5, -1$	-2
6	$-3, -4, -1, +6, -5, +1$	-6	38	$+3, -4, -1, +6, -5, +1$	0
7	$-3, -4, -1, +6, +5, -1$	$+2$	39	$+3, -4, -1, +6, +5, -1$	$+8$
8	$-3, -4, -1, +6, +5, +1$	$+4$	40	$+3, -4, -1, +6, +5, +1$	$+10$
9	$-3, -4, +1, -6, -5, -1$	-18	41	$+3, -4, +1, -6, -5, -1$	-12
10	$-3, -4, +1, -6, -5, +1$	-16	42	$+3, -4, +1, -6, -5, +1$	-10
11	$-3, -4, +1, -6, +5, -1$	-8	43	$+3, -4, +1, -6, +5, -1$	-2
12	$-3, -4, +1, -6, +5, +1$	-6	44	$+3, -4, +1, -6, +5, +1$	0
13	$-3, -4, +1, +6, -5, -1$	-6	45	$+3, -4, +1, +6, -5, -1$	0
14	$-3, -4, +1, +6, -5, +1$	-4	46	$+3, -4, +1, +6, -5, +1$	$+2$
15	$-3, -4, +1, +6, +5, -1$	$+4$	47	$+3, -4, +1, +6, +5, -1$	$+10$
16	$-3, -4, +1, +6, +5, +1$	$+6$	48	$+3, -4, +1, +6, +5, +1$	$+12$
17	$-3, +4, -1, -6, -5, -1$	-12	49	$+3, +4, -1, -6, -5, -1$	-6
18	$-3, +4, -1, -6, -5, +1$	-10	50	$+3, +4, -1, -6, -5, +1$	-4
19	$-3, +4, -1, -6, +5, -1$	-2	51	$+3, +4, -1, -6, +5, -1$	$+4$
20	$-3, +4, -1, -6, +5, +1$	0	52	$+3, +4, -1, -6, +5, +1$	$+6$
21	$-3, +4, -1, +6, -5, -1$	0	53	$+3, +4, -1, +6, -5, -1$	$+6$
22	$-3, +4, -1, +6, -5, +1$	$+2$	54	$+3, +4, -1, +6, -5, +1$	$+8$
23	$-3, +4, -1, +6, +5, -1$	$+10$	55	$+3, +4, -1, +6, +5, -1$	$+16$
24	$-3, +4, -1, +6, +5, +1$	$+12$	56	$+3, +4, -1, +6, +5, +1$	$+18$
25	$-3, +4, +1, -6, -5, -1$	-10	57	$+3, +4, +1, -6, -5, -1$	-4
26	$-3, +4, +1, -6, -5, +1$	-8	58	$+3, +4, +1, -6, -5, +1$	-2
27	$-3, +4, +1, -6, +5, -1$	0	59	$+3, +4, +1, -6, +5, -1$	$+6$
28	$-3, +4, +1, -6, +5, +1$	$+2$	60	$+3, +4, +1, -6, +5, +1$	$+8$
29	$-3, +4, +1, +6, -5, -1$	$+2$	61	$+3, +4, +1, +6, -5, -1$	$+8$
30	$-3, +4, +1, +6, -5, +1$	$+4$	62	$+3, +4, +1, +6, -5, +1$	$+10$
31	$-3, +4, +1, +6, +5, -1$	$+12$	63	$+3, +4, +1, +6, +5, -1$	$+18$
32	$-3, +4, +1, +6, +5, +1$	$+14$	64	$+3, +4, +1, +6, +5, +1$	$+20$

Table 6-4 DISTRIBUTION OF SAMPLE SUMS

x	$f(x)$	x	$f(x)$	x	$f(x)$
-20	$1/64$	-6	$4/64$	8	$4/64$
-18	$2/64$	-4	$4/64$	10	$4/64$
-16	$1/64$	-2	$5/64$	12	$3/64$
-14	$1/64$	0	$6/64$	14	$1/64$
-12	$3/64$	2	$5/64$	16	$1/64$
-10	$4/64$	4	$4/64$	18	$2/64$
-8	$4/64$	6	$4/64$	20	$1/64$

EXAMPLE 6-26 Three observations of radiation activity supposedly result from random draws from a "window" which has the mathematical form

$$f(x) = \begin{cases} x & 0 < x < 1 \\ 2 - x & 1 < x < 2 \\ 0 & \text{elsewhere} \end{cases}$$

Using the range of the three observations as a test variate and a two-sided 5 percent critical region, do the observations .4, 1.7, and .2 confirm the hypothesized physical situation?

The distribution of the range is obtained from the joint distribution of the smallest observation x_1 and the largest x_3 as $f(x_1, x_3)$. The joint occurrence of x_1 and x_3 has a different functional form for this probability depending upon the ranges of the variate. Thus

(1) $f(x_1, x_3)\, dx_1\, dx_3 = 6x_1 x_3 \left(\int_{x_1}^{x_3} x\, dx \right) dx_1\, dx_3$

$\qquad\qquad = 3x_1 x_3 (x_3{}^2 - x_1{}^2)\, dx_1\, dx_3 \qquad \text{if } x_1, x_3 < 1$

(2) $f(x_1, x_3)\, dx_1\, dx_3 = 6(2 - x_3)x_1 \left[\int_{x_1}^{1} x\, dx + \int_{1}^{x_3} (2 - x)\, dx \right] dx_1\, dx_3$

$\qquad\qquad = 3x_1 x_3 (4x_3 - x_1{}^2 - x_3{}^2 - 2)\, dx_1\, dx_3 \quad \text{if } x_1 < 1, x_3 > 1$

(3) $f(x_1, x_3)\, dx_1\, dx_3 = 6(2 - x_1)(2 - x_3) \left[\int_{x_1}^{x_3} (2 - x)\, dx \right] dx_1\, dx_2$

$\qquad\qquad = 3(2 - x_1)(2 - x_3)[(2 - x_1)^2 - (2 - x_3)^2]\, dx_1\, dx_3$

$\qquad\qquad\qquad\qquad\qquad\qquad\qquad\qquad\qquad \text{if } x_1, x_3 > 1$

If either case 1 or case 3 is integrated over the appropriate domain of the variables, the integral value is $\frac{1}{8}$, which is consistent with the probability that all three observations would occur either to the right or to the left of $x = 1$. This leaves a probability of $\frac{3}{4}$ for case 2, which is also consistent with the situation.

If the density function for the range R, $f(R)$ where $R = x_3 - x_1$, is now evaluated, it must be computed separately for all three cases. However for cases 1 and 3 the range is restricted between 0 and 1. Thus substituting $R = x_3 - x_1$ and $dR = dx_3$ we have, for case 1,

(1) $f(R)\, dR = 3R\, dR \int_{0}^{1-R} x_1(R + x_1)(R + 2x_1)\, dx_1$

$\qquad\qquad = \tfrac{3}{2}R(1 - R)^2\, dR \qquad R < 1$

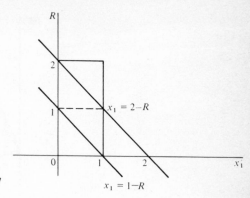

FIGURE 6-17

also, for case 3,

$$= 3 \, dR \int_{2}^{2-R} (2 - x_1)(2 - x_1 - R)[(2 - x_1)^2 - (2 - x_1 - R)^2] \, dx_1$$

$$= \tfrac{3}{2}R(1 - R)^2 \, dR \qquad R < 1$$

But for case 2,

(2) $f(R) \, dR = 3 \, dR \int_{1-R}^{1} x_1(2 - R - x_1)[2 - x_1{}^2 - (2 - R - x_1)^2] \, dx_1 \qquad R < 1$

$$= \left(6R^2 - 8R^3 + 3R^4 - \frac{3R^5}{10} \right) dR \qquad R < 1$$

or $f(R) \, dR = 3 \, dR \int_{0}^{2-R} x_1(2 - R - x_1)[2 - x_1{}^2 - (2 - R - x_1)^2] \, dx_1 \qquad R > 1$

$$= (2 - R)^3 - \tfrac{3}{10}(2 - R)^5 \qquad 1 < R \leq 2$$

The limits on case 2 may be obtained by examining the limits of x_1 for a fixed R from Fig. 6-17 where, from the relation $x_3 - x_1 = R$, we obtain the two lines $x_1 + R = 2$ and $X_1 + R = 1$ since in this case x_3 has two limits $x_3 = 1$ and $x_3 = 2$.

It is interesting to note that in case 3 the probability that R is between 0 and 1 is $\tfrac{11}{20}$ and only $\tfrac{9}{20}$ that it is between 1 and 2.

If the three parts of the probability density which contribute to R between 0 and 1 are added together, we obtain the final density of the random variable R:

$$f(R) = 3R - 5R^3 + 3R^4 - \tfrac{3}{10}R^5 \qquad 0 \leq R \leq 1$$

$$f(R) = (2 - R)^3 - \tfrac{3}{10}(2 - R)^5 \qquad 1 < R \leq 2$$

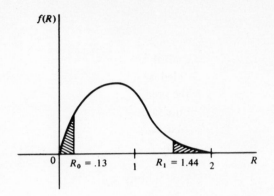

FIGURE 6-18

The graph of $f(R)$ is shown in Fig. 6-18 where R_0 and R_1 are to be determined so that the area to the left and right is .025. Approximately $R_0 = .13$ and $R_1 = 1.44$.

Since the observed range is $1.7 - .2 = 1.5$, the hypothesis is rejected at the 5 percent level and we assume the observations did not come from the triangular universe. ////

(*Suggested Problems 6-30 to 6-34*)

6-15 THE HYPERGEOMETRIC TEST FOR THE DIFFERENCE BETWEEN TWO PERCENTAGES

Having pursued research aimed at altering the relative frequency of some attribute A in the population considered, the investigator sets up an experiment to test the efficacy of the proposed procedure. For test purposes, two equivalent groups of individuals are secured. One of these, conventionally designated as the experimental group, will receive the treatment in question, and the other, designated as the control group, will not. After the treatment has had time to take its supposed effect, the two samples will be compared as to their relative frequencies of attribute A. For a *fixed* total sample size, maximum precision will be achieved by dividing the two groups equally. On the other hand, recruiting difficulties or the expense of treatment may restrict the size of the experimental group without appreciably affecting the size of the control group. In this situation, the total sample size is not definitely prescribed, and, provided that comparability can be maintained, the control group should be as large as practicable, for, consistent with any stipulated conditions, statistical precision increases with total sample size. Since the object of research here has been to produce a change in a stated direction, either increase or decrease as the case may be, the critical region should generally be one-sided. The statistical question to be decided is whether or not the two relative frequencies are significantly different, in the sense that they could not be reasonably considered as random specimens of the same population.

In the first sample, let us say, there are N_1 observations, and of these, a number a_1 exhibit the attribute A, whereas the remaining $N_1 - a_1$ observations do not. In the second sample, having N_2 observations, a number a_2 exhibit attribute A, and the remaining $N_2 - a_2$ observations do not. These facts are represented schematically in Table 6-5, wherein $N \equiv N_1 + N_2$ and $a \equiv a_1 + a_2$.

If both samples were drawn at random from the same infinite population, the common probability of attribute A would be some definite quantity p, but since its value is unknown, we resort to constructing a population in which the probability of A is a weighted average of the percentages observed in the two samples. That is, we simply combine the data, thereby obtaining a finite population consisting of $N = N_1 + N_2$ elements, of which the number $a = a_1 + a_2$ exhibit A. Imagine, now, the population of all distinct pairs of samples that could be formed from the combined data by arbitrarily allocating N_1 observations to the first sample and the remaining N_2 observations to the second. By construction, both samples are now being drawn from the same finite population of N observations, in the sense that initially all individuals have equal chances of being included in a particular sample. The population, of course, is exhausted by the two samples; nevertheless, it is intuitively plausible that in the long run, if an unlimited number of independent trials could be made, the probability of A would be the same in both. The truth of this conjecture can be established by a simple argument, based on mathematical expectation; but until this topic has been discussed (Chap. 7), the reader may accept the following statements without proof. In the first sample, the mean number of A's is $N_1 a/N$ and in the second, $N_2 a/N$. In each, therefore, the mean relative frequency is a/N.

Table 6-5 CONTINGENCY TABLE

Designation	A	Not A	Total
Sample 1	a_1	$N_1 - a_1$	N_1
Sample 2	a_2	$N_2 - a_2$	N_2
Total	a	$N - a$	N

In any selection, let x denote the number of type A elements assigned to sample 1. Then there are $a - x$ elements of type A in sample 2, and the percentages are x/N_1 and $(a - x)/N_2$, respectively. As a test variate, we should naturally consider employing the difference between the sample percentages, say

$$d = \frac{x}{N_1} - \frac{a - x}{N_2}$$

However, it is evident that d is uniquely determined by x, and since the latter is more convenient to use, we shall choose x as our test variate. By simplifying, we readily obtain

$$d = \frac{Nx - aN_1}{N_1 N_2} = \frac{N}{N_1 N_2}\left(x - \frac{aN_1}{N}\right)$$

whence it is clear that the two percentages will be equal ($d = 0$) when $x = aN_1/N$.

Depending on the working hypothesis, critical regions will be constructed (as indicated in Sec. 6-14) with the desired significance level. Since x is a hypergeometric variate, its distribution is

$$f(x) = \frac{C(a, x)C(N - a, N_1 - x)}{C(N, N_1)} \qquad x = 0, 1, \ldots, [a, N_1] \qquad (6\text{-}22)$$

Like many other contributions to statistics, the hypergeometric test is due to R. A. Fisher. The following example was based on a real experiment, but we have simplified it for purposes of illustration.

EXAMPLE 6-27 In a carefully controlled experiment on the value of elastic stockings in the treatment of thromboembolic diseases, case histories on 4,694 hospital patients were compiled. Half of this group wore elastic stockings during the period of treatment, and the other half (the control group) did not. The total sample was divided into 2,347 matched pairs chosen on the basis of nearly equal proneness to or affliction with thromboembolic diseases. By random selection, one member of each pair was allocated to the stocking group, and the other member, to the control group. A great deal of additional information was gained from the experiment, but to simplify the problem, let us confine our attention to the incidence of deaths. The issue, then, is whether we are justified in concluding, on the strength of the obtained evidence, that the stockings are a significant factor in reducing mortality. The null hypothesis is that the stockings have no effect. Letting x denote the number of deaths in the stocking group, we shall use a one-sided critical region which, in view of the working hypothesis, will be built up from the lower end of the scale.   Choose .05 as the level of significance.

As it turned out, eight deaths due to embolism occurred in the control group, but no deaths occurred in the stocking group.   The distribution of x, under the null hypothesis, is shown in Table 6-6.   The nearest we can approach .05 without overshooting is .035, and the corresponding boundary of the critical region will be $x = 1$.   Since the observed value $x = 0$ falls within the critical region, even with a more stringent level of significance than we initially planned to use, the null hypothesis is rejected, and we accept the working hypothesis that in the treatment of thromboembolic diseases, elastic stockings are a significant factor in reducing mortality.

Once again, the numerical labor of performing the test can be reduced a great deal by confining the computation to what is actually needed.   Since the critical region starts at the lower end, we need only compute $f(0)$, in this case, in order to find out whether the chosen level of significance is satisfied.

////

In the hypergeometric test, the magnitude test, and other tests commonly employed, the distribution obtained under the null hypothesis is actually a conditional distribution.   By holding to a constant level of significance in the conditional distributions, we ensure the same level in the marginal distribution; but the boundaries of the critical regions vary from one experiment to another, whereas they would remain fixed if the marginal distribution could be used. Admittedly, it would be preferable to base the test on the marginal distribution, but requirements of experimental control or difficulties in obtaining suitable data often impose such limitations on the availability of information that it is impossible to prescribe the marginal distribution of the test variate.   Therefore, the conditional distribution is accepted as the alternative to abandoning the problem entirely.

(Suggested Problems 6-35 and 6-36)

Table 6-6   DISTRIBUTION, UNDER NULL HYPOTHESIS, OF NUMBER OF DEATHS IN STOCKING GROUP

x	0	1	2	3	4	5	6	7	8
$f(x)$	.004	.031	.109	.219	.274	.219	.109	.031	.004
$F(x)$	.004	.035	.144	.363	.637	.856	.965	.996	1.000

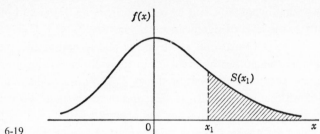

FIGURE 6-19

6-16   COLLECTIVE SIGNIFICANCE

The control of extraneous sources of variation in an experiment sometimes imposes such severe restrictions on the sample size that it is necessary to pool the results of several independent samples in order to obtain enough data for a satisfactory test of significance.   Under such circumstances, it rarely happens that the samples can be legitimately lumped together, but it is often possible to combine the respective probabilities into a single index of collective significance. Let us see how this can be done for a *continuous variate*.

Suppose that the criterion of significance can be regarded as the deviation (either algebraic or absolute) of a continuous variate x from an appropriate norm and that this variate has a density function $f(x)$.   We may assume that the significant values lie toward the right-hand extreme of the distribution of x, for this can be brought about by a change of variable.   Associated with any observed value x_1, the measure of significance $S(x_1)$ then consists of the area under the density curved beyond x_1 (Fig. 6-19).   It is readily seen that $S(x)$ itself is a random variable, and it follows from Sec. 6-5 that it is rectangular in the interval (0, 1).

If two independent tests yield significance measures S_1 and S_2, respectively, the probability of the joint event is the product $S_1 S_2$, which is the chance of drawing a deviation at least as great as x_1 on the first trial and a deviation at least as great as x_2 on the second trial.   What we want, therefore, is the chance that any pair of samples will yield a joint probability less than or equal to the actual product $S_1 S_2$.   Construct a rectangular coordinate system as in Fig. 6-20.   Under the null hypothesis, the points (S_1, S_2), representing the results of any pair of tests, are uniformly distributed over the unit square in the first quadrant.   The rectangular hyperbola $S_1 S_2 = \lambda$ defines the locus of all pairs of test results having equal likelihood λ.   The combined significance measure of any pair of tests for which $S_1 S_2 = \lambda$ is equal to the area of that portion of the

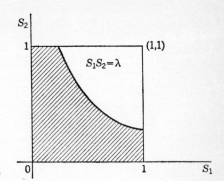

FIGURE 6-20

unit square which lies below the hyperbola, for this area represents the total probability of obtaining some pair of values for which the likelihood is equal to or less than λ. Denoting the combined significance by P, we have

$$P = 1 - \int_\lambda^1 \int_{\lambda/S_1}^1 dS_1 \, dS_2 = \lambda(1 - \ln \lambda) \qquad (6\text{-}23)$$

Similarly, for three independent tests the sample space is confined to a unit cube, within which all points have unit density. The likelihood contour is the hyperbolic surface $S_1 S_2 S_3 = \lambda$ and the combined significance measure is

$$P = 1 - \int_\lambda^1 \int_{\lambda/S_1}^1 \int_{\lambda/S_1S_2}^1 dS_1 \, dS_2 \, dS_3 = \lambda\left[1 - \ln \lambda + \frac{(\ln \lambda)^2}{2}\right] \qquad (6\text{-}24)$$

In general, for n independent tests, it is easy to show that the combined significance measure is

$$P = \lambda\left[1 - \ln \lambda + \frac{(\ln \lambda)^2}{2!} - \frac{(\ln \lambda)^3}{3!} + \cdots + (-1)^{n-1}\frac{(\ln \lambda)^{n-1}}{(n-1)!}\right] \qquad (6\text{-}25)$$

where λ equals the product of the S's.

Since very good tables of natural logarithms are available, P can be computed directly from the foregoing equation. However, Eq. (6-25) has a remarkable relationship to the distribution of chi-square with $2n$ degrees of freedom and, as a consequence, most of the arithmetic can be avoided by using tables of the chi-square integral. For $2n$ degrees of freedom [see Eq. (6-9)], the density function of chi-square is

$$g(\chi)^2 = \frac{1}{2^n \Gamma(n)} (\chi^2)^{n-1} e^{-\chi^2/2} \qquad (0, \infty) \qquad (6\text{-}26)$$

and it turns out that the corresponding probability integral is

$$\int_A^\infty g(\chi^2)\, d\chi^2 = e^{-A/2}\left[1 + \frac{A}{2} + \frac{(A/2)^2}{2!} + \frac{(A/2)^3}{3!} + \cdots + \frac{(A/2)^{n-1}}{(n-1)!}\right] \qquad (6\text{-}27)$$

Accordingly, if we are considering the combined significance of separate tests, we make the substitution $A = -2(\ln \lambda)$, and Eq. (6-27) becomes identical with Eq. (6-25). Therefore, the combined significance measure P can be evaluated from tables of the chi-square integral using $2n$ degrees of freedom and looking up under $-2(\ln \lambda)$. A more sophisticated way of establishing this connection with chi-square is outlined in Prob. 6-13.

EXAMPLE 6-28   Suppose that the test variate x is distributed as

$$f(x) = 12(4 + x^2)^{-5/2}$$

$(-\infty, \infty)$ and the significance measure is

$$S(u) = \int_u^\infty f(x)\, dx = \frac{1}{2}\left[1 - \frac{u(6 + u^2)}{(4 + u^2)^{3/2}}\right]$$

Three independent samples yield

$$x_1 = 1.42 \qquad x_2 = 1.61 \qquad x_3 = 1.79$$

$$S_1 = S(x_1) = .114 \qquad S_2 = .091 \qquad S_3 = .074$$

Thus

$$\lambda = S_1 S_2 S_3 = 7.68 \times 10^{-4} \qquad \ln \lambda = -7.172 \qquad (\ln \lambda)^2 = 51.438$$

and the collective significance is

$$P = 7.68 \times 10^{-4}(1 + 7.172 + 25.719)$$
$$= 7.68 \times 10^{-4}(33.891)$$
$$= 2.60 \times 10^{-2} = .026$$

Alternatively, we set chi-square equal to $-2(\ln \lambda) = 14.344$, and referring to standard tables for 6 degrees of freedom, the nearest tabulated value we find is 14.45, for which $P = .025$.   ////

The formula (6-25) derived for a continuous variate does not hold for a discrete variate. However, the general approach is much the same, as the following example illustrates.

EXAMPLE 6-29   Suppose a discrete variate x is distributed as $f(x) = (\tfrac{1}{2})^{x+1}$ ($x = 0, 1, 2, \ldots, \infty$) and the measure of significance is

$$S(u) = \sum_{u}^{\infty} f(x) = (\tfrac{1}{2})^{u}$$

Two independent samples yield

$$x_1 = 3 \qquad S_1 = S(x_1) = (\tfrac{1}{2})^3$$

$$x_2 = 4 \qquad S_2 = S(x_2) = (\tfrac{1}{2})^4$$

$$\text{Joint probability} = S_1 S_2 = (\tfrac{1}{2})^7$$

To evaluate the combined significance of the two samples, we determine the total probability that any pair of samples would yield a joint probability less than or equal to $(\tfrac{1}{2})^7$. The possible situations and corresponding probabilities are exhibited in Table 6-7. From this table we find that the combined significance is

$$P = 7(\tfrac{1}{2})^8 + (\tfrac{1}{2})^7 = \tfrac{9}{256} \approx .035 \qquad ////$$

(*Suggested Problems 6-37 to 6-39*)

Table 6-7   **POSSIBLE SAMPLE VALUES YIELDING $S_1 S_2 \leq (\tfrac{1}{2})^7$**

Values		Probabilities		Joint probabilities
x_1	x_2	x_1	x_2	
0	≥ 7	$f(0)$	$S(7)$	$f(0)S(7) = (\tfrac{1}{2})^8$
1	≥ 6	$f(1)$	$S(6)$	$f(1)S(6) = (\tfrac{1}{2})^8$
2	≥ 5	$f(2)$	$S(5)$	$f(2)S(5) = (\tfrac{1}{2})^8$
3	≥ 4	$f(3)$	$S(4)$	$f(3)S(4) = (\tfrac{1}{2})^8$
4	≥ 3	$f(4)$	$S(3)$	$f(4)S(3) = (\tfrac{1}{2})^8$
5	≥ 2	$f(5)$	$S(2)$	$f(5)S(2) = (\tfrac{1}{2})^8$
6	≥ 1	$f(6)$	$S(1)$	$f(6)S(1) = (\tfrac{1}{2})^8$
≥ 7	≥ 0	$S(7)$	$S(0)$	$S(7)S(0) = (\tfrac{1}{2})^7$

Total probability $P = 7(\tfrac{1}{2})^8 + (\tfrac{1}{2})^7 = \tfrac{9}{256}$

PROBLEMS

6-1   Expressed in appropriate monetary units, the distribution of yearly income in a certain population is given by $f(x) = (3/40,000)x(20 - x)^2(0, 20)$. (a) What proportion of the population have incomes below the mode? (b) What would the distribution be if everybody's income were exactly doubled?

                                                                              Ans. (a) $11/27$.

6-2   Sketch $f(x)$ and derive and sketch $g(y)$, the corresponding density function of y, for

    (a) $f(x) = (1/\pi\sqrt{1 - x^2})(-1, 1); y = \sin^{-1} x$. *Ans.* (a) $1/\pi(-\pi/2, \pi/2)$.

    (b) $f(x) = 1/\pi(-\pi/2, \pi/2); y = \tan x$.   *Ans.* (b) $1/\pi(1 + y^2)(-\infty, \infty)$.

    (c) $f(x) = (1/\sqrt{2\pi})x^{-2}\exp[-1/(2x^2)](-\infty, \infty); y = 1/x$.

6-3   Find the exact distribution of the variate w derived from a gamma variate x by the Wilson-Hilferty transformation. $w = [x/\beta(\alpha + 1)]^{1/3}$.

6-4   In a circle of radius R a chord AB is drawn (refer to Figs. 1-1a and 1-2a). The tangent CAD is constructed at the point A and the angle ϕ is taken as the angle DAB. Also, the perpendicular bisector OH of the chord is constructed and the point of intersection is denoted by X. (a) If the angle ϕ is rectangularly distributed over the range $(0, \pi)$ find the distribution of $x = OX$. (b) If x is rectangularly distributed over the range $(0, R)$, find the distribution of ϕ, given that the two supplementary values of this angle associated with a fixed value of x are equally likely. (c) Using part (a), find $P(x \leq R/2)$. (d) Using part (b), find $P(\pi/3 \leq \phi \leq 2\pi/3)$.

6-5   Find the distribution of $y = x^2$ given that $f(x) = (1/\pi)[1/(1 + x^2)](-\infty, \infty)$.

6-6   Find the distribution of $y = |x|$ given that $f(x) = 1/3(-1, 2)$.

6-7   A random voltage noise with probability density $f(V) = e^{-V}(0 \leq V \leq \infty)$ is present on the input to an amplifier which in turn subtracts a dc voltage of 2 from the noise and squares the result. What is the voltage distribution of the output?

6-8   Find the density function of the product $u = xy$, where x and y are distributed independently as follows:

    (a) $f_1(x) = 1 = f_2(y)$ $(0, 1)$ each   *Ans.* (a) $-\ln u\,(0, 1)$.

    (b) $f_1(x) = 6x(1 - x)$ $(0, 1); f_2(y) = 1$ $(0, 1)$

                                                            Ans. (b) $3(1 - u)^2(0, 1)$.

    (c) $f_1(x) = 6x(1 - x)$ $(0, 1); f_2(2) = 2y, (0, 1)$

    (d) $f_1(x) = 6x(1 - x)$ $(0, 1); f_2(y) = 6y(1 - y), (0, 1)$

6-9   Find the density function of the sum $u = x + y$, where x and y are independently distributed as follows:

    (a) $f_1(x) = ae^{-ax}(0, \infty); f_2(y) = be^{-by}(0, \infty)$ $0 < b < a$

    (b) $f_1(x) = xe^{-x}(0, \infty); f_2(y) = 4ye^{-2y}(0, \infty)$

    (c) $f_1(x) = 1/3, (1, 4); f_2(y) = 1/2(6, 8)$

6-10 Find the distribution of the quotient $u = x/y$, where x and y are distributed as follows:

    (a) x and y are independent unit normal variates.   *Ans.* (a) $1/\pi\,(1 + u^2)$.

    (b) x and y are independent observations from $f(z) = 1/z^2$ $(0, \infty)$.

6-11 Find the distribution of the function z, where the distribution of x and y is as given: (a) $f(x) = 1$ $(0, 1)$; y is distributed as $x_1 - x_2$, where x_1, x_2 are independent observations from $f(x)$; $z = |y|$. (b) z is the square root of the sum of two independent observations from $f(x) = e^{-x}$ $(0, \infty)$.

                                               Ans. (a) $2(1 - z)$ $(0, 1)$

6-12 The variate x has the Cauchy distribution $f(x) = 1/\pi(1 + x^2)$ $(-\infty, \infty)$. If two observations x_1, x_2 are taken at random and added together, find the distribution of their sum $u = x_1 + x_2$ and from this deduce the distribution of their average $\bar{x} = u/2$. Integrate as a rational fraction, making use of the identity

$$\frac{1}{(1 + x^2)[1 + (u - x)^2]} = \frac{1}{u^2(u^2 + 4)}\left[\frac{u^2}{1 + x^2} + \frac{2ux}{1 + x^2} + \frac{u^2}{1 + (x - u)^2}\right. $$
$$\left. - \frac{2u(x - u)}{1 + (x - u)^2}\right]$$

Indeterminate forms can be avoided by combining logarithms before substituting limits of integration.

6-13 The variate x is rectangularly distributed over the range $(0, 1)$. (a) Derive the distribution of $y = -2 \ln x$ and relate to chi-square. (b) Let $x_1, x_2, \ldots, x_n$ be random observations of x, define y_i as $-2 \ln x_i$ $(i = 1, 2, \ldots, n)$, and let z denote the sum $y_1 + y_2 + \cdots + y_n$. By immediate deduction from (a), state the distribution of z.

6-14 Find the distribution of the quotient $u = x/y$ of the independent variates x and y which are distributed as follows:

$$f_1(x) = \frac{2}{\sqrt{2\pi}}\, e^{-x^2/2} \qquad (0, \infty) \qquad f_2(y) = 2y^3 e^{-y^2} \qquad (0, \infty)$$

By comparing distributions, identify u as $|t|/\sqrt{2}$, where t has the t distribution with 4 degrees of freedom.

6-15 A variate x has the F distribution with a and b degrees of freedom and another unrelated variate y has the F distribution with b and a degrees of freedom. (a) If one had a table giving $P(x \geq c)$ for an arbitrary value $c > 0$, how could this be used to find $P(y \leq 1/c)$? (b) As a consequence of this property, show at once that in case $a = b$ the median of the F distribution is unity.

6-16 Establish the following properties of special cases of the chi-square, t, and F distributions:

    (a)   If x is distributed as chi-square with an even number of degrees of freedom $2n$, the density function can be integrated explicitly to yield the closed expression

$$P(x \geq 2u) = e^{-u}\left[1 + u + \frac{u^2}{2!} + \frac{u^3}{3!} + \cdots + \frac{u^{n-1}}{(n-1)!}\right]$$

(b) If x is distributed as t with n degrees of freedom, x^2 is distributed as F with 1 and n degrees of freedom.

(c) If x is distributed as F with 2 and n degrees of freedom, the density function can be integrated explicitly to yield

$$P(x \geq k) = \left(1 + \frac{2k}{n}\right)^{-n/2}$$

In Probs. 6-17 to 6-20, find the density function of z, where x and y are distributed as follows:

6-17 $f(x, y) = xe^{-x(1+y)}$, $(0, \infty)$ each, where $z = xy$.   *Ans.* e^{-z} $(0, \infty)$.

6-18 $f(x, y) = 6xy(2 - x - y)$, $(0, 1)$ each, where $z = x + y$.

6-19 $f(x, y) = \frac{1}{4}(1 + y + xy)$, $(0, 1)$ each, where $z = y(1 + x)$.

6-20 $f(x, y) = 120x(y - x)(1 - y)$, $(0 \leq x \leq y, 0 \leq y \leq 1)$, where $z = x/y$.

Find the distribution of the range R of random samples from the following distributions in Probs. 6-21 to 6-23.

6-21 n observations from $f(x) = 1$ $(0, 1)$.   Determine also the probability that the range of a sample of five will exceed .5.   *Ans.* Probability $= \frac{13}{16}$.

6-22 Four observations of the triangular variate x, where $f(x) = 2x$ $(0, 1)$.

6-23 Three observations of the beta variate x, where $f(x) = 6x(1 - x)$ $(0, 1)$.

6-24 (a) From an initial pair of variables x, y a new pair u, v is derived by the transformation $u = x + y$, $v = x/y$. Show that the Jacobian of u, v with respect to x, y is $-(x + y)/y^2$. By an independent derivation show that the Jacobian of x, y with respect to u, v is the reciprocal of the first Jacobian and can be expressed as $-u/(1 + v)^2$.

(b) Given $u = x + y$, $v = x/(x + y)$, find both Jacobians.

6-25 (a) The variates x, y are jointly distributed as follows: $f(x, y) = (2/\pi)$ $(1 - x^2 - y^2)$ $(0 \leq x^2 + y^2 \leq 1)$. Find the joint distribution of x and r and obtain the marginal distribution of r, where $r = (x^2 + y^2)^{1/2}$.

(b) Find the joint distribution of r and θ, where $\theta = \tan^{-1}(y/x)$ and $f(x, y)$ is as above.   *Ans.* (b) $(2/\pi) r (1 - r^2)$ $(0 \leq r \leq 1, 0 \leq \theta \leq 2\pi)$.

6-26 Given random observations x_1, x_2, x_3 from a unit normal population. Find (a) the joint distribution of y_1, y_2, y_3, (b) the marginal distribution of each, and (c) the distribution of $z = y_1^2 + y_2^2 + y_3^2$, where

$$y_1 = \frac{x_1 + x_2 + x_3}{\sqrt{3}} \qquad y_2 = \frac{x_1 - x_2}{\sqrt{2}} \qquad y_3 = \frac{2x_3 - x_1 - x_2}{\sqrt{6}}$$

6-27 Find the joint density function of u, v, where u, v are defined as follows:

(a) $u = y^2 - x^2$, $v = y - x$; $f_1(x) = 1$ $(0, 1)$; $f_2(y) = 1$ $(1, 2)$.

(b) $u = x + 2y$, $v = 2x + y$; $f_1(x) = e^{-x}$ $(0, \infty)$; $f_2(y) = 2e^{-2y}$ $(0, \infty)$.

(c) $u = xy$, $v = y/x$; x and y are independent observations from $f(x) = 1$ $(1, 2)$.   *Ans.* (c) $\frac{1}{2}v$ $(u < v < 1/u, \quad 0 < u < 1)$.

(d) $u = x + y$, $v = x/y$; x and y are independent observations from $f(x) = e^{-x}$ $(0, \infty)$. Are u and v independent?

6-28   Find the joint density function of u, v, where u, v are defined as follows:

(a) $u = x + y$, $v = x/(x + y)$; independent variates x and y are distributed as chi-square wih $2a$ and $2b$ degrees of freedom, respectively.

(b) $u = x + y$, $v = x - y$; $f_1(x) = (1\sqrt{2\pi})x^{1/2}e^{-x/2}$ $(0, \infty)$; $f_2(y) = (1/\sqrt{2\pi})y^{1/2}e^{-y/2}(0, \infty)$.

(c) $u = x + y$, $v = x - y$; x and y are independent observations from $f(x) = e^{-x}$ $(0, \infty)$.   *Ans.* (c) $\frac{1}{2}e^{-u}\,(-u < v < u,\ \ 0 < u < \infty)$.

6-29   Consider the distribution defined by

$$f(x, y) = \frac{\sin^{-1}[x/(x^2 + y^2)^{1/2}]}{\pi^2(x^2 + y^2)^{3/2}}, \qquad 1 \le x^2 + y^2 \le 4$$

(a)   Are the variables independent? (b) Transform into polar coordinates. (c) Show that $\iint f(r, \theta)r\,dr\,d\theta = 0$. (d) Find the marginal distribution of r and θ. (e) Are these variables independent?   *Ans.* (e)   Yes.

6-30   A man given an ordinary-looking die wonders about its fairness. In testing the null hypothesis that the probability of a 5 on a single toss is $\frac{1}{6}$ he finds that he must toss the die 12 times before he gets a 5. Is the hypothesis rejected at the .05 significance level?   *Ans.* No.

6-31   In normal use, a certain natural fiber is exposed to outdoor weathering. A synthetic fiber, intended as a substitute, was developed for greater breaking strength after exposure. Five strands of each fiber were chosen at random from typical lots, and each strand of natural fiber was paired at random with a strand of synthetic. The pairs were then exposed to weather at different places, which were far enough apart to justify the assumption of independence from pair to pair. Later tested for breaking strength, the two samples gave the following results:

Pair	1	2	3	4	5
Synthetic	18	15	25	13	17
Natural	10	16	11	6	8

(a)   At the least, what must the significance level α be in order that the above results would be judged significant by the sign test, using the appropriate one-sided critical region? State the critical region which corresponds.

(b)   How small could α be in order that the same data would be judged significant by the magnitude test? State the critical region which corresponds.

6-32   If the fiber experiment of Prob. 6-31 could be carried out under uniform weather conditions, pairing would become irrelevant and it would be correct to regard the observations as constituting two independent samples of five each. In that case, we could employ another form of the magnitude test in which the two sets of five are regarded as (conditionally) random selections from a combined set of 10, and the corresponding test variate is the difference D between the two sums of five observations. Using the data of Prob. 6-31, verify the following portion of the frequency table of D, wherein n denotes the number of ways (out of 252 possible samples) of obtaining a given value of D, and P denotes the probability that D will equal or exceed a stated value. Notice that the value of D actually obtained would be significant even if α were appreciably less than .05.

D	43	39	37	35	33	31	29
n	1	1	1	2	3	2	4
P	.004	.008	.012	.020	.032	.040	.056

6-33 The probability density function for the life of a certain electronic tube is hypothesized as

$$f(t) = e^{-t} \quad 0 < t < \infty$$

where t is measured in appropriate units. Five tubes are tested at random out of a lot, and the values are 1.3, .5, 2.6, 1.1, and .2. Since the consumer is particularly worried about tubes with very small life, he decides to test the hypothesis by examining the behavior of the smallest observation of a group of five. Using a 5 percent level of significance and a critical area on the left, do the data confirm or reject the hypothesis that $f(t) = e^{-t}$. *Ans.* Confirm.

6-34 Given the density function $f(x) = x$ (0, 1), $f(x) = 2 - x$ (1, 2), $f(0) = 0$ otherwise as the hypothesis. The test variate to check this hypothesis is to be the smaller of two observations taken on the population. Using a two-sided critical region of 10 percent, do the two observed readings .4 and .6 reject the original hypothesis? *Ans.* No.

6-35 In two random samples of five observations the *relative* frequencies of attribute A are respectively 0 and 1. (*a*) Using the hypergeometric test with a two-sided region, determine the probability of this result under the hypothesis that both samples are taken from the same population. (*b*) Repeat if the relative frequencies are .2 and .8, respectively. *Ans.* (*a*) $\frac{1}{126}$, (*b*) $\frac{26}{126}$.

6-36 Either level A_1 or level A_2 of an experimental vaccine is used on a group of patients, and the recovery period for the disease treated noted in each case as either short (B_1) or long (B_2). The results are shown in the following contingency table.

		B	
		B_1	B_2
A	A_1	2	18
	A_2	4	6

At a 5 percent level of significance is the null hypothesis tenable that the results are due to chance only or is there some association between the recovery period and the level of the dose?

6-37   In most applications of the F test the appropriate critical region occupies the right-hand portion of the distribution.   With 2 and 4 degrees of freedom, it is easy to show that under the null hypothesis the probability of attaining or exceeding any particular value of F, say $F = A$, is given by

$$S(A) \equiv \int_{A}^{\infty} f(F)\, dF = \frac{4}{(2 + A)^2}$$

Two independent samples using this test variate (F with 2 and 4 degrees of freedom) gave $F = 4$ and $F = 6$, respectively.   Determine whether or not the combined result is significant at the 5 percent level.   *Ans.* Yes.

6-38   With the test variate $z = x_i{}^3 / y_i{}^3$ in the fiber problem (Example 5-9) show that the significance measure under the working hypothesis that synthetic fiber is stronger than natural ($a < b$) is $S(z) = (z + 1)^{-1}$.   Test the collective significance of the data in Prob. 6-31, given that with 10 degrees of freedom and $\alpha = .05$ the critical value of chi-square is 18.307.   Actually the parameters a, b in the respective distributions of x_i and y_i should be multiplied by a common factor k_i which adjusts for local conditions; a common factor, however, would not affect the distribution of z.

6-39   The object of the series of fiber problems is to illustrate the increase of power which can be gained by narrowing the specification of the distribution either through direct information or by reducing the number of distinct parameters. Under the conditions of Prob. 6-32, the parameter k_i can be incorporated with a and b, yielding $f(x) = 3ax^2 e^{-ax^3}$ $(0, \infty)$ and $g(y) = 3by^2 e^{-by^3}$ $(0, \infty)$.   Still assuming two independent samples of five observations, show that the test variate $\sum x^3 / \sum y^3$ is distributed as F with 10 and 10 degrees of freedom under the null hypothesis.   Compute this ratio for the data in question and judge the comparative power of the latter test from the following fractiles $\alpha = .05$, $F = 2.97$, $\alpha = .025$, $F = 3.72$; $\alpha = .01$, $F = 4.85$.

7

MATHEMATICAL EXPECTATION

7-1 DEFINITION

The concept of mathematical expectation is of paramount importance both in the mathematical development and practical use of probability theory. The term "mathematical expectation" dates back to the pioneer work in the theory of probability, which received impetus from games of chance. The common experience of fortuitous variations in luck prompted gamblers to pose the question as to how much, in the long run, one might expect to win per game under stated rules of play, and mathematicians answered the question by defining "expected value."

Consider the distribution function $F(x)$ of any variate x, which may be either discrete or continuous or, in fact, such that its distribution function is continuous over part of its range but discontinuous at a series of discrete values, such as x_1 and x_2 in Fig. 7-1a. Simultaneously, consider a function $\phi(x)$ which is continuous over the entire region of definition of $F(x)$ (see Fig. 7-1b).

As one of the first steps in building the concept of the definite integral in the calculus, one divides the range of definition of the function into a series of parts Δx_i where the sum of the Δx_i equals the range of x. In this case, however, remembering that $F(x)$ is monotonic, let us divide the ordinate range

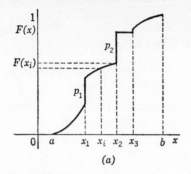 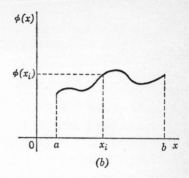

FIGURE 7-1

from 0 to 1 into n divisions where each part is $\Delta F(x_i)$. These divisions may be equal or not, as we please, but the sum of the $\Delta F(x_i)$ must again equal the range, which in this case is unity. Lines parallel to the x axis from two consecutive points on the vertical axis will ordinarily intersect the curve $F(x)$ in two points (see Fig. 7-1a). The abscissas of these two points may be the same or different, as one may see by examining the figure. However, if a point x_i is chosen any-where between the two abscissas, it will determine a value of $\phi(x)$, say $\phi(x_i)$. Under very general restrictions placed on $F(x)$, including the fact that it must be monotonic, the sum $\sum \phi(x_i) \Delta F(x_i)$ (where the summation goes from $i = 1$ to $i = n$) can be shown to approach a definite limit as the number of divisions $\Delta F(x_i)$ become infinite, irrespective of the choice of the size of each $\Delta F(x_i)$, provided they all approach zero as n becomes infinite. This unique limit is defined as the expected value of $\phi(x)$ and is denoted as the Stieltjes integral,

$$E[\phi(x)] = \int_a^b \phi(x)\, dF(x) \qquad (7\text{-}1)$$

It should be noted that in the case of the vertical jumps in the function $F(x)$, (such as at x_1 and x_2 in Fig. 7-1a) the function $\phi(x)$ remains constant along those jumps and the $\Delta F(x_i)$ sum to give the total rise in each case. The contributions to the integral at the points x_1 and x_2 would be $\phi(x_1)p_1 + \phi(x_2)p_2$. Thus, in the case of the distribution function of a discrete random variable, which must consist of a series of these jumps, the incremental value of the integral is dif-ferent from zero only at the jump points, since that portion of $F(x)$ which is parallel to the x axis, such as between x_2 and x_3 in Fig. 7-1a, will contribute nothing to the integral, as $\Delta F(x_i) = 0$ in such a case. Thus, the expected value can be represented by a pure summation as

$$E[\phi(x)] = \sum_{i=1}^n \phi(x_i) f(x_i) \qquad (7\text{-}2)$$

where the x_i's are the values of x for which the probability is defined.

In the case wherein $F(x)$ is differentiable, which implies a continuous variate having a density function $f(x)$, we have

$$E[\phi(x)] = \int_a^b \phi(x)f(x)\,dx \qquad (7\text{-}3)$$

The case shown in Fig. 7-1a is a combination of these two concepts, so that (except at two points) we have a function which is differentiable and, therefore, the integral can be computed as in Eq. (7-3); at the two points x_1 and x_2 the integral must be evaluated as in Eq. (7-2). The sum of these two integrals, of course, gives the expected value, and it must be remembered that $\int_a^b f(x)\,dx$ plus the heights p_1 and p_2 will give a value equal to unity.

7-2   AVERAGES

By the ordinary rule, the average of a set of N numbers is obtained by computing their sum, then dividing by N. Now since it is not necessary that these numbers all be different, let us suppose that there are n distinct values, $x_1, x_2, \ldots, x_n$, respectively, occurring $N_1, N_2, \ldots, N_n$ times, where $N_1 + N_2 + \cdots + N_n = N$. The average would then be given by

$$\text{Average} = \frac{\sum_{i=1}^n N_i x_i}{N} = \sum_n^{i=1} R_i x_i$$

where $R_i = N_i/N$. Recognizing R_i as the relative frequency of x_i, it may be seen that the average of a random variable could be defined in exactly the same way but with the relative frequency replaced by a probability in each case. Hence, we arrive at the following definitions:

1 For a discrete variate x with an exhaustive set of admissible values x_i ($i = 1, 2, \ldots, n$) and corresponding probabilities $f(x_i)$ we have

$$\text{Average of } x = \sum_{i=1}^n x_i f(x_i) \qquad (7\text{-}4)$$

2 For a continuous variate x with density function $f(x)$ defined in the range (a, b) we have

$$\text{Average of } x = \int_a^b x f(x)\,dx \qquad (7\text{-}5)$$

3 For a random variate x with a distribution function $F(x)$ defined over the region (a, b) we have in general, because of the definition of the Stieltjes integral defined in Sec. 7-1,

$$\text{Average of } x = \int_a^b x\,dF(x) \qquad (7\text{-}6)$$

where, of course, Eq. (7-6) includes (7-5) and (7-4) as special cases. In statistical terminology, the average as defined by Eqs. (7-4) to (7-6) is called the arithmetic mean, or briefly, the *mean*, and is often denoted by the Greek letter μ.

By comparing Eq. (7-6) with the definition of expected value [Eq. (7-1)] we see that for the special case of $\phi(x) = x$ the two are equivalent and, therefore,

$$E(x) = \int_a^b x \, dF(x) = \text{average of } x \qquad (7\text{-}7)$$

Actually, the expected value can always be thought of as some kind of average. Although the following specialization is not essential to the argument, it very well illustrates this concept. Let $\phi(x)$ be a continuous monotonic function and $f(x)$ the density function of the variable x over the range (a, b). In this case, the expected value is given by Eq. (7-3). Let y be a new variable equal to $\phi(x)$ so that it is now possible to compute the probability density of y as

$$g(y) = f[\phi^{-1}(y)] \left| \frac{d}{dy} [\phi^{-1}(y)] \right|$$

Therefore, if we substitute $y = \phi(x)$ in Eq. (7-3), replace $f(x) \, dx$ by its equal $g(y) \, dy$, and revise the limits of integration accordingly, we have

$$E[\phi(x)] = E(y) = \int_{\phi(a)}^{\phi(b)} yg(y) \, dy \qquad (7\text{-}8)$$

Equation (7-8) thus indicates that the expected value of $\phi(x)$ is the average value of $\phi(x)$ taken with respect to its own probability density.

Now let us return to an application of the idea which originally motivated the concept of expected value. From the probability standpoint, a game of chance can be regarded as a process of making a random draw x_i from a possible set of n distinct scores or n distinct payoffs $x_1, x_2, \ldots, x_n$. If the probability of any particular score (payoff) x_i is $f(x_i)$, where $\sum_i f(x_i) = 1$, then the *expected value* $E(x)$ of the scores (payoffs) is defined as the sum of products obtained by weighting each score (payoff) by its own chance of occurrence:

$$E(x) = \sum_1^n x_i f(x_i)$$

EXAMPLE 7-1 A player tosses a die, and if the numbers 1, 2, or 3 appear, he will receive 2 cents. If, however, a 4 or 5 turns up, he will receive 1 cent, whereas the occurrence of a 6 will cost him 2 cents. Thus, in ascending order, the payoffs are $-2, 1, 2$, and their respective probabilities are $\frac{1}{6}, \frac{1}{3}, \frac{1}{2}$. The

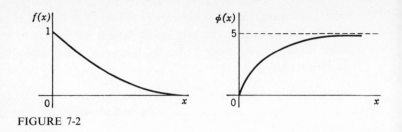

FIGURE 7-2

expected value of the game is obtained by multiplying each payoff by its chance of occurrence and then adding these products over all possibilities. Hence

$$E(x) = (-2)(\tfrac{1}{6}) + (1)(\tfrac{1}{3}) + (2)(\tfrac{1}{2}) = 1 \text{ cent}$$

If our gambler plays this game over and over, we should expect him to average 1 cent for each individual toss. ////

EXAMPLE 7-2 The profit realized by the utilization of a particular power-generating plant during August is a function of the rainfall during that month. The value of the profit is approximately $5(1 - e^{-x})$ million dollars, where x (itself a pure number) represents the total number of inches of rainfall during the month. Assuming the distribution of x is $f(x) = e^{-x}$ $(0, \infty)$, what is the expected value of rain for the power company?

Denoting the value of rain by $\phi(x)$, we have

$$E[\phi(x)] = \int_0^\infty 5(1 - e^{-x})e^{-x}\,dx = 5\int_0^\infty e^{-x}\,dx - 5\int_0^\infty e^{-2x}\,dx$$

$$= 5(1 - \tfrac{1}{2}) = 2\tfrac{1}{2} \text{ million dollars}$$

The leveling off of the profit function as indicated in Fig. 7-2 is due to the spilling of water over the dam in periods of heavy rainfall. The $2\tfrac{1}{2}$ million dollars is the amount that the company would receive on the average for the month of August if the average were to be computed for that month for a large number of years. ////

EXAMPLE 7-3 Find the expected value, hence the average, of the geometric variate x, distributed as follows:

$$f(x) = (1 - \alpha)\alpha^x \qquad 0 < \alpha < 1; x = 0, 1, 2, \ldots, \infty$$

By definition of the expected value

$$E(x) = \sum_0^\infty x(1-\alpha)\alpha^x = (1-\alpha)\sum_0^\infty x\alpha^x$$

To evaluate this summation, put

$$S_n = \sum_0^n x\alpha^x$$

Multiply by α and subtract:

$$S_n - \alpha S_n = \sum_0^n x\alpha^x - \sum_0^n x\alpha^{x+1} = \sum_1^n \alpha^j - n\alpha^{n+1}$$

whence

$$(1-\alpha)S_n = \frac{\alpha(1-\alpha^n)}{1-\alpha} - n\alpha^{n+1}$$

Now

$$\lim_{n\to\infty} n\alpha^{n+1} = 0$$

and so

$$E(x) = \lim_{n\to\infty}(1-\alpha)S_n = \frac{\alpha}{1-\alpha} \qquad \text{////}$$

EXAMPLE 7-4   The discrete variate x has the Poisson distribution

$$f(x) = e^{-\mu}\frac{\mu^x}{x!} \qquad \mu > 0; \; x = 0, 1, 2, \ldots, \infty$$

Find the average value of x.
   Since $0f(0) = 0$, we may write

$$E(x) \equiv \sum_0^\infty xf(x) = \sum_1^\infty xf(x)$$

Now if $x > 0$ we have

$$xf(x) = e^{-\mu}\frac{\mu^x}{(x-1)!} = \mu e^{-\mu}\frac{\mu^{x-1}}{(x-1)!}$$

Setting $t = x - 1$ and noting that the Maclaurin expansion for e^μ is $\sum_0^\infty \mu^t/t!$ we obtain

$$E(x) = \mu e^{-\mu}\sum_0^\infty \frac{\mu^t}{t!} = \mu e^{-\mu}e^\mu = \mu$$

As an aid in the geometric interpretation of the mean, this distribution is plotted for $\mu = 9$ in Fig. 7-3. $\qquad \text{////}$

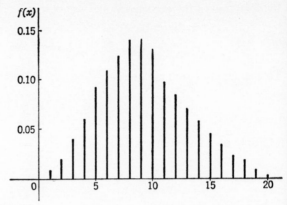

FIGURE 7-3
Poisson distribution for $\mu = 9$.

EXAMPLE 7-5　Find the average value of the continuous variate x, where

$$f(x) = 6x(1 - x) \qquad (0, 1)$$

It follows from the definition that

$$E(x) = \int_0^1 xf(x) = \int_0^1 6x^2(1 - x)\, dx = \tfrac{1}{2} \qquad ////$$

EXAMPLE 7-6　Find the average value of the function $y = x^2$ given that x is unit normal.　Using the distribution of x we have

$$E(y) = E(x^2) = \int_{-\infty}^{\infty} x^2 \frac{1}{\sqrt{2\pi}} e^{-x^2/2}\, dx$$

$$= \frac{2\sqrt{2}}{\sqrt{2\pi}} \Gamma(\tfrac{3}{2}) = 1$$

Alternatively, we may evaluate the expected value of y from its own distribution, namely,

$$g(y) = \frac{1}{\sqrt{2\pi}} y^{-1/2} e^{-y/2} \qquad (0, \infty)$$

Thus

$$E(y) = \int_0^{\infty} yg(y)\, dy = \frac{1}{\sqrt{2\pi}} \int_0^{\infty} y^{1/2} e^{-y/2}\, dy$$

$$= \frac{2\sqrt{2}}{\sqrt{2\pi}} \Gamma(\tfrac{3}{2}) = 1 \qquad ////$$

7-3 SOME ADDITIONAL PROPERTIES OF THE MEAN

Two rather obvious properties of the mean are worthy of explicit statement.

Property 1.   If $A \leq x \leq B$, then $A \leq E(x) \leq B$.

Property 2.   If $\alpha(x) = a_0 + a_1\alpha_1(x) + \cdots + a_n\alpha_n(x)$, then

$$E[\alpha(x)] = a_0 + a_1 E[\alpha_1(x)] + \cdots + a_n E[\alpha_n(x)]$$

The first property guarantees that the mean cannot fall outside the limits for which the variate is defined.   By way of proof, we notice that $\int_A^B A \, dF(x) \leq \int_A^B x \, dF(x) \leq \int_A^B B \, dF(x)$, hence $A \leq E(x) \leq B$.   The verbal translation of the second property is that if $\alpha(x)$ is a linear function of any number of components, then its expected value is given by the same linear function of the expected values of those components.   To prove this fact, we substitute $\alpha(x) = a_0 + \sum a_i \alpha_i(x)$ into the equation

$$E[\alpha(x)] = \int_A^B \alpha(x) \, dF(x)$$

and integrate term by term.

For some purposes it is convenient to consider an average value of a function $\alpha(x)$ over a portion of the total range of x.   Thus, if $A \leq x < B$ we define a *submean* $M(a, b)$ over the range a, b as

$$M(a, b) = \frac{\int_a^b \alpha(x) \, dF(x)}{\int_a^b dF(x)} \tag{7-9}$$

From this definition, it follows that a submean may be interpreted as an average taken with respect to a conditional distribution.   It is readily seen that properties 1 and 2 of regular means hold also for submeans.   As an application, suppose we divide the total range of x into n arbitrary intervals by marking off $n + 1$ points, $(x_0 \equiv A), x_1, x_2, \ldots, x_{n-1}, (x_n \equiv B)$.   Putting

$$P_i = \int_{x_{i-1}}^{x_i} dF(x) \quad \text{and} \quad M_i = \frac{1}{P_i}\int_{x_{i-1}}^{x_i} \alpha(x) \, dF(x)$$

we then have the identity

$$E[\alpha(n)] \equiv \sum_1^n P_i M_i \tag{7-10}$$

Formulas of this type have long been used in mechanics in connection with center of gravity.

EXAMPLE 7-7   Find the submean of a normal variate x over the range $\mu + \sigma \leq x < \infty$.   Consider, first, a unit normal variate u.   The corresponding range of u is 1, ∞ and the submean m is given by

$$m = \frac{\int_1^\infty (1/\sqrt{2\pi})ue^{-u^2/2}\,du}{\int_1^\infty (1/\sqrt{2\pi})e^{-u^2/2}\,du} = \frac{[-(1/\sqrt{2\pi})e^{-u^2/2}]_1^\infty}{R(1)}$$

$$= \frac{f(1)}{R(1)} = \frac{.2420}{.1587} = 1.525$$

Now applying property 2 we have $x = \mu + \sigma u$ whence

$$M = \mu + \sigma m = \mu + 1.525\sigma$$

More generally, if t is any positive number, the submean $m(t, \infty)$ of a unit normal variate in the range t, ∞ is

$$m(t, \infty) = \frac{f(t)}{R(t)}$$

and the corresponding submean $M(\mu + t\sigma, \infty)$ of any normal variate is

$$M(\mu + t\sigma, \infty) = \mu + \frac{\sigma f(t)}{R(t)}$$

where $f(t)$ and $R(t)$ refer to the unit normal distribution.   /////

(*Suggested Problems 7-1 to 7-9*)

7-4   VARIANCE

The expected value is one of several possible measures (called norms) commonly employed to indicate the general location of a distribution along the coordinate axis of the variate considered.   Two other measures often used for the same purpose are the median and the mode.   Once a suitable norm (measure of central tendency) is chosen, there arises the question of concentration, that is, the degree to which the norm represents the distribution as a whole.   Intuitively we feel that a measure of concentration should indicate how much area there is in the neighborhood of the norm; but upon reflection, it becomes obvious that unless a definite range about the norm is specified, more or less arbitrarily, what we are really considering is a probability function akin to $W(x)$, as defined for the unit normal distribution. Because an arbitrarily chosen range is of dubious merit, let us propose a measure which is itself an average derived from the entire distribution. As the intuitive concept of concentration applies equally to both sides of the norm, the function to be averaged should depend upon the magnitude of the difference between the norm and an individual value; moreover, since it should put a penalty upon large deviations in order to distinguish the immediate from the remote, the function should increase with the magnitude of the deviation. We thus arrive at a measure of departure from the norm, or a measure of variability, rather than a direct measure of concentration, but it will serve the same purpose. Initially we might consider using merely the average absolute deviation from the norm, and, in fact, this measure has gained a limited acceptance; but its exceedingly poor combinative properties render it useless for most purposes, practical and theoretical alike. The simplest measure which satisfies both intuitive and analytical requirements is the mean-square deviation from the norm; and when the mean μ is chosen as the norm, this measure of departure is called the *variance* and is denoted by the symbol σ^2. The positive square root of the variance is known as the *standard deviation* and is denoted by the symbol σ. Thus,

$$\text{Variance} \equiv \sigma^2 \equiv E[(x - \mu)^2] \qquad \text{Standard deviation} \equiv \sigma \equiv \sqrt{\sigma^2} \qquad (7\text{-}11)$$

By expanding the quadratic $(x - \mu)^2$ and making use of property 2 (Sec. 7-3), with $E(x)$ replaced by its equal μ, we obtain the following identity, which often simplifies the determination of σ^2:

$$\sigma^2 \equiv E(x^2) - \mu^2 \qquad (7\text{-}12)$$

EXAMPLE 7-8 Find the mean, variance, and standard deviation of the variate x where $f(x) = (x/\beta^2)e^{-x/\beta}$ $(0, \infty; \beta > 0)$. The mean is given by

$$\mu = \int_0^\infty xf(x)\, dx = \beta\,\Gamma(3) = 2\beta$$

and the variance is

$$\sigma^2 = E(x^2) - \mu^2 = \int_0^\infty x^2 f(x)\, dx - 4\beta^2$$

$$= \beta^2\,\Gamma(4) - 4\beta^2 = 2\beta^2$$

Thus the standard deviation is $\sigma = \beta\sqrt{2}$. ////

EXAMPLE 7-9 The variance of the Poisson distribution: Use the identity $x^2 \equiv x(x-1) + x$. Then

$$E(x^2) = e^{-\mu} \sum_{x=0}^{\infty} x(x-1) \frac{\mu^x}{x!} + e^{-\mu} \sum_{x=0}^{\infty} x \frac{\mu^x}{x!} = e^{-\mu}\left(\mu^2 \sum_{y=0}^{\infty} \frac{\mu^y}{y!} + \mu \sum_{v=0}^{\infty} \frac{\mu^v}{v!}\right)$$

$$= e^{-\mu}(\mu^2 e^\mu + \mu e^\mu) = \mu^2 + \mu$$

$$\sigma^2 = E(x^2) - [E(x)]^2 = (\mu^2 + \mu) - \mu^2 = \mu$$

Thus the mean and variance of a Poisson distribution are both equal to μ. ////

EXAMPLE 7-10 The mean and variance of the binomial distribution: Put $m = n - 1$; then

$$\mu = E(x) = \sum_{x=0}^{n} \frac{xn!}{x!(n-x)!} p^x q^{n-x} = np \sum_{u=0}^{m} \frac{m!}{u!(m-u)!} p^u q^{m-u} = np$$

Put $x^2 \equiv x(x-1) + x$; $L = n - 2$; then

$$E(x^2) = \sum_{x=0}^{n} \frac{x(x-1)n!}{x!(n-x)!} p^x q^{n-x} + \sum_{x=0}^{n} \frac{xn!}{x!(n-x)!} p^x q^{n-x}$$

$$= n(n-1)p^2 \sum_{v=0}^{L} \frac{L!}{v!(L-v)!} p^v q^{L-v} + np$$

$$= n(n-1)p^2 + np = n^2 p^2 + np - np^2$$

$$\sigma^2 = E(x^2) - \mu^2 = np - np^2 = np(1-p) = npq$$

Therefore, $\mu = np$ and $\sigma^2 = npq$. ////

EXAMPLE 7-11 The mean and variance of the normal distribution: For the present write the normal distribution as

$$f(x) = \frac{1}{\beta\sqrt{2\pi}} e^{-(x-\alpha)^2/2\beta^2} \qquad (-\infty, \infty; \beta > 0)$$

Also, put $u = x - \alpha$; then,

$$\mu = E(x) = \int_{-\infty}^{\infty} xf(x)\,dx = \int_{-\infty}^{\infty} (u + \alpha)f(x)\,dx$$

$$= \frac{1}{\beta\sqrt{2\pi}} \int_{-\infty}^{\infty} ue^{-u^2/2\beta^2}\,du + \alpha \int_{-\infty}^{\infty} f(x)\,dx = 0 + \alpha = \alpha$$

Thus $\mu = \alpha$. The variance is

$$\sigma^2 = \int_{-\infty}^{\infty} (x - \mu)^2 f(x)\, dx = \frac{2}{\beta\sqrt{2\pi}} \int_0^{\infty} u^2 e^{-u^2/2\beta^2}\, du = \frac{2\beta^2}{\sqrt{\pi}} \Gamma(\tfrac{3}{2}) = \beta^2$$

Since the parameters α, β of the normal distribution are respectively equal to the mean μ and standard deviation σ, the conventional notation for these parameters is appropriately chosen.   ////

Often the zero point and unit of measurement of a variate are chosen arbitrarily; hence it is useful to establish general formulas for the mean and variance of a linear function $y = a + bx$ $(b \neq 0)$. Denote the respective means of x, y by μ_x, μ_y and the variances by σ_x^2, σ_y^2. By property 2, we have at once

$$\mu_y = E(a + bx) = a + b\mu_x \qquad (7\text{-}13)$$

Thus $y - \mu_y = b(x - \mu_x)$ whence (again by property 2)

$$\sigma_y^2 = b^2 E[(x - \mu_x)^2] = b^2 \sigma_x^2 \qquad \sigma_y = |b| \sigma_x \qquad (7\text{-}14)$$

The special transformation $z = (x - \mu)/\sigma$ is known as a reduction to *standard scores;* the mean of any such transformed variate is zero and its standard deviation is unity. When this process is applied to a normal variate, the result is a unit normal variate.

EXAMPLE 7-12   Although most of us are aware of the difficulty of forecasting the weather, comparatively few people are familiar with the problem of devising a scoring scheme for weather forecasts. Limiting attention to the simplest case, let us consider the two-category forecast of rain or no rain during a 24-hour period at a given observation station and assume that trial dates are far enough apart to be regarded as independent. As a point of departure, let us examine a scheme wherein a forecaster will score one point if the trial date is correctly classified as to rain or no rain, and zero otherwise; no distinction being made as to relative importance, the total score will equal the total number of correct forecasts. In the utter absence of skill, if r days out of N trials were selected at random and arbitrarily labeled rain, and the remaining N-r days were labeled no rain, what would be the expected value of the score? Let $p < \tfrac{1}{2}$ denote the probability of rain, and indicate forecast and observed events by the respective subscripts f, 0. Making use of Table 7-1, which refers to repeated independent samples, we find that the expected score on a single trial is $q - r(q - p)/N$ and the expected total score is $Nq - r(q - p)$. Hence, if a pessimistic individual felt that his skill were negligible but still wanted to maximize his score, he could do

Table 7-1

Event	(R_f, R_0)	$(R_f, \bar{R}_0)$	$(\bar{R}_f, R_0)$	$(\bar{R}_f, \bar{R}_0)$
Probability	pr/N	qr/N	$p(N-r)/N$	$q(N-r)/N$

this by setting $r = 0$, thus obtaining an expected total score of Nq by forecasting no rain on every trial. This way of beating the system can be thwarted by a more judicious choice of scores. Assigning weights w, $-x$, $-y$, z to the four possible events (in the order tabulated), we obtain, as the expected score under the same conditions of randomness,

$$r(wp + yp - zq - xq) + N(zq - yp)$$

Since we want this to vanish for all choices of r, we set the coefficients of r and N equal to zero, obtaining $x = wp/q$, $y = zq/p$.

On the other hand, consider a skilled forecaster; let $\lambda = P(R_0 | R_f)$ and $\rho = P(R_f)$. Then we readily find that the respective joint probabilities are $\rho\lambda$, $\rho(1 - \lambda)$, $p - \rho\lambda$, $q - \rho(1 - \lambda)$, and the expected score on a single trial is

$$S = (\lambda - p)\rho\left(\frac{w}{q} + \frac{z}{p}\right)$$

showing that a competent forecaster's score is jointly proportional to his marginal probability of giving a forecast of rain and the excess of the conditional probability of the occurrence of rain when forecast over the unconditional probability of rain. In case of perfect accuracy, $\lambda = 1$ and $\rho = p$, whence $S = pw + qz$.

We have yet to specify w and z. To this end, consider the variance of the score per trial of a random forecast having the same marginal distribution as the actual rain; that is, $\lambda = p = \rho$. This variance is

$$\sigma^2 = \frac{p^2 w^2}{q} + \frac{q^2 z^2}{p}$$

We now propose to choose w and z in such a way that the expected score per trial will be unity in the case of perfect prediction and further that the variance of the scores will be as small as possible in the random case. We thus put $pw + qz = 1$ and minimize σ^2 subject to this restriction:

$$\sigma^2 = \frac{p^2 w^2}{q} + \frac{(1 - pw)^2}{p}$$

$$\frac{d\sigma^2}{dw} = 0 \quad \text{when } w = \frac{q}{p(p + q)} = \frac{q}{p} \quad \text{Min } \sigma^2 = q + p = 1$$

Table 7-2

Event	(R_f, R_0)	$(R_f, \bar{R}_0)$	$(\bar{R}_f, R_0)$	$(\bar{R}_f, \bar{R}_0)$
Score	q/p	-1	-1	p/q

Correspondingly, $z = (1 - pw)/q = p/q$; $x = wp/q = 1$; $y = zq/p = 1$. Our scoring scheme thus is given by Table 7-2.

The skilled forecaster's expected score per trial is $S = (\lambda - p)\rho/pq$; this becomes unity in the case of perfect skill, and the random forecast will have zero mean and unit variance.   ////

(*Suggested Problems 7-10 to 7-13*)

7-5 UNBIASED ESTIMATOR OF VARIANCE

A statistic b is said to be an unbiased estimator of a parameter β if $E(b) = \beta$. Let us explore this notion in connection with the problem of estimating the population variance from a sample.

In addition to matters directly related to variability, the variance enters into several questions of fairly general interest which concern the sample mean—what confidence limits can be attached to it, whether an observed sample mean supports a prior hypothesis as to the population mean, or whether two or more sample means could be considered representative of a common population mean. Even assuming a normal distribution for the sample means, numerical values of the probabilities in question can not be obtained so long as they involve the unknown parameter σ^2. However, introducing a suitable estimate s^2 for σ^2 and making a simple change of variables, it turns out, in the normal case, that a parameter-free variate results, and the questions can be answered. As a consequence of the approximate normality of sample means from ordinary nonnormal populations, the sampling distributions derived from the normal population are sufficiently close to the true distributions for most practical purposes. Both by theoretical and empirical methods, the effects of nonnormality on the standard tests involving the mean have been investigated by top-notch men in the field, and although the conclusions are necessarily limited to the general classes of nonnormality actually examined, the results indicate surprisingly little disturbance of the conventional significance levels.

In devising an estimator for σ^2, we are confronted with choosing between conflicting principles. Applying the method of maximum likelihood to a random sample $x_1, x_2, \ldots, x_n$ from a normal population with both parameters μ,

σ unknown, we find that the estimator for μ is the sample mean $\bar{x}$, the estimator for σ^2 is $\sum_i (x_i - \bar{x})^2/n$, and that for σ is the square root of the latter quantity. On the other hand, it is readily shown that this maximum likelihood estimator for σ^2 is biased, for taking expectations of both sides of the identity,

$$\sum_i (x_i - \mu)^2 \equiv \sum_i (x_i - \bar{x})^2 + n(\bar{x} - \mu)^2 \qquad (7\text{-}15)$$

and using a result which will be proved in Sec. 7-9, we obtain

$$n\sigma^2 = E\left[\sum_i (x_i - \bar{x})^2\right] + \sigma^2$$

whence
$$E\left[\sum_i (x_i - \bar{x})^2\right] = (n - 1)\sigma^2 \qquad (7\text{-}16)$$

Consequently, the expected value of the maximum likelihood estimator is $\sigma^2(n - 1)/n$ rather than σ^2. If we were sure we had a normal population in a particular instance, we might be influenced more strongly in favor of the maximum likelihood estimator, despite its bias. But we have at our immediate disposal the alternative estimator,

$$s^2 \equiv \frac{\sum_i (x_i - \bar{x})^2}{n - 1} \qquad (7\text{-}17)$$

which is unbiased for all populations such that the variance itself is definable. Therefore, in the interests of generality, and also because standard tests have already been developed from this point of view, we adopt $s^2 \equiv \sum_i (x_i - \bar{x})^2/(n - 1)$ as our estimator of σ^2, and its square root s as our estimator of σ. It should be pointed out, however, that there is no general rule for obtaining an unbiased estimator of the standard deviation, for the mean of a square root is not in general equal to the square root of the mean of the initial variate. Thus s is usually a biased estimator of σ, and here again our choice between conflicting aims is dictated by the relative breadth of applicability of our options.

By fairly heavy algebra, which we omit, it can be shown that the variance of s^2 in repeated random samples of size n from an infinite population is

$$\text{Variance } (s^2) \equiv E(s^2 - \sigma^2)^2 = \frac{1}{n}\left[\mu_4 - \frac{(n - 3)\sigma^4}{n - 1}\right] \qquad (7\text{-}18)$$

where $\mu_4 \equiv E(x - \mu)^4$. For a normal variate, $\mu_4 = 3\sigma^4$, whence the variance of s^2 in random samples from a normal population is $2\sigma^4/(n - 1)$.

Assuming an adequate reason for hypothesizing a normal distribution, the actual process of fitting a normal curve to a particular sample of n independent observations consists of estimating the two parameters μ and σ from the data. The fitted density function then becomes

$$f(x) = \frac{1}{s\sqrt{2\pi}} e^{-(x - \bar{x})^2/2s^2} \qquad (-\infty, \infty)$$

The extent to which this density function actually agrees with the observed relative frequencies depends upon how well the sample itself is typical of the normal distribution. In this connection, it is important to remember that if the sample were, in fact, drawn at random from a normal universe the relative frequencies would exhibit this randomness in the form of irregular deviations about the fitted curve. We shall see how to decide whether the normal hypothesis is tenable in Chap. 10.

EXAMPLE 7-13   The annual amounts of rainfall in the Susquehanna valley during the years 1860–1951 are recorded to the nearest hundredth of an inch in Table 7-3.   Fit a normal curve to this sample.

Table 7-3   SUSQUEHANNA ANNUAL RAINFALL, 1860-1951†

Amount (x)	Freq.	Cum. freq.	Amount (x)	Freq.	Cum. freq.	Amount (x)	Freq.	Cum. freq.
28.04	1	1	38.08	1	33	40.75	1	64
31.21	1	2	38.15	1	34	40.78	1	65
31.63	1	3	38.18	1	35	41.12	1	66
32.11	1	4	38.43	1	36	41.18	1	67
33.02	1	5	38.51	1	37	41.23	1	68
34.25	1	6	38.65	1	38	41.45	1	69
34.34	1	7	38.81	1	39	41.72	1	70
34.58	1	8	38.86	1	40	41.80	1	71
35.11	1	9	38.89	1	41	41.85	1	72
35.43	1	10	38.91	1	42	41.90	1	73
35.76	1	11	39.08	1	43	41.96	1	74
36.35	1	12	39.13	1	44	42.03	1	75
36.39	1	13	39.55	1	45	42.09	1	76
36.62	2	15	39.56	1	46	42.26	1	77
36.63	1	16	39.74	1	47	42.85	1	78
36.65	1	17	39.82	1	48	43.17	1	79
36.68	1	18	39.99	1	49	43.38	1	80
36.70	1	19	40.06	1	50	43.44	1	81
36.86	1	20	40.08	1	51	44.34	1	82
37.00	1	21	40.14	2	53	44.46	1	83
37.13	1	22	40.15	1	54	44.56	1	84
37.15	1	23	40.23	1	55	44.76	1	85
37.33	1	24	40.40	1	56	44.82	1	86
37.57	1	25	40.41	1	57	45.06	1	87
37.72	1	26	40.48	1	58	45.89	1	88
37.77	1	27	40.54	1	59	46.51	1	89
37.79	2	29	40.57	1	60	47.31	1	90
37.96	1	30	40.60	1	61	48.22	1	91
38.04	1	31	40.62	1	62	48.61	1	92
38.07	1	32	40.67	1	63			

† Courtesy of the Pennsylvania Water and Power Company

Assuming that the annual amounts of rainfall are independently distributed, which is in fact fairly close to the truth, we compute the two statistics $\bar{x}$ and s^2.   Here we have

$$N = 92 \qquad \sum x = 3633.24 \qquad \sum x^2 = 144700.7034$$

$$\bar{x} = 39.492 \qquad \sum (x - \bar{x})^2 = \sum x^2 - \bar{x} \sum x = 1216.7893$$

$$s^2 = \frac{\sum (x - \bar{x})^2}{N - 1} = \frac{1216.7893}{91} = 13.3713$$

$$s = 3.657$$

Therefore the fitted normal distribution is

$$f(x) = \frac{1}{s\sqrt{2\pi}} e^{-(x-\bar{x})^2/2s^2} \qquad (-\infty, \infty)$$

$$= \frac{1}{(3.657)\sqrt{2\pi}} e^{-(x-39.492)^2/26.743} \qquad (-\infty, \infty) \qquad ////$$

7-6   PROBABILITY OF EXTREME VALUES

The intimate connection between the standard deviation and the intuitive concept of concentration is brought out by the fact that, in terms of standard deviation, an upper bound can be set on the probability of extreme deviations from the mean.   A very general relation of this sort, applicable to any distribution for which μ and σ are definable, is known as the Tchebysheff inequality; a stronger inequality, but one restricted to unimodal distributions, was previously established by Gauss.

The Tchebysheff Inequality.   For any distribution with finite mean and variance,

Tchebysheff inequality:

$$P(|x - \mu| \geq t\sigma) \leq \frac{1}{t^2} \qquad (7\text{-}19)$$

where $t > 0$.

Proof.   Let P denote the probability that $|x - \mu| \geq t\sigma$.   Let A represent the class of values of x for which $|x - \mu| \geq t\sigma$ and B, the complementary class, for which $|x - \mu| < t\sigma$.   Denote the submean of $(x - \mu)^2$ by α^2 in class A and by β^2 in class B; thus $\alpha^2 \geq t^2\sigma^2$.   By definition, $\sigma^2 = P\alpha^2 + (1 - P)\beta^2$ whence

$$P = \frac{\sigma^2 - (1 - P)\beta^2}{\alpha^2} \leq \frac{\sigma^2 - (1 - P)\beta^2}{t^2\sigma^2} \leq \frac{1}{t^2}$$

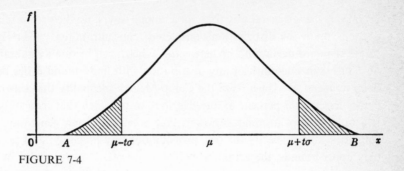

FIGURE 7-4

Note that this proposition yields no information unless $t > 1$, for $P \leq 1$ in any event. Note further that P is definitely less than $1/t^2$ if the probability of class B is other than zero; also P is less than $1/t^2$ if class A contains any area beyond its inside limit $|x - \mu| = t\sigma$. Hence $P < 1/t^2$ for all variates having a density function, and the same is true for most discrete variates as well.

The Gauss Inequality. For a unimodal density function, let m denote the mode, and put $\lambda = (\mu - m)/\sigma$. Then when $t > |\lambda|$,

Gauss inequality:

$$P(|x - \mu| \geq t\sigma) \leq \frac{4(1 + \lambda^2)}{9(t - |\lambda|)^2} \qquad (7\text{-}20)$$

Thus if $\lambda = 0$,

Camp-Meidell inequality:

$$P \leq \frac{1}{2.25t^2} \qquad (7\text{-}21)$$

The latter relation is commonly called the *Camp-Meidell* inequality. For an indication of the proof of the Gauss inequality, the reader is referred to H. Cramer.[1] The probability P is represented graphically as the shaded portion of Fig. 7-4.

From the quantity $(|x - \mu|)/\sigma$ in these inequalities, it is seen that the probability of exceeding a stated magnitude of deviation from the mean tends to decrease as σ decreases or, what amounts to the same thing, the probability of keeping within preassigned limits tends to increase with decreasing σ;

---

[1] H. Cramer, "Mathematical Methods of Statistics," pp. 183, 256, Princeton University Press, Princeton, N.J., 1946.

accordingly, in a general way (but not a unique way, since the precise relationship depends upon the distribution considered) concentration is inversely related to the standard deviation. Whereas the Tchebysheff inequality guarantees that no more than one-ninth of any distribution with finite μ and σ lies beyond 3σ from its mean, we learn from the Camp-Meidell inequality that this region includes less than 5 percent of any distribution to which that inequality applies. It should be kept in mind, however, that because of their generality, these inequalities cannot specify the true probabilities very precisely; and as they yield only upper bounds, the actual probabilities are often much lower. With reference to significance tests, such overestimates of the tail areas (outlying portions of the curve) ordinarily would be on the conservative side, inasmuch as the evidence in support of an innovation would then be deemed inconclusive. Even when the mean and mode do not coincide, the Camp-Meidell inequality frequently leads to an overestimate of tail areas of unimodal distributions. In Example 7-8, for instance, where the mean is exactly twice the mode, the true probability is .039 that a random point will deviate from μ by as much as 2σ, and .014 that it will deviate by as much as 3σ, whereas the corresponding bounds, as given by the Camp-Meidell inequality, are .111 and .049, respectively.

(Suggested Problems 7-14 and 7-15)

7-7 MULTIVARIATE FUNCTIONS

The rule for finding the expected value of a function $\phi(x)$ can be extended to continuous functions of any number of variables. Thus, if the joint-distribution function $F(x_1, x_2, \ldots, x_n)$ of several random variables $x_1, x_2, \ldots, x_n$ is defined over the region of space R_x, and $\phi(x_1, x_2, \ldots, x_n)$ is any continuous function of the x's, then the expected value of this function is given by

$$E[\phi(x_1, x_2, \ldots, x_n)] = \iint \cdots \int_{R_x} \phi(x_1, x_2, \ldots, x_n)\, dF(x_1, x_2, \ldots, x_n)$$

$$(7\text{-}22)$$

For a set of discrete variates, this reduces to a multiple summation, whereas for continuous variates having a joint density function, the equation becomes

$$E[\phi(x_1, x_2, \ldots, x_n)] = \iint \cdots \int_{R_x} \phi(x_1, x_2, \ldots, x_n) f(x_1, x_2, \ldots, x_n)$$

$$dx_1\, dx_2 \ldots dx_n \qquad (7\text{-}23)$$

EXAMPLE 7-14 Find the expected value of the function $z = xy$ given that x and y are jointly distributed as follows:

$$f(x, y) = xe^{-x(1+y)} \qquad (0, \infty \text{ each})$$

By Eq. (7-23) we have

$$E(z) = \int_0^\infty \int_0^\infty xyf(x, y)\, dx\, dy = \int_0^\infty e^{-x}\left(\int_0^\infty x^2 y e^{-xy}\, dy\right) dx$$

$$= \int_0^\infty e^{-x}\, dx = 1 \qquad\qquad\qquad ////$$

As an important special case of Eq. (7-23) we notice that the expected value of any one of the variates x_i of a multivariate system $x_1, x_2, \ldots, x_n$ can be evaluated directly from the joint distribution without first obtaining the marginal distribution. Thus,

$$E(x_i) = \iint \cdots \int_{R_x} x_i\, dF(x_1, x_2, \ldots, x_n) \qquad (7\text{-}24)$$

Two consequences of the equation for the expected value of a function are sufficiently important to be stated as theorems. The first concerns the sum of any number of variates, whether dependent or independent, and the second concerns the product of independent variates only.

Theorem 7-1 If $y = x_1 + x_2 + \cdots + x_n$, then

$$E(y) = E(x_1) + E(x_2) + \cdots + E(x_n)$$

PROOF From the equation

$$E(y) = \iint \cdots \int_{R_x} (x_1 + x_2 + \cdots + x_n)\, dF(x_1, x_2, \ldots, x_n)$$

we obtain by expansion

$$E(y) = \sum_{i=1}^n \iint \cdots \int_{R_x} x_i\, dF(x_1, x_2, \ldots, x_n) = \sum_{i=1}^n E(x_i) \qquad ////$$

Corollary If $z = a_0 + a_1 x_1 + a_2 x_2 + \cdots + a_n x_n$ where the a's are constants, then

$$E(z) = a_0 + a_1 E(x_1) + a_2 E(x_2) + \cdots + a_n E(x_n)$$

The proof is similar to that of Theorem 7-1. $\qquad\qquad\qquad ////$

Theorem 7-2   If $y = x_1 x_2 \cdots x_n$, where $x_1, x_2, \ldots, x_n$ are *independent*, then

$$E(y) = E(x_1)E(x_2) \cdots E(x_n)$$

PROOF   If the variates are independent, their joint-probability increment $dF(x_1, x_2, \ldots, x_n)$ factors into the product of the increments of their respective distribution functions

$$dF(x_1, x_2, \ldots, x_n) = dF_1(x_1)\, dF_2(x_2) \cdots dF_n(x_n)$$

Also, the respective regions of integration are independent.   Consequently, the whole expression defining $E(y)$ may be factored, and the result is

$$E(y) = \int_{R_1} x_1\, dF_1(x_1) \int_{R_2} x_2\, dF_2(x_2) \cdots \int_{R_n} x_n\, dF_n(x_n)$$

$$= E(x_1)E(x_2) \cdots E(x_n)$$

where, of course, R_i represents the region of definition of $dF_i(x_i)$.   ////

Corollary   If y is the product of pure functions of *independent* variates, for instance $y = \phi_1(x_1)\phi_2(x_2) \cdots \phi_n(x_n)$, then

$$E(y) = E[\phi_1(x_1)]E[\phi_2(x_2)] \cdots E[\phi_n(x_n)]$$

The proof is similar to that of Theorem 7-2.   ////

EXAMPLE 7-15   If $z = (x + y)^2$ then

$$E(z) = E[(x + y)^2] = E(x^2 + 2xy + y^2)$$

Hence, by Theorem 7-1

$$E(z) = E(x^2) + 2E(xy) + E(y^2)$$

This result holds for any variates whatever.   If the variates x and y are *independent*, the cross-product term can be factored by virtue of Theorem 7-2 to yield $E(xy) = E(x)E(y)$.   Hence, for independent variates,

$$E(z) = E(x^2) + 2E(x)E(y) + E(y^2)$$

The reader's attention is called to the fact that, for brevity, an expression like $E[(x + y)^2]$ is usually written as $E(x + y)^2$.   ////

EXAMPLE 7-16   From an infinite population in which the mean value of x is μ, a random sample of n observations $x_1, x_2, \ldots, x_n$ is drawn and the sample average $\bar{x} = (x_1 + x_2 + \cdots + x_n)/n$ is computed.   If this experiment could be repeated for an indefinitely large number of independent trials, find the mean value of $\bar{x}$.   From the corollary to Theorem 7-1, the mean value of $\bar{x}$ in repeated trials is

$$E(\bar{x}) = \frac{1}{n} E(x_1) + \frac{1}{n} E(x_2) + \cdots + \frac{1}{n} E(x_n)$$

By construction, x_1, x_2, $\ldots$, x_n constitute independent random variables all having the same distribution as x, the variate being sampled.   Hence, they all have the same expected value μ, and

$$E(\bar{x}) = \frac{n\mu}{n} = \mu \qquad (7\text{-}25)$$

////

From the foregoing propositions, we may deduce that the variance is a more comprehensive modulus of variability than its definition at first suggests, for it actually measures the squared difference of all possible pairs of values. Now the mean-square difference of all possible pairs x_1, x_2 can be obtained by regarding x_1, x_2 as independent variates, each distributed as x.   Then

$$E(x_1 - x_2)^2 = E(x_1{}^2) - 2E(x_1 x_2) + E(x_2{}^2) = 2[E(x^2) - \mu^2] = 2\sigma^2$$

showing that the variance is equal to half the mean square of all possible differences.

7-8   COVARIANCE AND CORRELATION

Two or more variates are dependent when their joint-distribution function is not simply the product of their marginal-distribution functions.   Dependence in this sense is so loosely defined that it can range from the borderline of utter independence to the opposite limit, which is strict mathematical determinism. Except when the latter extreme is approached, the existence of useful relationships between variates is often difficult to detect observationally, and the mathematical representation of the relationship, once demonstrated to exist, is an art in itself.

Standard methods of studying the interrelationships between variates make constant use of the average value of a cross product of the type $(x_1 - \xi_1)(x_2 - \xi_2)$, where $\xi_1 = E(x_1)$, $\xi_2 = E(x_2)$.   This quantity

$$E[(x_1 - \xi_1)(x_2 - \xi_2)]$$

denoted by the symbol σ_{12}, is called the *covariance* of (or between) x_1 and x_2,

Covariance:

$$\sigma_{12} \equiv E[(x_1 - \xi_1)(x_2 - \xi_2)] \equiv E(x_1 x_2) - \xi_1 \xi_2 \qquad (7\text{-}26)$$

As a special case of this notation, σ_{11} would represent the variance of x_1, although the symbol σ_1^2 would ordinarily be employed. If x_1 and x_2 are independent, $\sigma_{12} = \xi_1 \xi_2 - \xi_1 \xi_2 = 0$, but the converse does not hold; two variates with zero covariance need not be independent.

Dimensionally, the covariance involves the product of the units of the two variates, whereas the variance involves the square of the unit of one of them. Therefore, a dimensionless parameter ρ_{12} could be constructed thus:

Coefficient of correlation:

$$\rho_{12} \equiv \frac{\sigma_{12}}{\sqrt{\sigma_{11}\sigma_{22}}} \equiv \frac{\sigma_{12}}{\sigma_1 \sigma_2} \qquad (7\text{-}27)$$

Such a parameter is, in fact, widely used, and it is called the *coefficient of linear correlation* between x_1 and x_2. The linear correlation between two variates is simply the covariance of their corresponding standard scores, and ρ ranges from -1 to $+1$ (see Prob. 7-29).

7-9 THE VARIANCE OF A LINEAR FUNCTION

From the viewpoint of general use, an important type of statistical function is a linear function of several variates. The variance of such a function can be expressed in terms of the variances and covariances of the component variates. Consider the linear function

$$y = a_0 + a_1 x_1 + a_2 x_2 + \cdots + a_n x_n$$

where the a's are constants, and put $\xi_i = E(x_i)$, $\mu = E(y)$. Then

$$\mu = a_0 + \sum_1^n a_i \xi_i \qquad y - \mu = \sum_1^n a_i(x_i - \xi_i)$$

and the variance σ^2 of y is

$$\sigma^2 = E(y - \mu)^2 = E\left[\sum_1^n a_i(x_i - \xi_i)\right]^2$$

But

$$\left[\sum_1^n a_i(x_i - \xi_i)\right]^2 \equiv \sum_{i=1}^n \sum_{j=1}^n a_i a_j (x_i - \xi_i)(x_j - \xi_j)$$

Therefore,

Variance of a linear function:

$$\sigma^2 = \sum_1^n \sum_1^n a_i a_j E[(x_i - \xi_i)(x_j - \xi_j)] = \sum_1^n \sum_1^n a_i a_j \sigma_{ij} \qquad (7\text{-}28)$$

If the x's are independent, their covariances vanish, and the variance of y reduces to $\sum_1^n a_i^2 \sigma_{ii}$, which is usually written as

Variance of a linear function of independent variates:

$$\sigma^2 = \sum_1^n a_i^2 \sigma_i^2 \qquad (7\text{-}29)$$

since $\sigma_{ii} \equiv \sigma_i^2$.

EXAMPLE 7-17 A sample of n independent observations $x_1, x_2, \ldots, x_n$ of the same variate x is drawn at random, and the sample mean $\bar{x} = (x_1 + x_2 + \cdots + x_n)/n$ is computed. Find the variance of $\bar{x}$ in repeated independent trials of this experiment, if the variance of x is σ^2. In repeated independent trials the observations $x_1, x_2, \ldots, x_n$ may be regarded as independent variates, all distributed as x. Hence, denoting the variance of $\bar{x}$ by $\sigma_{\bar{x}}^2$, we have

Variance of $\bar{x}$:

$$\sigma_{\bar{x}}^2 = \sum_1^n a_i^2 \sigma_i^2 = \frac{n\sigma^2}{n^2} = \frac{\sigma^2}{n} \qquad (7\text{-}30)$$

Therefore, once the mean and variance μ, σ^2 are given for any variate x, the mean and variance μ, σ^2/n of $\bar{x}$ may be inferred directly. ////

(Suggested Problems 7-16 to 7-31)

7-10 APPLICATIONS OF EXPECTED-VALUE THEORY

Applications of probability to small- and large-scale military and industrial operations often involve the consideration of pertinent random variables. As the appropriate mathematical models contain more and more of these random variables, the actual solutions become very complicated. It is the purpose of this section to illustrate the use of "expected value" in a group of problems for which the mathematical models are relatively simple, so that the manipulations can be easily performed. It must be remembered that the point of departure between the method of using the distribution functions to compute an average and the elementary technique of addition and division is very important and is worth considerable thought on the part of the reader.

EXAMPLE 7-18   A player is to be rewarded with an amount of money equal to x where he chooses x according to the following mechanism: He makes a random draw from a population with density

$$f(x) = \begin{cases} 1 & 0 < x \le 1 \\ 0 & \text{elsewhere} \end{cases}$$

If he wishes, he may keep this value of x.   If not, he may discard it permanently and make another random draw.   The procedure may be repeated for a total of three draws.   What is the best strategy to follow if he wishes to maximize the expected reward, and if he must accept the third draw?

The strategy obviously involves the value of λ below which value he will discard the obtained value of x and above which he will accept it for each of the three opportunities.   Let λ_1 be the dividing value for draw 1, λ_2 that for draw 2, and the third attempt must be accepted whatever its value.   Also let $P_1 = \int_{\lambda_1}^{1} dx = (1 - \lambda_1)$ be the probability that our player keeps the first draw and $P_2 = \int_{\lambda_2}^{1} dx = (1 - \lambda_2)$ be the probability that he keeps the second draw, given that he rejects the first.   Clearly the expected monetary value of the first draw is $(1 + \lambda_1)/2$, of the second draw, $(1 + \lambda_2)/2$, and of the third draw, $\frac{1}{2}$.   Thus the expected value E of this game is given by

$$E = (1 - \lambda_1)\frac{1 + \lambda_1}{2} + \lambda_1(1 - \lambda_2)\frac{1 + \lambda_2}{2} + \frac{\lambda_1\lambda_2}{2} = \frac{1}{2}(1 - \lambda_1{}^2)$$

$$+ \frac{\lambda_1}{2}(1 - \lambda_2{}^2) + \frac{\lambda_1\lambda_2}{2} \qquad (7\text{-}31)$$

For a maximum it is necessary that

$$\frac{\partial E}{\partial \lambda_1} = 0 = -\lambda_1 + \frac{1 - \lambda_2{}^2}{2} + \frac{\lambda_2}{2}$$

$$\frac{\partial E}{\partial \lambda_2} = 0 = -\lambda_1\lambda_2 + \frac{\lambda_1}{2}$$

whence $\lambda_2 = \frac{1}{2}$ and $\lambda_1 = \frac{5}{8}$.   These values substituted in Eq. (7-31). give an expected value of $.695 \approx .7$.   Since

$$\frac{\partial^2 E}{\partial \lambda_1{}^2} = -1 \qquad \frac{\partial^2 E}{\partial \lambda_2{}^2} = -\lambda_1 \qquad \frac{\partial^2 E}{\partial \lambda_1\,\partial \lambda_2} = (-\lambda_2 + \frac{1}{2}) \qquad (7\text{-}32)$$

the solution $\lambda_2 = \frac{1}{2}$, $\lambda_1 = \frac{5}{8}$ is a relative maximum.   ////

EXAMPLE 7-19   A problem often found in the literature to illustrate the concept of the best distribution of effort deals with a secretary looking for a piece of correspondence which may be in any one of N file folders.   The time to

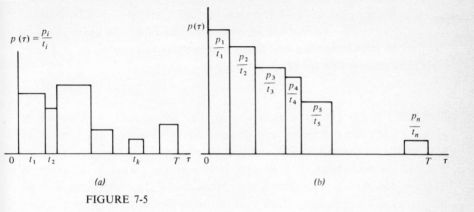

$$p(\tau) = \frac{p_i}{t_i}$$

(a)

(b)

FIGURE 7-5

search the ith folder is t_i, and the probability of finding the letter there is p_i. What is the order in which the folders should be searched in order to minimize the expected search time?

Assuming any searching sequence is adopted and that the correspondence is equally likely to be anywhere inside a folder, the probability of finding the correspondence between τ and $\tau + d\tau$ is $p(\tau)\,d\tau = p_i/t_i\,d\tau$ if τ falls in the ith file, or $\sum_{k=1}^{i-1} t_k < \tau < \sum_{k=1}^{i} t_k$. Thus $p(\tau)$ for any ordering is shown in Fig. 7-5a, where $\sum_{i=1}^{N} p_i = 1$ and $\sum_{i=1}^{N} t_i = T$, the total time to search all folders. The average time to find the correspondence is then given by

$$\bar{\tau} = \int_0^T \tau p(\tau)\,d\tau \qquad (7\text{-}33)$$

This integral (7-33) will be a minimum if the files are ordered so that the blocks are monotonically decreasing along the axis (see Fig. 7-5b), and therefore the secretary should arrange the search (and thus number the files) in decreasing order of p_i/t_i. If we compute the value of $\bar{\tau}$ on the basis that the files are so ordered in sequence, then

$$\bar{\tau} = \frac{p_1 t_1}{2} + p_2\left(t_1 + \frac{t_2}{2}\right) + p_3\left(t_1 + t_2 + \frac{t_3}{2}\right) + \cdots + p_N\left(t_1 + t_2 + \cdots + \frac{t_N}{2}\right)$$

$$= \frac{p_1 t_1}{2} + \cdots + \frac{p_N t_N}{2} + p_2 t_1 + p_3(t_1 + t_2) + \cdots + p_N(t_1 + \cdots + t_{N-1})$$

$$(7\text{-}34)$$

where

$$\frac{p_1}{t_1} \geq \frac{p_2}{t_2} \geq \frac{p_3}{t_3} \cdots \geq \frac{p_N}{t_N}$$

From the expression 7-34, it is clear that the interchanging of the order of any two files in the sequence will increase the value of $\bar{\tau}$. To prove this, one has only to interchange the two values and subtract this new value of $\bar{\tau}$ from (7-34) and observe that the result is negative. ////

EXAMPLE 7-20 Let us consider again the random-walk problem where player A has z units of money and player B has $a - z$ units. The probability of A winning a particular game and thus obtaining any of B's units of money is p and the probability of B winning is $(1 - p) = q$. What is the expected length of the game? The game is over when one of the competitors wins all the money of his opponent, which is of course equivalent to the walk reaching one boundary or the other.

If D_z is equal to the expected number of plays if the random walk starts at z, we have the following equation by considering the situation after one game has been played.

$$D_z = p(D_{z+1} + 1) + q(D_{z-1} + 1) = pD_{z+1} + qD_{z-1} + 1 \qquad (7\text{-}35)$$

where $0 < z < a$ and $D_0 = D_a = 0$.

The homogeneous equation is

$$D_z = pD_{z+1} + qD_{z-1}$$

which had a solution (see Sec. 3-5).

$$D_z = A + B\left(\frac{q}{p}\right)^z \qquad (7\text{-}36)$$

The particular solution of the difference equation would, upon inspection, be a constant C, but since a constant is a part of the solution of the homogeneous equation, the C must be multiplied by z, or $D_z = Cz$.

Substituting this value in the original difference equation we have

$$Cz = pC(z + 1) + qC(z - 1) + 1$$
$$= Cz + C(p - q) + 1$$

or

$$C = \frac{-1}{p - q} = \frac{1}{q - p}$$

Thus the general solution of Eq. (7-35) is of the form

$$D_z = \frac{z}{q - p} + A + B\left(\frac{q}{p}\right)^z$$

Using the condition that $D_0 = D_a = 0$, the proper value of A and B may be determined, giving for the final value of D_z

$$D_z = \frac{z}{q-p} - \frac{a}{q-p} \frac{1-(q/p)^z}{1-(q/p)^a} \qquad (7\text{-}37)$$

as $p \to \frac{1}{2}$, and thus $q = (1-p) \to \frac{1}{2}$, we have for D_z using L'Hospital's Rule

$$D_z = z(a-z) \qquad (7\text{-}38)$$

If each of our contestants has 10 units of money, the expected length of a fair contest would be 100 games. There is a wide spread of possible lengths about this expected value since the 100 is merely the μ of the distribution.   ////

EXAMPLE 7-21   A seasonal article which must be ordered in advance and stocked by a department store sells for $100 per unit and costs the store $50 per unit irrespective of disposal; however, any article not sold during the season must be sold at a sacrifice to a special dealer for $35 per unit. Given that the distribution of customer orders for the item is $f(z) = e^{-9}9^z/z!$ ($z = 0, 1, 2, \ldots, \infty$) and that the number of orders during any season is a random draw from this distribution, how many units should be stocked in order to maximize the expected value of the profit?

Letting x denote an arbitrary number of units stocked, we shall express the expected value of the profit as a function of x, say $\bar{P}(x)$, and then determine x so as to maximize this function. Depending on the magnitude of the number of orders z relative to x, the profit function assumes two distinct forms. In range A $(0 \le z \le x)$, we have the profit function P_A based on the analysis that we sell z units at $100 each, sell the remaining $x - z$ units at $35 each, and at the same time we have to pay for x units at $50 each. In range B $(z > x)$, we have the profit function P_B based on the fact that we have only x units available for sale regardless of the actual demand, where each unit is sold for $100 but costs the store $50. Hence the two profit functions are

$$P_A = 100z + 35(x-z) - 50x = 65z - 15x \qquad 0 \le z \le x$$
$$P_B = 50x \qquad z > x$$

The overall expected value of the profit $\bar{P}(x)$, for any value of x, is given by

$$\bar{P}(x) = \sum_{z=0}^{\infty} P(x)f(z) = \sum_{z=0}^{x} P_A f(z) + \sum_{z=x+1}^{\infty} P_B f(z)$$
$$= \sum_{0}^{x} (65z - 15x)f(z) + \sum_{x+1}^{\infty} 50xf(z)$$

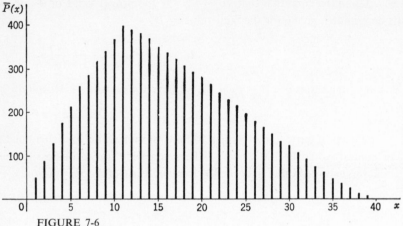

FIGURE 7-6
Expected profit $\bar{P}(x)$ vs. amount stocked x.

Now

$$\sum_{x+1}^{\infty} f(z) = 1 - \sum_{0}^{x} f(z)$$

Hence

$$\bar{P}(x) = \sum_{0}^{x} (65z - 15x)f(z) + 50x\left[1 - \sum_{0}^{x} f(z)\right] = 50x + 65\sum_{0}^{x} (z - x)f(z)$$

An obvious method of arriving at the maximum profit would be to evaluate this function at successive points until the maximum is located. For illustrative purposes we have carried out this calculation, with the results shown in Fig. 7-6; the maximum is reached $x = 11$. On the other hand, a method of solution which relies so heavily on computation is both time-consuming and inflexible. Therefore, we present the following method of analysis because of its greater efficiency and mathematical generality.

To determine the optimum value of x, let us investigate whether we would increase or decrease the expected profit by increasing x by one unit. By substitution,

$$\bar{P}(x + 1) = 50(x + 1) + 65\sum_{0}^{x+1} (z - x - 1)f(z)$$

$$= 50(x + 1) + 65\sum_{0}^{x} (z - x - 1)f(z)$$

since $(z - x - 1) = 0$ when $z = x + 1$.   Subtracting $\bar{P}(x)$ from $\bar{P}(x + 1)$, we find that the difference is

$$\Delta\bar{P}(x) = \bar{P}(x + 1) - \bar{P}(x) = 50 + 65 \sum_{0}^{x} [(z - x - 1) - (z - x)]f(z)$$

$$= 50 - 65 \sum_{0}^{x} f(z) = 50 - 65F(x)$$

where $F(x)$ is the value of the distribution function of z at $z = x$.   From the latter equation it is clear that $\Delta\bar{P}(x)$ will be positive as long as $F(x) < {}^{50}\!/_{65}$ and will be negative when $F(x) > {}^{50}\!/_{65}$, whereas if $F(x) = {}^{50}\!/_{65}$, $\bar{P}(x) = \bar{P}(x + 1)$. Therefore, the *maximum value* of the *profit* will be obtained if we choose x to be the *smallest* value for which $\Delta\bar{P}(x)$ is zero or negative.   For in case $F(x) = {}^{50}\!/_{65}$ for an integral value of x, we get the same expected profit with x as with $x + 1$, and the smaller number has the advantage of smaller investment.   Moreover, in case $F(x)$ cannot equal ${}^{50}\!/_{65}$ for an integral value of x, the smallest value of x such that $\Delta\bar{P}(x)$ is negative has the property that the next larger value of x would yield a smaller profit, in view of this negative difference, and the next smaller value would also yield a smaller profit, because for the latter the difference is positive.

In the particular case considered here, the distribution is the well-known Poisson distribution, and extensive tables have been prepared by E. C. Molina of the Bell Telephone Laboratories.   Adapting Molina's tables, we find the following values (Table 7-4) near ${}^{50}\!/_{65} \approx .769$.   Thus, the maximum value of the expected profit will be obtained by choosing $x = 11$.   This solution, of course, agrees with Fig. 7-6.

We know, however, from the solution of Example 7-4 that the average number of orders is 9, and from experience the manager would be aware of this fact even though he might not know the form of the distribution.   Thus, he might pick 9 as the appropriate number to stock.   The expected profit would be in that case $373 as opposed to the maximum of $389, and the difference is about 4 percent of the smaller figure.   It happens that there are four values of x (10, 11, 12, 13) which would yield a higher profit than $x = 9$ even though the curve

Table 7-4

x	9	10	11
$F(x)$	.587	.706	.803
$\Delta\bar{P}(x)$	11.8	4.1	−2.2

is fairly flat near the average order level. The 4 percent additional profit in most businesses is well worth going after, but it brings out a point which must be considered carefully. What is the company's policy or what is its object in being in business? In this case we have maximized the profit; but obviously one might wish to maximize the return on investment, which would give a different answer, and one certainly cannot maximize both at the same time. A mathematical model of the operation, therefore, is not the only part of a statistical problem which is important; there is also the formulation of the objective to be accomplished by the operation. Thus if one is going to pick the best strategy to use against a set of random causes, one has to have a measure of effectiveness which, if maximized, will convince one that the procedure is correct.   ////

EXAMPLE 7-22 A company has a sole purchaser for its product. If this purchaser on a particular day does not obtain the number of items which he requests, he will merely buy them on the open market; but his failure to obtain the necessary number from this company does not affect his future course of action. The margin of profit on the item is m dollars and, since the item is perishable, the loss is n dollars for any items not sold to this single purchaser. This purchaser will buy either 100, 200, or 300 items on a given day, with probabilities p_1, p_2, $1 - p_1 - p_2$, respectively. What should be the strategy of the company in order to maximize its profits?

Let x be the number of items the company produces. It is clear that x will not be greater than 300, since this is the maximum number which can be sold. The profit will be given by different mathematical expressions, depending on whether x is less than 100, between 100 and 200, or between 200 and 300. The following is the expected value of the profit (P) for each of these ranges of x.

$$0 \leq x \leq 100 \qquad E(P) = mx$$

$$100 < x \leq 200 \qquad E(P) = p_1[100m - n(x - 100)] + (1 - p_1)mx$$
$$= 100m + (x - 100)[m - p_1(m + n)]$$

$$200 < x \leq 300 \qquad E(P) = p_1[100m - n(x - 100)] + p_2[200m - n(x - 200)]$$
$$+ (1 - p_1 - p_2)mx$$
$$= 200m - 100(m + n)p_1$$
$$+ (x - 200)[m - (m + n)(p_1 + p_2)]$$

In a simple case like this, one would guess without any analysis that there are only three strategies capable of maximizing the profit. The company would produce 100, 200, or 300 items. This is clearly shown in Fig. 7-7, where the

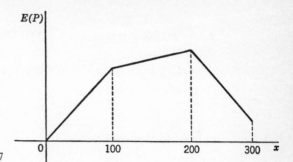

FIGURE 7-7

expected value of the profit is plotted against x. The straight line running from $x = 0$ to $x = 100$ always has a positive slope, but the slopes of the other two segments may be negative. Therefore, at least 100 items should be produced. In examining the equation for the expected value of the profit in the range from $x = 100$ to $x = 200$ we see that the slope is positive if $p_1 < m/(m + n)$ which means that, if this inequality holds, 200 items should be produced as the profit increases continuously from $x = 100$ to $x = 200$. If, however, the slope is negative, the company would, of course, make only the 100 items. If we now examine the equation for the expected profit in the final range, that is, from 200 to 300, we find that the slope is positive for this segment if $p_1 + p_2 < m/(m + n)$, which again gives the condition that the company should move to the 300-item level. It is obviously impossible for the slope to be negative in the middle range and positive in the final range. Therefore, the best strategy out of a theoretically infinite number of possible strategies is determined by comparing the probabilities p_1 and $p_1 + p_2$ with the ratio $m/(m + n)$ of marginal profit per item to the sum of marginal profit per item and loss per item.   ////

EXAMPLE 7-23   An item to be sold during the Christmas season brings a net profit of m dollars for each unit sold and a net loss of n dollars for each unit left unsold when the season ends. If the distribution of order volume v may be approximated by the density function $f(v)$ $(0 < v < V)$, determine the optimum number of units x to be stocked in order to maximize the expected value $\bar{P}$ of the profit P.

Two forms of the profit function exist, depending upon whether v is less than x or greater than x. Thus

$$P(x) = mv - n(x - v) = (m + n)v - nx \qquad v \le x$$
$$P(x) = mx \qquad\qquad\qquad\qquad\qquad v > x$$

Hence, the expected value of the profit is

$$\bar{P}(x) = \int_0^x [(m + n)v - nx] f(v) \, dv + \int_x^V mx f(v) \, dv$$

or, rewriting $\int_x^V mx f(v) \, dv$ as $mx[1 - \int_0^x f(v) \, dv]$, the expected value becomes

$$\bar{P}(x) = mx + (m + n) \int_0^x (v - x) f(v) \, dv$$

Differentiating this function with respect to x we obtain

$$\frac{\partial \bar{P}(x)}{\partial x} = m + (m + n)(x - x) f(x) - (m + n) \int_0^x f(v) \, dv = m - (m + n) F(x)$$

Hence, the maximum average profit is reached if x is so chosen that

$$F(x) = \frac{m}{m + n} = \frac{\text{unit profit}}{\text{unit profit} + \text{unit loss}}$$

If $m = \$1$ and $n = \$2$, then x must be chosen so that the area under $f(v)$ to its left will equal $\frac{1}{3}$.

It might be interesting to ask the question, at this point, as to what would happen if the loss of sales due to lack of the item might also produce a loss to the company because of a good-will factor. This loss of future profit could be taken into account during the current year by assigning a net loss to the company for each item for which there is a demand that cannot be supplied. From a practical viewpoint, the loss per item might be proportionately greater in case there were many items which could not be supplied. However, for illustrative purposes, let us consider that the loss to the company is a constant for each item that cannot be supplied. The integral for the expected value of the profit would then be

$$\bar{P}(x) = \int_0^x [(m + n)v - nx] f(v) \, dv + \int_x^V [mx - a(v - x)] f(v) \, dv$$

where a is the loss to the company for each item not available to a customer. Differentiating this function with respect to x and setting equal to zero, we obtain

$$F(x) = \frac{m + a}{m + a + n}$$

Therefore, the effect of the good-will factor in determining the value of x for the best strategy is to add the loss per item due to good will to the margin of profit per item in order to determine an effective margin of profit. It is clear from this that as a increases it tends to move x farther and farther toward its upper limit. If we had applied a good-will factor to the problem in Example 7-21 the effect would be exactly the same, that is, the margin of profit would be augmented by the numerical value of the good-will factor.   ////

EXAMPLE 7-24   In Example 7-23 let us suppose that the probability density of sales can be represented by

$$f(v) = \frac{1}{2b} \qquad s - b \le v \le s + b$$

This is a situation in which a sales forecast indicates that the amount to be sold is s, but previous history indicates that there is an error in estimating the sales and that this error is equally likely to assume any value from $-b$ to b. What is the additional profit that would be realized by stocking the number which maximizes the profit rather than stocking the amount s which is the estimated sale of the item?

The expected value of the profit (P), if we stock s, is given by

$$E(P_s) = \frac{1}{2b} \int_{s-b}^{s} [(m + n)x - ns] \, dx + \frac{1}{2b} \int_{s}^{s+b} ms \, dx$$

By carrying out the indicated integration and substituting the appropriate limits, we obtain for the expected value of the profit

$$E(P_s) = ms - \frac{b(m + n)}{4}$$

By choosing n sufficiently large in this formula, it is seen that the expected profit may become negative. The possibility of this situation is enhanced when b is almost as large as s. The expression for the maximum expected value of the profit is given by

$$E(P_{max}) = \frac{1}{2b} \int_{s-b}^{y} [mx - n(s - x)] \, dx + \frac{1}{2b} \int_{y}^{s+b} ms \, dx$$

where y is the solution (see Example 7-23) of the equation

$$\frac{1}{2b} \int_{s-b}^{y} dx = \frac{m}{m + n}$$

By carrying out the indicated integration and solving the result for y, we have

$$y = \frac{m(s + b) + n(s - b)}{m + n}$$

Evaluating the maximum expected value of the profit by substituting the correct value of y in the integral for $E(P_{max})$, we obtain

$$E(P_{max}) = ms - \frac{bmn}{m + n}$$

Even for $b = s$, this expression still remains positive for any n. The net gain in profit by stocking the value y rather than the value s is now obtained by subtracting the two expected profits as

$$E(P_{max}) - E(P_s) = \frac{b}{4} \frac{(m - n)^2}{m + n}$$

This expression is equal to zero when $m = n$ but for $m \neq n$ the gain is always positive. The numerical unit of this calculated difference is necessarily the total gain in dollars, since b has the unit of number of items and both m and n are in dollars. This calculated difference is independent of s but, of course, in most practical cases the range of error of forecasting is a function of the value forecasted. It we assume a constant percentage error of estimate in the sales, then $b = ks$ (where k, in this case, must lie between 0 and 1) and the percentage gain in profit would be given by

$$\frac{100(ks/4)(m - n)^2/(m + n)}{sm - ks(m + n)/4} = \frac{100k(m - n)^2}{(m + n)[4m - k(m + n)]} \approx 25k \frac{(1 - K)^2}{1 + K}$$

The last approximate value is obtained under the assumption that $k(m + n)$ is small compared with $4m$ and with the notation that K is the ratio of n to m. As an illustration, if we lose \$2 for every \$1 (the margin of profit) so that $K = 2$ and the range of error of the estimate of the sales forecast is ± 25 percent, then the percentage gain in profit is approximately 2 percent. If, however, $K = 3$ in this instance, the gain is about 6 percent.   ////

EXAMPLE 7-25 Raw material must be ordered in the spring for sale in a finished product in the fall. The cost of raw material is c_1 dollars per item. If this raw material is not used, it has a scrap value which is c_2 dollars where $c_2 < c_1$. The raw material may be processed during the summer at a cost of $g_1(x)$ dollars per item (that is, the unit cost) where x is the number to be manufactured. The loss incurred on those completed products which are not sold can be again expressed as the fact that the scrap value is c_3 dollars per item. It is here assumed that the scrap value is less than the cost incurred to produce the item. On the other hand, it is possible to produce W items during the fall season (thus at the time that the demand is known) at a higher rate of $g_2(W)$ dollars per item so that one never has to suffer the loss on manufactured goods not sold but merely on the raw material. There is, however, a maximum total number (A) which can be produced during the season of sale itself because of limited plant capacity. With a selling price of s dollars per item, what is the maximum amount of raw material to be ordered and how much should be processed during the summer in order to maximize the profit, assuming the probability of sales can be represented by a density function $f(v)$ ($0 < v < \infty$)?

Assuming that the total number of pieces of raw material is made up of the x pieces which are to be manufactured during the summer and y pieces which are to be produced during the season if the demand is demonstrated, we have for the expected value of the profit the following expression, where it must be remembered that the function representing the profit differs in the three ranges of the volume of sales, namely, $0 < v < x$, $x < v < x + y$, and $x + y < v < \infty$.

$$\text{Expected profit} = E(P) = \int_0^x [sv - c_1(x + y) - xg_1(x) + c_2 y + c_3(x - v)]f(v)\, dv$$

$$+ \int_x^{x+y} [sv - c_1(x + y) - xg_1(x) - (v - x)g_2(v - x) + c_2(x + y - v)]f(v)\, dv$$

$$+ \int_{x+y}^{\infty} [s(x + y) - c_1(x + y) - xg_1(x) - yg_2(y)]f(v)\, dv$$

It is tentatively assumed at this point in the analysis that there is no restriction on the amount y which can be manufactured during the season, and the integral for this expected profit can be differentiated first with respect to x and then with respect to y.

For this situation, the derivatives with respect to the limits on the integrals cancel each other in both equations because of the continuity conditions that exist from one range of v to the next. The only time that differentiation with respect to the limits produces nonvanishing terms is when a setup charge or its equivalent is introduced at a particular volume of sales so that it occurs in one integral and not the other, which is not true here. Therefore we have

$$\frac{\partial E(P)}{\partial x} = \int_0^x \left[-c_1 - \frac{d}{dx}[xg_1(x)] + c_3 \right] f(v)\, dv$$

$$+ \int_x^{x+y} \left\{ -c_1 - \frac{d}{dx}[xg_1(x)] - \frac{d}{dx}[(v - x)g_2(v - x)] + c_2 \right\} f(v)\, dv$$

$$+ \int_{x+y}^{\infty} \left\{ s - c_1 - \frac{d}{dx}[xg_1(x)] \right\} f(v)\, dv$$

and

$$\frac{\partial E(P)}{\partial y} = \int_0^x [-c_1 + c_2]f(v)\, dv + \int_x^{x+y} [-c_1 + c_2]f(v)\, dv$$

$$+ \int_{x+y}^{\infty} \left\{ s - c_1 - \frac{d}{dy}[yg_2(y)] \right\} f(v)\, dv$$

These equations can now be simplified by considering the integral from 0 to x of $f(v) \, dv$ in $\partial E(P)/\partial x$ as $1 - $ (integral from x to $x + y$ plus the integral from $x + y$ to ∞) and the integral from 0 to x plus the integral from x to $x + y$ of $f(v) \, dv$ as $1 - $ (integral from $x + y$ to ∞) in $\partial E(P)/\partial y$ to give

$$\frac{\partial E(P)}{\partial x} = -c_1 + c_3 - \frac{d}{dx}[xg_1(x)] + (c_2 - c_3) \int_x^{x+y} f(v) \, dv$$

$$- \int_x^{x+y} \frac{d}{dx}[(v - x)g_2(v - x)]f(v) \, dv + (s - c_3) \int_{x+y}^{\infty} f(v) \, dv$$

$$\frac{\partial E(P)}{\partial y} = -c_1 + c_2 + \left\{ s - c_2 - \frac{d}{dy}[yg_2(y)] \right\} \int_{x+y}^{\infty} f(v) \, dv$$

The integral involving $d[(v - x)g_2(v - x)]/dx$ must be kept intact, since this term involves v which is the variable of integration, whereas the other coefficients of $f(v)$ in the several integrals are constants.

Following the usual technique, one would set these two derivatives equal to zero in order to obtain a stationary point as the minimum. This would generate two equations in the two unknowns x and y, which could then be solved by any of several methods of successive approximations.

It must be noted that in this problem there are very definite restrictions which have to be satisfied, and a stationary point may not occur within these limitations. The total area under the probability density $f(v)$ must equal unity, x and y must each be equal to or greater than zero, and y cannot be greater than A. Such minimization with lateral restrictions is, of course, often found in practical applications. In order to analyze the difficulties and to illustrate the solutions of a problem in this category, we now simplify the above expressions by assuming $g_1(x)$ and $g_2(W)$ to be constants c_4 and c_5, respectively. Then

$$\frac{\partial E(P)}{\partial x} = (-c_1 + c_3 + c_4) + (c_2 - c_3 + c_5) \int_x^{x+y} f(v) \, dv + (s - c_3) \int_{x+y}^{\infty} f(v) \, dv$$

$$\frac{\partial E(P)}{\partial y} = -c_1 + c_2 + [s - c_2 - c_5] \int_{x+y}^{\infty} f(v) \, dv$$

If the expression for $\partial E(P)/\partial y$ is set equal to zero we have

$$\int_{x+y}^{\infty} f(v) \, dv = \frac{c_1 - c_2}{s - c_2 - c_5}$$

Since $c_1 > c_2$ and $s > c_1 + c_5$ by the conditions imposed upon the problem, this integral is always positive and less than unity. It therefore always uniquely determines the value of $x + y$ if the function $f(v)$ is given, and thus the stationary point, as far as $\partial E(P)/\partial y$ is concerned, is also uniquely determined.

Substituting this value for the integral in $\partial E(P)/\partial y$, we have

$$\frac{\partial E(P)}{\partial x} = (-c_1 + c_3 + c_4) + (c_2 - c_3 + c_5) \int_x^{x+y} f(v)\, dv + \frac{(s - c_3)(c_1 - c_2)}{s - c_2 - c_5}$$

which gives the slope of the $E(P)$ surface in the x direction.

For a stationary point to exist, it is necessary that $\partial E(P)/\partial x$ change sign. This is not always the case, as one can see by letting $c_3 = c_5$ and $c_2 = 0$, which are possible values of these constants. The sign of the derivative is then always positive, which indicates that the profit is increasing with x, so that we should not plan to produce any of the product after the season starts. Since the value of $x + y$ is known from the integral and $y = 0$, we have determined the amount of x to order. The exact opposite conclusion would be the case if, for some values of the constants, the derivative were always negative with, therefore, $x = 0$ and thus $y = x + y$. This, of course, neglects the limitation on the total amount of y possible.   More generally, the fact that $\int_x^{x+y} f(v)\, dv$ is restricted to lie between 0 and $1 - \int_{x+y}^{\infty} f(v)\, dv$ means that there are many situations where $\partial E(P)/\partial x$ will not change sign in the possible region of the integral, and the above argument will again necessarily follow.   For some ranges of the constants, the partial derivative $\partial E(P)/\partial x$ can be set equal to zero and a value for $\int_x^{x+y} f(v)\, dv$ determined which is within its range of possibility.   In this case, we have a unique stationary point and can calculate the values of x and y.

After x and y are determined, we must consider the restriction that $y \le A$. If y comes out less than or equal to A, we have the final solution to the problem. If, however, y is greater than A, it indicates that y should be as large as possible, namely, A.   We cannot, however, assign the excess of y over A to x, since this may not produce a maximum profit, but $E(P)$ must be again set up as a function of x alone as

$$E(P) = \int_0^z [sv - c_1(x + A) - xg_1(x) + c_2 A + c_3(x - v)] f(v)\, dv$$

$$+ \int_x^{x+A} [sv - c_1(x + A) - xg_1(x) - (v - x)g_2(v - x) + c_2(x + A - v)] f(v)\, dv$$

$$+ \int_{x+A}^{\infty} [s(x + A) - c_1(x + A) - xg_1(x) - Ag_2(A)] f(v)\, dv$$

This function can now be differentiated with respect to x and the derivative set equal to zero.   Since the conditions of the problem were such that a profit is possible, and the range of x is between 0 and ∞, this process will yield a stationary point and therefore constitute a solution to the problem.   ////

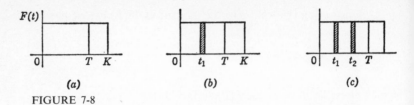

FIGURE 7-8

EXAMPLE 7-26 In a building with n lamps lighted at the same time daily a bulb is replaced immediately when it burns out. The costs for replacing one bulb consist of the purchasing cost a_1, the cost a_2 of changing the bulb (i.e., taking out the old and putting a new one in the lamp), and all the costs a_3 for bringing the bulb, ladder, and other necessary things to the required place. Another alternative is to change all the bulbs at the same time periodically with time period T, but still change a bulb when it burns out. When all the bulbs are to be changed together, the costs a_1 and a_2 are the same per bulb, but instead of a_3, we now have a total cost a_4 for bringing all the bulbs and other materials to the place. Determine T such that the average cost per unit of time the bulbs are lit becomes a minimum. The lifetime of a bulb is unknown but we shall assume that its probability density is $f(t) = 1/K$ $(0 < t < K)$.

Given the probability density, it is possible to compute the expected number of burnouts per socket between $t = 0$ and $t = T$, where T is arbitrary and is to be determined later. Obviously, when a bulb fails and is replaced by another, this new replacement bulb has the same rectangular distribution for its life as the burned-out bulb, but starting with the moment of replacement. By reference to Fig. 7-8 the probabilities $P(0)$, $P(1)$, and $P(2)$ of exactly 0, 1, 2 burnouts, respectively, before time T is reached are as follows:

$$P(0) = 1 - \int_0^T \frac{dt}{K} = 1 - \frac{T}{K}$$

$$P(1) = \int_0^T \frac{dt_1}{K} \left(1 - \int_{t_1}^T \frac{dt}{K} \right) = \int_0^T \frac{dt_1}{K} - \int_0^T \frac{T - t_1}{K^2} \, dt_1 = \frac{T}{K} - \frac{T^2}{2K^2}$$

$$P(2) = \int_0^T \frac{dt_1}{K} \int_{t_1}^T \frac{dt_2}{K} \left(1 - \int_{t_2}^T \frac{dt}{K} \right) = \int_0^T \frac{dt_1}{K} \left(\int_{t_1}^T \frac{dt_2}{K} - \int_{t_1}^T \frac{T - t_2}{K^2} \, dt_2 \right)$$

$$= \frac{T^2}{2K^2} - \frac{T^3}{6K^3} = \frac{T^2}{K^2 \times 2!} - \frac{T^3}{K^3 \times 3!}$$

An examination of the form of these integrals indicates that the probability of exactly n burnouts in any one socket would be

$$P(n) = \frac{T^n}{K^n n!} - \frac{T^{n+1}}{K^{n+1}(n + 1)!}$$

From the definition of expected value, the average number of burnouts per socket is

$$E(n) = \sum_{n=0}^{n=\infty} nP(n) = \sum_{0}^{\infty} n\left[\frac{T^n}{K^n n!} - \frac{T^{n+1}}{K^{n+1}(n+1)!}\right]$$

$$= \frac{T}{K}\sum_{1}^{\infty}\frac{T^{n-1}}{K^{n-1}(n-1)!} - \sum_{0}^{\infty}\frac{[(n+1)-1]T^{n+1}}{K^{n+1}(n+1)!}$$

$$= \frac{T}{K}e^{T/K} - \frac{T}{K}\sum_{0}^{\infty}\frac{T^n}{K^n n!} + \sum_{0}^{\infty}\frac{T^{n+1}}{K^{n+1}(n+1)!}$$

$$E(n) = \frac{T}{K}e^{T/K} - \frac{T}{K}e^{T/K} + (e^{T/K} - 1) = e^{T/K} - 1$$

It is interesting to note that the average number of burnouts per socket over the entire range K of the bulb life obtained by setting $T = K$ is $(e - 1) \approx 1.72$. Referring to the costs given in the statement of the problem, the average cost per unit of time the bulbs are lighted is

$$\text{Cost} = C = \frac{n(a_1 + a_2 + a_3)(e^{T/K} - 1) + n(a_1 + a_2) + a_4}{T}$$

assuming that they were all changed at the specific cycle time T. The derivative of this cost with respect to T is equal to zero if a stationary point exists and thus

$$\frac{dC}{dT}$$

$$= \frac{(T/K)n(a_1 + a_2 + a_3)e^{T/K} - [n(a_1 + a_2 + a_3)(e^{T/K} - 1) + n(a_1 + a_2) + a_4]}{T^2} = 0$$

If $T/K = x$ and $[n(a_1 + a_2) + a_4]/n(a_1 + a_2 + a_3) = A$, the equation for this stationary point becomes

$$xe^x - (e^x - 1) - A = 0$$

or

$$e^x(1 - x) = 1 - A$$

Since A is less than 1 (in order for this problem to have any meaning the cost of replacing the bulbs all at one time must be less than changing them one at a time), this transcendental equation has a solution for $x < 1$ which determines the period T at which they should all be replaced in order to make this cost a minimum.   ////

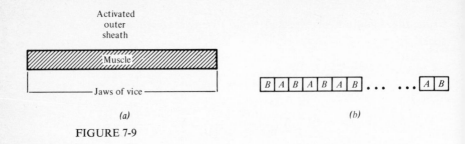

(a)

(b)

FIGURE 7-9

EXAMPLE 7-27   A problem which illustrates the use of probabilistic analyses and expected-value theory in building a mathematical model deals with a very simplified picture of the way a piece of animal muscle behaves when held with slight tension in the jaws of a vice.   The muscle in its entirety is activated by applying an electric current to the outer sheath (see Fig. 7-9a).   We shall assume that the muscle is made up of a very large number of long fibers of vary-ing length.   Each fiber consists of sections A which can be activated, connected by sections B which we shall assume in this simplified model are inert and in-elastic (see Fig. 7-9b).   When any section of type A becomes activated, the entire fiber shrinks slightly, creating a tension as both ends are fixed.   We shall also assume that a section of type A contracts instantaneously upon being activated and then gradually expands to its original length so that the tension it creates falls off as $T = ae^{-\beta(t-t_0)}$ $(t > t_0)$, where t_0 is the instant when it was activated, and where a is the tension created instantaneously at the moment of activation.   We shall also assume that the probability that any section A is activated at time t is given by the density function $f(t) = \alpha e^{-\alpha t}$, where t is measured from the instant of the activation of the outer sheath of muscle.   All activations of the sections A are independent and at random and of course an individual section is activated only once.

Thus as the numerous A's are activated one after another the tension builds up in an individual, long fiber and eventually (as the A's have all been activated and relaxed) the tension in that fiber drops again to zero.   What is the equation of the tension of the entire muscle as a function of α, β, and time t?

Let us start with a single fiber consisting of N A's and consider the amount of tension at a time T after the original activation of the sheath.   The prob-ability that a particular A will have been activated by T is

$$\int_0^T \alpha e^{-\alpha t}\, dt = 1 - e^{-\alpha T}$$

and the probability that it will not have been activated is

$$\int_T^\infty e^{-\alpha t}\, dt = e^{-\alpha T}$$

Thus the probability (at time T) that n A's are under tension out of a total of N possible A's is

$$P(n) = C(N, n)(1 - e^{-\alpha T})^n (e^{-\alpha T})^{N-n}$$

The conditional distribution (given that n A's have been activated) that the first one was tensioned between t_1 and $t_1 + dt_1$, the second between t_2 and $t_2 + dt_2$, ..., etc., is, since the total volume is unity,

$$\frac{n!\alpha e^{-\alpha t_1}\,dt_1\alpha e^{-\alpha t_2}\,dt_2\cdots\alpha e^{-\alpha t_n}\,dt_n}{n!\int_0^T dt_n\int_0^{t_n} dt_{n-1}\cdots\int_0^{t_3} dt_2\int_0^{t_2} dt_1 e^{-\alpha(t_1+t_2+\cdots+t_n)}}$$

$$= \frac{n!\alpha e^{-\alpha t_1}\,dt_1\alpha e^{-\alpha t_2}\,dt_2\cdots\alpha e^{-\alpha t_n}\,dt_n}{(1 - e^{-\alpha T})^n}$$

where $t_1 < t_2 < t_3 < \cdots < t_n = T$. If the n units are tensioned at $t_1, t_2, \ldots, t_n$, the total pull at time T is given by the sum of the tensions of each one separately, if we assume linearity. Thus the expected tension E_c would be, by definition,

$$E_c = a \int \cdots \int (e^{-\beta(T-t_1)} + \cdots + e^{-\beta(T-t_n)}) \frac{n!\alpha e^{-\alpha t_1}\,dt_1\alpha e^{-\alpha t_2}\,dt_2\cdots\alpha e^{-\alpha t_n}\,dt_n}{(1 - e^{-\alpha T})^n}$$

There are n integrals to evaluate in the expression for E_c, but they are all of the same form. A typical one involving t_k as the activation time would be

$$\frac{an!e^{-\beta T}}{(1 - e^{-\alpha T})^n}\iint_R \cdots \int \alpha e^{-\alpha t_1}\,dt_1 \cdots \alpha e^{-(\alpha-\beta)t_k}\,dt_k \cdots \alpha e^{-\alpha t_n}\,dt_n$$

Because the integral involving dt_k has $e^{-(\alpha-\beta)t_k}$ whereas the others are $e^{-\alpha t_i}\,dt_i$, this integration should be performed last; thus the multiple integral with appropriate limits would be

$$\frac{e^{-\beta T}n!a}{(1 - e^{-\alpha T})^n}\int_0^T \alpha e^{-(\alpha-\beta)t_k}\,dt_k\left(\int_0^{t_k}\alpha e^{-\alpha t_1}\,dt_1\int_{t_1}^{t_k}\alpha e^{-\alpha t_2}\,dt_2\cdots\int_{t_{k-2}}^{t_k}\alpha e^{-\alpha t_{k-1}}\,dt_{k-1}\right)$$

$$\cdot\left(\int_{t_k}^T\alpha e^{-\alpha t_{k+1}}\,dt_{k+1}\int_{t_{k+1}}^T\alpha e^{-\alpha t_{k+2}}\,dt_{k+2}\cdots\int_{t_{n-1}}^T\alpha e^{-\alpha t_n}\,dt_n\right)$$

$$= \frac{e^{-\beta T}n!a}{(1 - e^{-\alpha T})^n}\int_0^T\alpha e^{-(\alpha-\beta)t_k}\,dt_k\cdot\frac{(1 - e^{-\alpha t_k})^{k-1}}{(k-1)!}\frac{(e^{-\alpha t_k} - e^{-\alpha T})^{n-k}}{(n-k)!}$$

This expression can be summed for all k from 1 to n. Realizing that t_k is a dummy variable of integration and does not affect that summation, we now have for E_c

$$E_c = \frac{ae^{-\beta T}n!}{(1 - e^{-\alpha T})^n}\int_0^T\alpha e^{-(\alpha-\beta)t}\sum_{k=1}^n\frac{(1 - e^{-\alpha t})^{k-1}}{(k-1)!}\frac{(e^{\alpha t} - e^{-\alpha T})^{n-k}}{(n-k)!}\,dt$$

or

$$E_c = \frac{ae^{-\beta T}n}{(1 - e^{-\alpha T})^n} \int_0^T \alpha e^{-(\alpha - \beta)t} \sum_{k=1}^n C(n-1, n-k)(1 - e^{-\alpha t})^{k-1}(e^{-\alpha t} - e^{-\alpha T})^{n-k} \, dt$$

$$= a \frac{e^{-\beta T}n(1 - e^{-\alpha T})^{n-1}}{(1 - e^{-\alpha T})^n} \int_0^T \alpha e^{-(\alpha - \beta)t} \, dt$$

and finally

$$E_c = \frac{ae^{-\beta T}n\alpha}{\alpha - \beta} \frac{1 - e^{-(\alpha - \beta)T}}{1 - e^{-\alpha T}}$$

where E_c is the conditional expected value of the tension, given that exactly n elements of type A are activated before T.

Thus by multiplying by the probability of exactly n activations and summing over all n from 0 to N, we have for the total tension in this one fiber

$$E = \sum_{n=0}^N \frac{ae^{-\beta T}n\alpha}{\alpha - \beta} \frac{1 - e^{-(\alpha - \beta)T}}{1 - e^{-\alpha T}} C(N, n)(1 - e^{-\alpha T})^n(e^{-\alpha T})^{N-n}$$

Except for a multiplying constant independent of n, E is equivalent to the expected value of a binomial distribution (N, p) with $p = 1 - e^{-\alpha T}$ and $q = 1 - p = e^{-\alpha T}$ so that

$$E = \frac{a\alpha N}{\alpha - \beta} (e^{-\beta T} - e^{-\alpha T})$$

Since, also, N occurs only as a coefficient, if we take the expected value of this expression E for any distribution of fiber lengths $f(N)$ there results

$$E = N_0 \frac{a\alpha}{\alpha - \beta} (e^{-\beta T} - e^{-\alpha T})$$

where N_0 is the average length of all fibers in the sheath. When $\alpha = \beta$, E becomes $N_0 a\alpha T e^{-\alpha T}$ by using L'Hospital's rule.   ////

(*Suggested Problems 7-32 to 7-39*)

PROBLEMS

7-1   A player throws a die and is awarded twice the number thrown (in dollars). How much would you pay to play?   *Ans.* $7.

7-2   A group giving a supper spends A dollars for enough food for 200 people. If the number of people who attend is equally likely to fall anywhere in the range 150 to 200, what should the group charge per meal in order to have an expected 10 percent profit?   *Ans.* $.0062A$.

7-3 Given $f(x) = 2e^{-2x}$ ($0 \leq x \leq \infty$), find $E(y)$, where $y = e^x$, and check by using the distribution of y. *Ans.* 2.

7-4 Given $y = x^3$ and $v = (x - \mu)^4$, find $E(y)$ and $E(v)$, where x is normal.

7-5 Given $f(x) = \frac{3}{4}x(2 - x)$ (0, 1), find $E(y)$, where $y = 1/x$. *Ans.* $\frac{3}{2}$.

7-6 Given $f(x) = 4e^{-4x}$ (0, ∞), find $E(y)$, where $y = e^{3x}$, and check by using the distribution of y. *Ans.* 4.

7-7 Maxwell demonstrated that for a fixed absolute temperature the speed x of individual molecules of a given gas is distributed as follows:

$$f(x) = kx^2 e^{-3x^2/2w^2} \qquad 0 < x < \infty$$

where w is a physical parameter and k is a mathematical parameter so adjusted that the area is unity. Determine k in terms of w, and then find the mean molecular kinetic energy of the gas, letting m represent the mass of an individual molecule.

7-8 (a) If c is an arbitrary constant, show that the quantity $E[(x - c)^2]$ is minimized by choosing $c = \mu$. (b) If x has a density function, show that the quantity $E(|x - c|)$ is minimized by choosing c as the median.

7-9 From records accumulated over a period of 20 years, the number of significant rainfalls during the first 6 weeks of the summer has a Poisson distribution with $\mu = 5$. The dollar value of a certain agricultural index is given by $100(e^n - 1)$, where n is the number of significant rainfalls. What is the expected value of the index? *Ans.* $538,400.

7-10 Four independent observations x_1, x_2, x_3, x_4 are taken on the discrete variate x, where $f(x) = \frac{1}{2}$ (-1, 1). State the mean and variance of $y = (x_1 + x_2 + x_3 + x_4)/4$ by obtaining the distribution of y and computing the mean and variance from it. *Ans.* $\mu = 0, \sigma^2 = \frac{1}{4}$.

7-11 Find the mean and variance of x, where

$$f(x) = \frac{1}{b - a} \qquad (a, b)$$

7-12 Find the mean and variance of the following distributions: (a) gamma, (b) beta, (c) F distribution. (d) Give an approximate formula for the variance of the F distribution for large values of both parameters.

7-13 A random variable can take on the values -2, -1, and 2 with the respective probabilities .125, .250, .625. Find the mean and the standard deviation of the distribution by direct calculation. *Ans.* $\mu = .750, \sigma^2 = 2.69$.

7-14 Test the Camp-Meidell inequality with $t = 1, 2, 3$ for the distribution

$$f(x) = 30x^2(1 - x)^2 \qquad (0, 1)$$

7-15 The variate x is distributed as follows:

$$f(x) = k[1 - (x/c)^2]^{(c^2 - 3)/2} \qquad (-c, c)$$

where $c > 1$. (a) Determine the value of k and find the mean and variance of x. (b) Sketch the curve for $c^2 < 3$, $c^2 = 3$, and $c^2 = 9$. (c) Using an appropriate inequality, give an upper limit for $P(|x| > 3)$ when $c^2 > 9$.

7-16 Given a random sample of two observations x_1 and x_2 from any population for which the variance is finite, show that the statistic

$$\tfrac{1}{2}(x_1 - x_2)^2$$

is an unbiased estimate of the population variance.

7-17 (a) Independent variates are distributed as follows:

$$f_1(x) = 20x^3(1 - x) \quad (0, 1) \qquad f_2(y) = 2y \quad (0, 1)$$

Find the expected value of $f(x, y) = y/x^3 + x/y$.

  (b) Find the expected value of $(x + y)^3$, where x and y are independent unit normal variates.   *Ans.* (a) 8.

7-18 The independent variates x_1, x_2 are normally distributed with respective means $2, -3$ and variance $4, 9$.   Write the distribution of the function $y = 2x_1 + x_2 - 1$.

7-19 Given x_1 and x_2 as two independent normal variates with unknown means μ_1 and μ_2 but standard deviations 4 and 9, respectively.   A new variate is found as follows:

$$y = 1 + 2x_1 + 3x_2$$

Four simultaneous values of x_1 and x_2 are made and thus four values of y are computed by the above formula.   These four values are averaged and give a value of 50.   What are the 90-percent confidence limits for this average of 50?   *Ans.* 26.7 to 73.3.

7-20 Given that x_1 and x_2 are two independent and unbiased measures of the same parameter but have different standard deviations which are σ_1 and σ_2, respectively.   A linear combination z of the two measures is as follows, where p is a constant:

$$z = px_1 + (1 - p)x_2$$

Show that z is an unbiased estimate and find p if the variance of z is to be made as small as possible.   *Ans.* $p = \sigma_2^2/(\sigma_1^2 + \sigma_2^2)$.

7-21 For any variate for which the mean μ is finite, prove the identities (a) $E(x - \mu) = 0$, (b) $E(x - a)^2 = E(x - \mu)^2 + (\mu - a)^2$, where $a = $ constant.   Use the substitution $(x - a) = (x - \mu) + (\mu - a)$.   For any sample of N observations x_1, x_2, ..., x_N put $\bar{x} = (1/N)\sum_i x_i$ and prove analogous properties of $\bar{x}$: (c) $\sum_i (x_i - \bar{x}) = 0$, (d) $\sum_i (x_i - a)^2 = \sum_i (x_i - \bar{x})^2 + N(\bar{x} - a)^2$.   As a special case of (d) notice that $\sum_i (x_i - \mu)^2 = \sum_i (x_i - \bar{x})^2 + N(\bar{x} - \mu)^2$.

7-22 Derive a formula for the covariance of two linear functions $y_1 = a_0 + a_1x_1 + \cdots + a_nx_n$, $y_2 = b_0 + b_1x_1 + \cdots + b_nx_n$ of the same variates $x_1, x_2, ..., x_n$.

7-23 In any sample of N independent observations from an infinite population with mean μ and variance σ^2, let x_i, x_j denote any pair of observations in the same sample and $\bar{x}$, the corresponding mean of the sample.   Prove the following statements by expanding the expressions algebraically and taking expected values term by term.

  (a) $E[(x_i - \bar{x})(x_j - \bar{x})] = \dfrac{-\sigma^2}{N} \qquad i \neq j$

  (b) $E\left[\displaystyle\sum_1^N (x_i - \bar{x})^2\right] = (N - 1)\sigma^2$

7-24 The covariance of x and y is σ_{xy} and the respective means μ_x and μ_y. From the bivariate distribution, a random sample of N independent pairs of observations $(x_1, y_1), (x_2, y_2), \ldots, (x_N, y_N)$ is drawn, and the corresponding sample means $\bar{x} = \sum x_i/N$, $\bar{y} = \sum y_i/N$ are computed. Prove that (a) $E(\bar{x}\bar{y}) = (1/N)\sigma_{xy} + \mu_x\mu_y$; (b) $E[\sum_i(x_i - \bar{x})(y_i - \bar{y})] = (N-1)\sigma_{xy}$.

7-25 From an infinite population with mean μ and variance σ^2, k random samples are drawn. These contain, respectively, $n_1, n_2, \ldots, n_k$ observations. A record of the corresponding sample means $\bar{x}_1, \bar{x}_2, \ldots, \bar{x}_k$ is kept, but the individual observations are not put on record. At a later date, there arises the problem of estimating μ and σ^2 from the available information. Show that the statistics $\bar{x}$ and s^2 have the properties that $E(\bar{x}) = \mu$, $E(s^2) = \sigma^2$, where $\bar{x}$ and s^2 are defined as follows:

$$\bar{x} = \frac{\sum_1^k n_i \bar{x}_i}{\sum_1^k n_i} \qquad s^2 = \frac{\sum_1^k n_i(\bar{x}_i - \bar{x})^2}{k-1}$$

7-26 From independent normal variates x_1, x_2 (respective means ξ_1, ξ_2; variances $\sigma_1{}^2$, $\sigma_2{}^2$) two linear functions $u_1 = c_0 + c_1 x_1 + c_2 x_2$ and $u_2 = k_0 + k_1 x_1 + k_2 x_2$ are derived. (a) Show that u_1 and u_2 will be independent if and only if their covariance is zero. (b) Express this condition as a relation between the coefficients.

    Hint: Solve for $x_1 - \xi_1$ and $x_2 - \xi_2$ in terms of $u_1 - m_1$ and $u_2 - m_2$, where $m_1 = E(u_1)$, $m_2 = E(u_2)$; then consider the quadratic.

7-27 The respective means of independent normal variates x_1, x_2 are $N_1 p$, $N_2 p$ and the variances are $N_1 pq$, $N_2 pq$. Show that the variates $u_1 = (x_1 + x_2)/(N_1 + N_2)$, $u_2 = (x_1/N_1 - x_2/N_2)/\sqrt{pq(1/N_1 + 1/N_2)}$ are independent and that u_2 is unit normal.

7-28 Given n independent normal variates $x_1, x_2, \ldots, x_n$ with common mean and variance, set $\bar{x} = \sum_i x_i/n$ and $v_i = x_i - \bar{x}$. The general term of a variable change known as MacMahon's canonical transformation is

$$z_k = \frac{(k+1)v_k + v_{k+1} + v_{k+2} + \cdots + v_{n-1}}{\sigma\sqrt{k(k+1)}}$$

From what general proposition does it follow that z_k is normally distributed? By taking expected values, prove that z_k has zero mean and unit variance and hence is unit normal.

7-29 Let a, b denote arbitrary real constants and x_1, x_2 any pair of variates for which $E(x_1 - a)^2$ and $E(x_2 - b)^2$ exist. Observing that $E[(x_1 - a) - \lambda(x_2 - b)]^2 \geq 0$ for all real λ, show that $[E(x_1 - a)(x_2 - b)]^2 \leq [E(x_1 - a)^2][E(x_2 - b)^2]$. Hence, deduce that $-1 \leq \rho \leq 1$. Also infer that $\rho^2 = 1$ when and only when one variate is an exact linear function of the other. [*Hint:* Put $\lambda = [E(x_1 - a)(x_2 - b)]/E(x_2 - b)^2$.]

7-30 The variates $x_1, x_2, \ldots, x_{2n}$ all have a common variance σ^2, and for any pair of variates x_i, x_j $(i \neq j)$ the correlation coefficient has the same value ρ. Determine the correlation between the two sums $S_1 = x_1 + x_2 + \cdots + x_n$; $S_2 = x_{n+1} + x_{n+2} + \cdots + x_{2n}$. *Ans.* $np/(1 + (n-1)\rho)$.

7-31 Expressed as deviations from the curve of seasonal trend, the daily means of temperature at a certain locality may be regarded as identically distributed over a period of a month, with mean and variance 0 and σ^2, respectively. The covariance of any pair of daily mean deviations is given by $\sigma^2 \phi_k$, where ϕ_k is known as the autocorrelation coefficient and depends only on the number of days k between the two dates considered ($\phi_0 = 1$; $|\phi_k| \leq 1$). If τ represents the average of n consecutive daily mean deviations $t_1, t_2, \ldots, t_n$ find the variance of τ.

7-32 A seasonal item which must be stocked in advance by a retailer costs \$1 per unit and sells for \$2 per unit. Unsold items at the end of the season are a complete loss. The probability of k customer orders is $(\frac{1}{2})^{k+1}$ ($k = 0, 1, \ldots$). Find the expected profit as a function of the number of items stocked. Note that

$$\sum_{k=0}^{\infty} (\frac{1}{2})^k = 2 \qquad \sum_{k=0}^{n} (\frac{1}{2})^k = 2[1 - (\frac{1}{2})^{n+1}]$$

$$\sum_{k=0}^{n} k(\frac{1}{2})^k = 2[1 - (\frac{1}{2})^n] - n(\frac{1}{2})^n$$

Ans. $2 - x - 2(\frac{1}{2})^x$.

7-33 The cost of operating a hydroelectric plant for a specific period consists of a fixed charge of K_1 dollars independent of the amount of electricity generated and a cost of K_2 dollars per kilowatthour for generation. All the power generated can be sold at a price which is K_3 dollars per kilowatthour. The possible amount of generation is related to the rainfall in the area, which has a certain probability density, thus permitting the computation of the probability density of the distribution of kilowatthours generated as $f(x) = a^2 x e^{-ax}$ ($0 < x < \infty$). What is the expected value of the profit? By finding the probability density of the profit, determine the probability of going into bankruptcy at the end of the period if only K_4 dollars are available as reserves at the beginning of the period with no borrowing capacity ($K_4 < K_1$). Plot the distribution of the profit.

7-34 A random variable x has the density function

$$f(x) = \begin{cases} 2x & 0 \leq x \leq 1 \\ 0 & \text{otherwise} \end{cases}$$

A random draw is made from this population which determines a payoff equal to its value. This payoff may be refused in favor of one equal to the square of the value of a second random observation. (*a*) What should your strategy be as to the acceptance and rejection of the first observation in order to maximize your expected value? (*b*) What is this maximum expected value?

7-35 A business consists of renting out the use of 10 washing machines. Assume that under present conditions, any given machine is likely to be out of commission one day out of five and also that breakdowns occur at random. What is the probability of more than one machine being out of order on a given day? Assuming the average loss is c dollars per day for a machine out of order, write an expression for the expected loss due to having machines out of service and evaluate by making the substitution $u = k - 1$ and recognizing the binomial expansion. *Ans.* 2c.

7-36 A manufacturer wishes to estimate the mean μ of a certain attribute x of his product by computing the sample mean $\bar{x}$ of a set of n independent observations of x. The cost C of the sampling process is composed of one part proportional to the sample size and one part proportional to the magnitude of the error of the estimate:

$$C = \sqrt{\frac{2}{\pi}} \, n + 16|\bar{x} - \mu|$$

If x is assumed to be normally distributed with unit variance, find the value of n which minimizes the expected value of the cost. *Ans.* 4.

7-37 The selling price of a certain seasonal item is $100 per thousand whereas its cost is $50 per thousand. Unsold items have no scrap value. The number of orders x (in thousands) is a random variable with density function

$$f(x) = \begin{cases} \frac{3}{4} & 0 < x < 1 \\ \frac{1}{4} & 1 < x < 2 \\ 0 & \text{otherwise} \end{cases}$$

Let s (in thousands) be the number stocked. What should s be in order to maximize profit?

7-38 The sales volume forecast (in hundreds) for a certain item is

$$f(V) = \left(\frac{3}{2,500}\right) V^2 (10 - V) \qquad 0 < V < 10$$

The unit raw material cost is $3 and the unit manufacturing cost is $2. Items are to be manufactured after orders have been received, provided raw material is available. Raw material may be purchased only at the beginning of the period and may be scrapped at a unit price of $1. The profit on each item sold is m; the object, as usual, is to maximize profit. Find an expression relating the number of units x of raw material to be ordered and the unit profit m.

7-39 The sales volume of a given concern is given by the following probability density function:

$$f(v \mid \overline{V}) = \frac{4}{\overline{V}^2} \, v e^{-2v/\overline{V}} \qquad 0 < v < \infty$$

which has an expected value of $\overline{V}$, that is, $\int_0^\infty v f(v \mid \overline{V}) \, dv = \overline{V}$. This expected value $\overline{V}$ depends upon the selling price per unit, s, and also upon the total amount of money spent on advertising, t.   Assume that in the range of application the relation is

$$\bar{V} = M \frac{1 + bt}{1 + t} e^{-as}$$

where M, a, and b are constant.   Also assume that the cost c per item is independent of the volume, since this item is purely one of assembly.   The margin of profit m per item is, of course, reduced by the amount spent on production and advertising, so that $m = s - c - (t/v)$.   Find the expected value of the product mv and then maximize this with respect to t and s in order to determine the best selling price and amount of money to be spent in advertising.

8

GENERATING FUNCTIONS

8-1 MOMENTS

In Secs. 7-2 and 7-4, two of the important characterizing parameters of a distribution function were defined, namely: the mean μ and the variance σ^2. In a mechanical system, these are equivalent to the center of gravity and moment of inertia, and they are usually sufficient to define the response characteristics of such a system. A probabilistic system, when subjected to a sampling process, has a behavior which depends upon a more refined knowledge of the distribution function. This suggests using more characterizing parameters for the particular distribution under consideration; because of theoretical considerations discussed in this chapter, these parameters are chosen as the expected values of higher powers of x. We define the rth moment v_r as the expected value of the rth power and thus

$$v_r = E(x^r) = \int_{-\infty}^{\infty} x^r \, dF(x) \qquad (8\text{-}1)$$

which becomes

$$v_r = \int_a^b x^r f(x) \, dx$$

in the case $F(x)$ is continuous and differentiable in $[a, b]$ or

$$v_r = \sum_{i=1}^{N} x_i^r f(x_i)$$

in the case $F(x)$ is a discrete variable.   Note that $v_1 = \mu$.

Since these moments are affected by the chosen origin of the variate x, it is preferable to consider the moments (μ_r) about the center of gravity or mean μ, thus making the characterizing parameters more comparable from one distribution function to another.   These parameters, called *central moments*, are defined as

$$\mu_r = E(x - \mu)^r = \int_{-\infty}^{\infty} (x - \mu)^r \, dF(x) \qquad (8\text{-}2)$$

where $\mu_1 = 0$ and $\mu_2 = \sigma^2$.   Again these may be written as ordinary integrals or summations as in the case of the v_r.   If the factor $(x - \mu)^r$ is expanded by the binomial theorem and the result integrated term by term, μ_r can be expressed in terms of $v_1, v_2, \ldots, v_r$.   For example,

$$\mu_3 = \int_{-\infty}^{\infty} (x - \mu)^3 \, dF(x) = \int_{-\infty}^{\infty} x^3 \, dF(x) - 3\mu \int_{-\infty}^{\infty} x^2 \, dF(x)$$

$$+ \, 3\mu^2 \int_{-\infty}^{\infty} x \, dF(x) - \mu^3 \int_{-\infty}^{\infty} dF(x)$$

$$= v_3 - 3\mu v_2 + 2\mu^3 \qquad (8\text{-}3)$$

When nothing is known about the distribution function except the moments, sometimes a partial characterization can be based on the first four moments.   From the previous discussions (Secs. 7-2 and 7-4), the information given by μ and σ^2 is apparent.   Two additional nondimensional parameters are often useful, namely, $\alpha_1 = \mu_3/\sigma^3$ and $\alpha_2 = \mu_4/\sigma^4$.   The constant α_1 measures the amount of skewness of the distribution, and one can compare this value with that of known curves with given α_1's.   The same is true of μ_4/σ^4 which is called the *kurtosis* of the distribution and has a value of 3 for a gaussian density function; for this reason, kurtosis -3 is termed the *excess*.

(Suggested Problems 8-1 to 8-3)

8-2   CHARACTERIZATION OF A DISTRIBUTION FUNCTION

Let us consider the fundamental question of how we might obtain the equation of any given distribution function in terms of a set of parameters for which estimates can be obtained from a random sample of data drawn from the population in question. In order to simplify the argument, only distribution functions which are continuous will be considered so that their derivatives (or, equivalently, their density functions) will be the subject of this section.

The probability density function can be expressed as a function of the random variable x and a certain set of parameters $\lambda_1, \lambda_2, \ldots, \lambda_N$ which will be defined in such a way that they can be approximated by calculations carried out on a finite sample. Thus, the unknown function may be written as $f(x, \lambda_1, \lambda_2, \ldots, \lambda_N)$, where the λ's will vary, of course, from one density function to another. Our objective is to arrive at a definition of the λ's. As a means to this end, we consider calculations on a hypothetical sample of N observations. Ordinarily, a clear distinction must be made between a population constant, termed a parameter, and an estimate of it, termed a statistic, computed from a sample of observations. For the purpose of defining the λ's, however, the role of the sample is only to provide a vehicle for passing to the limit as $N \to \infty$. Hence, to denote the constants being defined, we shall use the symbols $\lambda_1, \lambda_2, \ldots, \lambda_N$ that apply to the limit as $N \to \infty$.

Suppose that a random sample from a particular population yields observations $x_1, x_2, \ldots, x_N$. The parameters (i.e. the λ's) must be symmetric functions of the N observations, since the sequential order of the x's cannot affect the functional form of the distribution. However, at the most, we can define only N λ's in terms of the N x's. From algebra we know that the basic N symmetric functions of the $x_1, x_2, \ldots, x_N$ upon which all other symmetric functions depend are as follows:

$$S_0 = x_1^{\,0} + x_2^{\,0} + \cdots + x_N^{\,0} = N$$
$$S_1 = x_1^{\,1} + x_2^{\,1} + \cdots + x_N^{\,1}$$
$$S_2 = x_1^{\,2} + x_2^{\,2} + \cdots + x_N^{\,2} \qquad (8\text{-}4)$$
$$\cdots\cdots\cdots\cdots\cdots\cdots\cdots\cdots\cdots$$
$$S_N = x_1^{\,N} + x_2^{\,N} + \cdots + x_N^{\,N}$$

These functions $S_0, S_1, S_2, \ldots, S_N$ are usually referred to as the power sums.

For reasons that will emerge as we proceed, we choose to define the N λ's by approximating an exponential function involving these N λ's by a polynomial in the S's as

$$S_0 \exp\!\left(\lambda_1 y + \frac{\lambda_2 y^2}{2!} + \cdots + \frac{\lambda_N y^N}{N!}\right) \approx S_0 + S_1 y + \frac{S_2 y^2}{2!} + \cdots + \frac{S_N y^N}{N!} \qquad (8\text{-}5)$$

It can be only an approximation, unless $N \to \infty$, since the exponential has an infinite number of terms when it is expanded in powers of y. However, we shall define the N λ's in such a manner that the first N derivatives of the two functions are identical at $y = 0$. Notice that the two functions are already equal at $y = 0$.

The equality of the first N derivatives of the two functions means that the coefficients of equal powers of y up to the Nth are identical on the two sides of Eq. (8-5). Remembering that if $e^u = W$, then $W \dfrac{du}{dy} = \dfrac{dW}{dy}$, we have

$$\left(S_0 + S_1 y + \frac{S_2 y^2}{2!} + \cdots + \frac{S_N y^N}{N!}\right)\left(\lambda_1 + \frac{\lambda_2 y}{1!} + \frac{\lambda_3 y^2}{2!} + \cdots + \frac{\lambda_N y^{N-1}}{(N-1)!}\right) \approx$$

$$S_1 + \frac{S_2 y}{1!} + \frac{S_3 y^2}{2!} + \cdots + \frac{S_N y^{N-1}}{(N-1)!}$$

When we equate the coefficients of equal powers of y, it immediately gives

$$S_0 \lambda_1 = S_1 \qquad S_1 \lambda_1 + S_0 \lambda_2 = S_2$$

$$\frac{S_2 \lambda_3}{2} + S_1 \lambda_2 + \frac{\lambda_1 S_2}{2} = \frac{S_3}{2} \qquad \text{etc.}$$

and therefore

$$\lambda_1 = \frac{S_1}{S_0} \qquad \lambda_2 = \frac{S_2 S_0 - S_1{}^2}{S_0{}^2} \qquad \lambda_3 = \frac{S_3 S_0{}^2 - 3 S_1 S_2 S_0 + 2 S_1{}^3}{S_0{}^3} \qquad \text{etc.}$$

$$(8\text{-}6)$$

which shows that all the λ's are uniquely determined in terms of the N fundamental symmetric functions or power sums. The reason for defining the λ's by means of (8-5) will now become apparent.

If we examine the right-hand side of Eq. (8-5) we see that, for large N, the following approximation for the first N terms is valid as given by

$$S_0 + S_1 y + S_2 \frac{y^2}{2!} + \cdots + \frac{S_{N2} y^N}{N!} = (1 + 1 + \cdots + 1) + (x_1 + x_2 + \cdots + x_N)\frac{y}{1!}$$

$$+ (x_1{}^2 + x_2{}^2 + \cdots + x_N{}^2)\frac{y^2}{2!} + \cdots + (x_1{}^N + \cdots + x_N{}^N)\frac{y^N}{N!}$$

$$\approx e^{x_1 y} + e^{x_2 y} + \cdots + e^{x_N y} = \sum_{j=1}^{N} e^{x_j y}$$

This fact may now be utilized in simplifying Eq. (8-5), and if at the same time y is replaced by iy, where $i = \sqrt{-1}$, we have

$$\exp\left[\lambda_1(iy) + \lambda_2 \frac{(iy)^2}{2!} + \cdots + \lambda_N \frac{(iy)^N}{N!}\right] \approx \frac{1}{N}\sum_{j=1}^{N} e^{ix_j y}$$

since $S_0 = N$. The identity is now in iy instead of y.

If the total number of observations is large, the number of measurements having a particular value x_j may be greater than 1 since the measuring device eventually fails to distinguish values whose differences are very small. Let the number with value x_j be $N(x_j)$ so that $N(x_j)/N = \phi(x_j)$ is the percentage of the observations which do have x_j for a value. Thus Eq. (8-5) now becomes

$$\exp\left[\lambda_1(iy) + \lambda_2 \frac{(iy)^2}{2!} + \cdots + \lambda_N \frac{(iy)^N}{N!}\right] \approx \sum_j \phi(x_j)e^{ix_j}$$

Since our interest centers on the behavior of the summation on the right-hand side as $N \to \infty$, this summation may be replaced by an integral where $\phi(x_j) = f(x)\,dx$ and thus where $f(x)$ is the density of the observations at x. Of course, the usual continuity conditions have to be satisfied in order that the integral exist. The exponential involving the λ's now has an infinite number of terms.

Thus we have finally under certain general conditions,

$$\exp\left[\lambda_1(iy) + \lambda_2 \frac{(iy)^2}{2!} + \cdots\right] = \int_{-\infty}^{\infty} f(x)e^{ixy}\,dx \qquad (8\text{-}7)$$

When $y = 0$, we have $\int_{-\infty}^{\infty} f(x)\,dx = 1$ which is certainly a condition that the density function must satisfy.

From the definition of the S_i, one observes that as $N \to \infty$, the λ's become functions of the moments v_i of the particular density function so that from (8-6) we have

$$\lambda_1 = v_1 = \mu \qquad \lambda_2 = v_2 - v_1{}^2 = \sigma^2 \qquad \lambda_3 = v_3 - 3v_1 v_2 + 2v_1{}^3 \qquad \text{etc.} $$
$$(8\text{-}8)$$

The λ's are usually referred to as the "cumulants" and sometimes also as "semi-invariants."

8-3 CHARACTERISTIC FUNCTIONS

The right-hand side of Eq. (8-7) is the definition of the Fourier transform of $f(x)$ or, equivalently, the characteristic function of $f(x)$ so that the left side of Eq. (8-1) is necessarily the value of this transform. The reader is referred to a book on advanced calculus where the following Fourier theorem, which relates the transform of $f(x)$ to $f(x)$, is developed as follows: If

$$g(y) = \int_{-\infty}^{\infty} e^{ixy}f(x)\,dx$$

then
$$f(x) = \frac{1}{2\pi}\int_{-\infty}^{\infty} e^{-ixy}g(y)\,dy \qquad (8\text{-}9)$$

for such functions $g(y)$ and $f(x)$ for which the integrals exist.

Since Eq. (8-7) is of the same form as the first part of Eq. (8-9), we may now solve for $f(x)$ explicitly as

$$f(x) = \frac{1}{2\pi} \int_{-\infty}^{\infty} e^{-ixy} \exp\left[\lambda_1(iy) + \frac{\lambda_2(iy)^2}{2!} + \frac{\lambda_3(iy)^3}{3!} + \cdots\right] dy \qquad (8\text{-}10)$$

Thus, if the λ's are all known in Eq. (8-10) the $f(x)$ is uniquely determined by the evaluation of this one integral, at least theoretically. The importance of this conclusion involves the concept that if all the moments of a particular distribution function are known and thus all exist, these moments uniquely determine the density function itself. Any density function having a finite domain obviously has the property that all the moments do exist. In the case of an infinite domain, the moments $[\int_{-\infty}^{\infty} x^n(f(x) \, dx$, for any $n]$ do exist provided $x^n f(x) \to 0$, as $x \to \pm\infty$, rapidly enough so that the integral is finite.

If we set all the λ's except the first one, λ_1, equal to zero, we have as a solution of Eq. (8-10) only a point $x = \lambda_1$ which occurs with probability 1. However, we arrive at a very interesting case if we set all the λ's except the first two equal to zero or $\lambda_r = 0$ $(r > 2)$. Then Eq. (8-10) becomes

$$f(x) = \frac{1}{2\pi} \int_{-\infty}^{\infty} e^{-ixy} e^{i\lambda_1 y + \lambda_2(iy)^2/2} \, dy$$

Since

$$e^{i(\lambda_1 - x)y} = \cos(\lambda_1 - x)y + i\sin(\lambda_1 - x)y$$

we may break the integral for $f(x)$ into two parts. The part involving the sine integral is equal to zero since it is the integral of an odd function over the entire domain, and thus Eq. (8-10) in this case is

$$f(x) = \frac{1}{2\pi} \int_{-\infty}^{\infty} e^{-\lambda_2 y^2/2} \cos[(\lambda_1 - x)y] \, dy \qquad (8\text{-}11)$$

The integral represented by Eq. (8-11) is a common one and is found in most integral tables under the form

$$\int_{0}^{\infty} e^{-ax^2} \cos bx \, dx$$

which gives the value of (8-11) as

$$f(x) = \frac{1}{\sqrt{2\pi\lambda_2}} \exp \frac{-(x - \lambda_1)^2}{2\lambda_2} \qquad (8\text{-}12)$$

which is a normal curve (gaussian) with $\sigma = \sqrt{\lambda_2}$ and $\mu = \lambda_1$. This shows that the only continuous curve which is determined by the first two cumulants is the normal curve.

We may generalize the definition of characteristic function to the case where $F(x)$ is not differentiable as

$$g(y) = \int_{-\infty}^{\infty} e^{ixy} \, dF(x)$$

which again simplifies in the case where x is a discrete variable to the summation

$$g(y) = \sum_{j=1}^{N} e^{ix_j y} f(x_j)$$

Any $g(y)$ can, of course, be continuous, even if x itself is a discrete variable. It usually is continuous but nevertheless $g(y)$ is not always differentiable everywhere, particularly at the origin.

The characteristic function or Fourier transform of the Poisson distribution is by definition

$$g(y) = \sum_{x=0}^{\infty} e^{ixy} \frac{e^{-\mu}\mu^x}{x!} = \sum_{x=0}^{\infty} \frac{e^{-\mu}(\mu e^{iy})^x}{x!}$$

$$= e^{\mu[(iy)/1! + (iy)^2/2! + \cdots]} \tag{8-13}$$

If we compare this result with the left-hand side of Eq. (8-7), we see that the two are identical if $\lambda_1 = \lambda_2 = \lambda_3 \cdots, = \lambda_N = \cdots = \mu$ in Eq. (8-7) or equivalently the Poisson distribution has the unique property that all the cumulants are the same and equal to the mean.

Actually the characteristic function always exists even if the moments of the distribution function do not exist. If one examines the integral $\int_{-\infty}^{\infty} e^{ixy} \, dF(x)$, it is apparent that $\int_{-\infty}^{\infty} dF(x)$ is finite and thus exists, and that multiplying it by e^{ixy} does not change this same convergence property since $|\cos xy + i \sin xy| = \sqrt{\cos^2 xy + \sin^2 xy} = 1$. Theoretically, of course, if the characteristic function $g(y)$ is known, the distribution function is also known by the inversion process as

$$\frac{dF(x)}{dx} = \frac{1}{2\pi} \int_{-\infty}^{\infty} e^{-ixy} g(y) \, dy$$

Thus, $g(y)$ actually characterized our function $F(x)$ so that two distribution functions having the same characteristic function are the same.

EXAMPLE 8-1   Find the characteristic function of the Cauchy distribution

$$f(x) = \frac{1}{\pi} \frac{1}{1 + x^2} \qquad (-\infty, \infty)$$

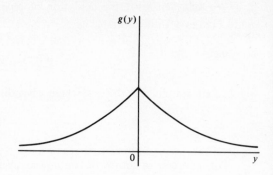

FIGURE 8-1

By definition

$$g(y) = \frac{1}{\pi} \int_{-\infty}^{\infty} \frac{e^{ixy} \, dx}{1 + x^2} = \frac{1}{\pi} \int_{-\infty}^{\infty} \frac{\cos xy \, dx}{1 + x^2}$$

$$= e^{-|y|} \quad -\infty < y < \infty$$

The Cauchy curve has the property that neither the mean nor the standard deviation exists; in fact, none of the moments defined as

$$\frac{1}{\pi} \int_{-\infty}^{\infty} \frac{x^r \, dx}{1 + x^2}$$

are finite for $r > 0$. This is now evident from the characteristic function that has a graph as shown in Fig. 8-1. The $g(y)$ does not have a unique derivative at the origin so that we cannot expand $g(y)$ as a power series in y around that point, which is the basic assumption of Eq. (8-7).

However, if we were given $g(y) = e^{-|y|}$, then

$$f(x) = \frac{1}{2\pi} \left(\int_{-\infty}^{0} e^{+yx} e^{-ixy} \, dy + \int_{0}^{\infty} e^{-yx} e^{-ixy} \, dy \right)$$

$$= \frac{1}{\pi} \int_{0}^{\infty} e^{-yx} \cos xy \, dy = \frac{1}{\pi} \frac{1}{1 + x^2} \qquad ////$$

8-4   MOMENT-GENERATING FUNCTIONS

The information concerning the moments of any distribution function taken about a chosen origin is all contained in a single function, as one might surmise from the discussion of the characteristic functions.

The expected value of $e^{\theta x}$ is defined as the moment-generating function and has the property that, if this expected value is differentiated r times with respect to θ and this derivative evaluated at $\theta = 0$, the value so obtained is the rth moment about the origin. If $M(\theta : x)$ is defined as the moment-generating function of the random variable X, then

$$M(\theta : x) = \int_{-\infty}^{\infty} e^{\theta x}\, dF(x) \quad \text{and} \quad \left.\frac{d^r M(\theta : x)}{d\theta^r}\right|_{\theta=0} = \int_{-\infty}^{\infty} x^r e^{\theta x}\, dF(x)\bigg|_{\theta=0} = v_r$$

$$(8\text{-}14)$$

In the case of a continuous and discrete random variable, we have, respectively,

$$M(\theta : x) = \int_{a}^{b} e^{\theta x} f(x)\, dx \quad \text{and} \quad M(\theta : x) = \sum_{j=1}^{N} e^{\theta x_j} f(x_j) \qquad (8\text{-}15)$$

For the integral representing $M(\theta : x)$ [Eq. (8-14)] to exist and give the moments of the distribution function upon differentiation, it is necessary that $\int_{-\infty}^{\infty} e^{\theta x}\, dF(x)$ converge for all values of θ in the neighborhood of the origin (that is, $|\theta| < \delta$) so that all the derivatives at $\theta = 0$ exist also.

From the previous discussion on cumulants, it is clear that any two distribution functions having the same moment-generating function must be equal. Furthermore, if the moment-generating function of any unknown distribution can be found by any means and if we can identify this generating function as one that we have observed in past experience, we obviously now know that unknown distribution function. This uniqueness property permits the solution of many problems arising in the applications of probability theory which otherwise would be intractable.

If we again consider Eq. (8-7) and retrace our steps by replacing iy by θ on both sides of the equality, we have

$$e^{\lambda_1 \theta/1! + \lambda_2 \theta^2/2! + \lambda_3 \theta^3/3! + \cdots} = \int_{-\infty}^{\infty} e^{\theta x} f(x)\, dx = M(\theta : x) \qquad (8\text{-}16)$$

In this case we have assumed all the moments exist so that an expansion around $\theta = 0$ on the left is legitimate for values of $|\theta| < \delta$.

By taking the natural logarithm of both sides of (8-16) we observe that

$$\lambda_1 \frac{\theta}{1!} + \lambda_2\left(\frac{\theta^2}{2!}\right) + \lambda_3\left(\frac{\theta^3}{3!}\right) + \cdots = \ln M(\theta : x) \qquad (8\text{-}17)$$

Thus, if all the moments of a distribution function exist and if the logarithm of the moment-generating function is computed, this logarithmic function may be expanded in powers of θ in an ordinary Taylor's series. The coefficient of $\theta^k/k!$ is now the cumulant λ_k. The first few cumulants are as follows:

$$\lambda_1 = \mu \qquad \lambda_2 = \sigma^2 \qquad \lambda_3 = \mu_3 \qquad \lambda_4 = \mu_4 - 3\sigma^4$$

EXAMPLE 8-2   The moment-generating function of the gamma variate x (parameters α, β) is

$$M(\theta : x) = \int_0^\infty e^{\theta x} f(x) \, dx = \int_0^\infty \frac{e^{\theta x} x^\alpha e^{-x/\beta}}{\beta^{(\alpha+1)} \Gamma(\alpha + 1)} \, dx$$

$$= \frac{1}{\beta^{\alpha+1} \Gamma(\alpha + 1)} \int_0^\infty x^\alpha e^{-(1-\theta\beta)x/\beta} \, dx = \frac{1}{(1 - \theta\beta)^{\alpha+1}}$$

The series expansion of this function exists for every value of θ in the interval $(-h, h)$ provided that $h < 1/\beta$; hence the moment-generating function uniquely determines the distribution.   ////

EXAMPLE 8-3   Given that $M(\theta : x) = (1 - \theta\beta)^{-(\alpha+1)}$ find the distribution of x. Since the moment-generating function uniquely determines the distribution, this must be a gamma distribution with parameters (α, β). Hence

$$f(x) = \frac{1}{\beta^{\alpha+1} \Gamma(\alpha + 1)} x^\alpha e^{-x/\beta} \qquad (0, \infty) \qquad ////$$

EXAMPLE 8-4   From the moment-generating function find $E(x)$, where

$$f(x) = (1 - \alpha)\alpha^x \qquad x = 0, 1, 2, \ldots, \infty; 0 < \alpha < 1$$

The moment-generating function of this geometric distribution is

$$M(\theta : x) = \sum_{x=0}^\infty e^{\theta x} f(x) = (1 - \alpha) \sum_{x=0}^\infty (\alpha e^\theta)^x = \frac{1 - \alpha}{1 - \alpha e^\theta}$$

From this function, we obtain $E(x)$ by differentiation:

$$E(x) = \frac{dM(\theta : x)}{d\theta}\bigg|_{\theta=0} = \frac{(1 - \alpha)\alpha e^\theta}{(1 - \alpha e^\theta)^2}\bigg|_0 = \frac{\alpha}{1 - \alpha}$$

This result checks with Example 7-3.   ////

EXAMPLE 8-5 Given that $M(\theta : x) = (1 - \alpha)/(1 - \alpha e^{\theta})$ and $x = 0, 1, 2, \ldots, \infty$, find $f(x)$. This could be done by direct comparison with Example 8-4 but the information that x is confined to integral values is sufficient to determine $f(x)$ from $M(\theta : x)$ without reference to the previous result. Expanding $M(\theta : x)$ in a series, we have

$$M(\theta : x) = (1 - \alpha) \sum_{x=0}^{\infty} \alpha^x e^{x\theta}$$

But by definition

$$M(\theta : x) = \sum_{x=0}^{\infty} f(x) e^{x\theta}$$

Therefore, equating general terms,

$$f(x) = (1 - \alpha)\alpha^x \qquad x = 0, 1, 2, \ldots, \infty \qquad\qquad ////$$

EXAMPLE 8-6 Find the moment-generating function of the binomial, and also find the first three cumulants. From the definition,

$$M(\theta : x) = E(e^{\theta x}) = \sum_{0}^{n} e^{\theta x} C(n, x) p^x q^{n-x} = \sum_{0}^{n} C(n, x)(pe^{\theta})^x q^{n-x}$$

$$= (q + pe^{\theta})^n \tag{8-18}$$

Taking the logarithm of this moment-generating function in order to obtain the cumulants, we have

$$\ln M(\theta : x) = n \ln(q + pe^{\theta})$$

$$= n \ln\left(1 + p\theta + p\frac{\theta^2}{2} + p\frac{\theta^3}{6} + \cdots\right)$$

The expansion for the $\ln(1 + x)$ around $x = 0$ is

$$\ln(1 + x) = x - \frac{x^2}{2} + \frac{x^3}{3} - \frac{x^4}{4} + \cdots \quad |x| < 1$$

and in this case

$$x = p\left(\theta + \frac{\theta^2}{2} + \frac{\theta^3}{6} + \cdots\right)$$

where θ can be made as small as necessary to guarantee convergence. Hence

$$\ln M(\theta : x) = n\left[p\left(\theta + \frac{\theta^2}{2} + \frac{\theta^3}{6} + \cdots\right) - \frac{p^2}{2}\left(\theta + \frac{\theta^2}{2} + \frac{\theta^3}{6} + \cdots\right)^2 \right.$$
$$\left. + \frac{p^3}{3}\left(\theta + \frac{\theta^2}{2} + \frac{\theta^3}{6} + \cdots\right)^3 + \cdots\right]$$
$$= n\left[p\left(\frac{\theta}{1!}\right) + (p - p^2)\left(\frac{\theta^2}{2!}\right) + (p + 2p^3 - 3p^2)\frac{\theta^3}{3!} + \cdots\right]$$

thus $\lambda_1 = np$, $\lambda_2 = npq$, $\lambda_3 = np(1 - 3p + 2p^2)$.   ////

EXAMPLE 8-7 Find the moment-generating function (MGF), the mean, and the variance of the discrete distribution $f(-1) = \frac{1}{8}$, $f(0) = \frac{1}{4}$, $f(2) = \frac{3}{8}$, and $f(5) = \frac{1}{4}$. From the definition $M(\theta : x) = \frac{1}{8}e^{-\theta} + \frac{1}{4} + \frac{3}{8}e^{2\theta} + \frac{1}{4}e^{5\theta}$ and

$$\frac{dM(\theta : x)}{d\theta} = -\frac{1}{8}e^{-\theta} + \frac{6}{8}e^{2\theta} + \frac{5}{4}e^{5\theta} \quad \text{and}$$

$$\frac{d^2 M(\theta : x)}{d\theta^2} = \frac{1}{8}e^{-\theta} + \frac{12}{8}e^{2\theta} + \frac{25}{4}e^{5\theta}$$

whence $\qquad\qquad\qquad v_1 = \mu = \frac{15}{8} \quad \text{and} \quad v_2 = \frac{63}{8}$

so that $\qquad \sigma^2 = \mu_2 = \frac{63}{8} - \left(\frac{15}{8}\right)^2 = \frac{279}{64} \approx 4.4 \quad \text{and} \quad \sigma \approx 2.1$ ////

EXAMPLE 8-8 Find the MGF of (a) the Poisson distribution and (b) the normal distribution.

From the definition

(a) $\qquad M(\theta : x) = \sum_0^\infty \frac{e^{\theta x}e^{-\lambda}\lambda^x}{x!} = e^{-\lambda}\sum_0^\infty \frac{(e^\theta \lambda)^x}{x!}$

Since the summation is just the expansion of the exponential, we have

$$M(\theta : x) = e^{-\lambda}e^{\lambda e^\theta} \qquad (8\text{-}19)$$

It is again interesting to note that the cumulants are all equal as

$$\ln M(\theta : x) = -\lambda + \lambda e^\theta = \lambda\left(\frac{\theta}{1!} + \frac{\theta^2}{2!} + \cdots\right) = \lambda\left(\frac{\theta}{1!}\right) + \lambda\left(\frac{\theta^2}{2!}\right) + \lambda\left(\frac{\theta^3}{3!}\right) + \cdots$$

(b) $\qquad M(\theta : x) = \int_{-\infty}^\infty e^{\theta x} \frac{1}{\sigma\sqrt{2\pi}} e^{-(x-\mu)^2/2\sigma^2}\, dx$

Since

$$\frac{(x-\mu)^2}{2\sigma^2} - \theta x = \frac{x^2 - 2x(\mu + \sigma^2\theta) + (\mu + \sigma^2\theta)^2 - (\mu + \sigma^2\theta)^2 + \mu^2}{2\sigma^2}$$

$$= \frac{[x - (\mu + \sigma^2\theta)]^2 - 2\mu\sigma^2\theta - \sigma^4\theta^2}{2\sigma^2}$$

the integral becomes

$$M(\theta : x) = e^{+\mu\theta}e^{+\theta^2\sigma^2/2}\frac{1}{\sqrt{2\pi}}\int_{-\infty}^{\infty} e^{-y^2/2}\,dy = e^{+\mu\theta}e^{+\theta^2\sigma^2/2} \qquad (8\text{-}20)$$

with the substitution

$$y = \frac{x - (\mu + \sigma^2\theta)}{\sigma} \qquad ////$$

EXAMPLE 8-9 An observation x_1 is drawn at random from a population with mean μ, standard deviation σ, skewness α_3, and kurtosis α_4. A second observation x_2 is now obtained through the relationship $x_2 = kx_1 + y_1$, where y_1 is a random variable independent of x_1 and where k is positive and a constant. The population of x_2's so generated is identical to that of the x_1's. It is thus apparent that x_1 and x_2 are correlated with $\rho = k$. What are the four characterizing parameters of the y's, namely: μ_y, σ_y, α_{3y}, and α_{4y}? On the basis of the hypothesis of the equality of the distributions of the x_1's and x_2's, we have

$$E(x_2) = E(kx_1 + y_1) \qquad \text{or} \qquad \mu = k\mu + E(y_1) \qquad \text{and} \qquad E(y_1) = (1-k)\mu$$

so that, in general,

$$E(x_2 - \mu)^n = E[k(x_1 - \mu) + (y_1 - (1-k)\mu)]^n$$

Since the expected value of the cross-product term on the right-hand side of the equation is zero, when $n = 2$, we have

$$\sigma^2 = k^2\sigma^2 + \sigma_y^2 \qquad \text{and} \qquad \sigma_y^2 = (1 - k^2)\sigma^2$$

When $n = 3$, since again the cross-product terms are zero, there results

$$\mu_3 = k^3\mu_3 + \mu_{3y} \qquad \text{and} \qquad \mu_{3y} = (1 - k^3)\mu_3$$

and

$$\alpha_{3y} = \frac{\mu_{3y}}{\sigma_y^{3/2}} = \frac{(1 - k^3)\mu_3}{(1 - k^2)^{3/2}\sigma^{3/2}}$$

$$= \frac{(1 - k^3)\alpha_3}{(1 - k^2)^{3/2}}$$

where $k < 1$. When $n = 4$, since one of the cross-product terms is now different from zero, we have

$$\mu_4 = k^4\mu_4 + 6k^2\sigma^2(1 - k^2)\sigma^2 + \mu_{4y}$$

and

$$\mu_{4y} = (1 - k^4)\mu_4 - 6k^2(1 - k^2)\sigma^4$$

so that

$$\alpha_{4y} = \frac{\mu_{4y}}{\sigma_y^4} = \frac{(1 - k^4)\sigma^4\alpha_4 - 6k^2(1 - k^2)\sigma^4}{(1 - k^2)^2\sigma^4}$$

$$= \frac{(1 + k^2)\alpha_4 - 6k^2}{1 - k^2} \qquad k < 1$$

Obviously, $\alpha_{3y} = \alpha_3$ and $\alpha_{4y} = \alpha_4$ when $k = 0$, and both these parameters become very large as $k \to 1$. ////

EXAMPLE 8-10   Given the n observations $x_1, x_2, \ldots, x_n$ developed as follows: $x_1, x_2 = kx_1 + y_1. x_3 = kx_2 + y_2, \ldots, x_i = kx_{i-1} + y_{i-1}, \ldots, x_n = kx_{n-1} + y_{n-1}$, where the y's are all independent observations from the same population and also x_{i-1} and y_{i-1} are independent. Here again, as in Example 8-9, the process is stationary in that all the x's belong to the same population which in this case we shall assume has $\mu = 0$ and standard deviation σ. Therefore, from the previous example, $\mu_y = 0$ and $\sigma_y^2 = (1 - k^2)\sigma^2$. What is the standard deviation $\sigma_{\bar{x}}$ of the mean $\bar{x}$ of a sample of n observations developed by means of the above procedure?

Each x_i can be written as

$$x_i = k^{i-1}x_1 + k^{i-2}y_1 + \cdots + ky_{i-2} + y_{i-1}$$

so that $\bar{x}$ can be expressed by

$$\bar{x} = \frac{1}{n}(x_1 + x_2 + \cdots + x_n)$$

$$= \frac{1}{n}[(1 + k + \cdots + k^{n-1})x_1 + (1 + k + \cdots + k^{n-2})y_1 + \cdots$$

$$+ (1 + k)y_{n-2} + y_{n-1}]$$

$$= \frac{1}{(n)(1 - k)}[(1 - k^n)x_1 + (1 - k^{n-1})y_1 + \cdots + (1 - k^2)y_{n-2}$$

$$+ (1 - k)y_{n-1}]$$

and then $E(\bar{x}) = 0$ since $E(x_1) = E(y_i) = 0$.   Thus

$$E(\bar{x})^2 = \sigma_{\bar{x}}^2 = \frac{1}{n^2(1-k)^2} E[(1-k^n)^2 x_1^2 + (1-k^{n-1})^2 y_1^2 + \cdots$$
$$+ (1-k)^2 y_{n-1}^2]$$

since the $E(x_1 y_i) = E(y_i y_j) = 0$ and, therefore,

$$\sigma_{\bar{x}}^2 = \frac{\sigma^2}{n^2(1-k)^2} [n(1-k^2) - 2k(1-k^n)] \qquad k < 1$$

by substituting $E(x_1^2) = \sigma^2$ and $E(y_i^2) = (1-k^2)\sigma^2$ and summing the geometric series.   If n is even moderately large,

$$\sigma_{\bar{x}} \approx \frac{\sigma}{\sqrt{n}} \sqrt{\frac{1+k}{1-k}}$$

which compares with $\sigma/\sqrt{n}$ for an uncorrelated series of x's.   ////

EXAMPLE 8-11   A communication system consists of N terminal outlets all divided into k subgroups of m outlets each ($N = km$).   Each subgroup of m outlets is connected to one control unit.   Each one of the N units has an instantaneous probability of p_1 of operating when called upon to function. This probability is independent of the state of any of the other outlets or control units.   Each control unit has a probability of p_2 of operating upon demand; this probability is also independent of the state of all the other equipment. However, if a control unit is not functioning, whether or not any of the m outlets connected to it could or could not operate is unimportant since all m units are now unusable.   Find the moment-generating function of the distribution of the number of terminal units that would be usable at a random time.

Let the probability that there are x control units operating be $P(x)$, the probability that there are y usable terminals operating be $P(y)$, their joint occurrence be $P(x, y)$, and the conditional probability that y units are usable, given that x control units are intact, be $P(y|x)$.   Then

$$P(x, y) = P(x)P(y|x) = C(k, x)p_1^x q_1^{k-x} C(mx, y)p_2^y q_2^{mx-y}$$

where the domain is $y \leq mx, 0 \leq x \leq k$.   Thus

$$P(y) = \sum_{\text{over } x} C(k, x)p_1^x q_1^{k-x} C(mx, y)p_2^y q_2^{mx-y}$$

Because of the domain of definition, the summation over x will have different limits, depending upon the value of y which is held fixed during the summation, and $P(y)$ is thus not a simple probability distribution. If, however, we define $C(a, b) = 0$ for $b > a$, then

$$P(y) = \sum_{x=0}^{x=k} C(k, x)p_1{}^x q_1{}^{k-x} C(mx, y)p_2{}^y q_2{}^{mx-y}$$

since all the terms will be zero except the correct ones in the summation.

The moment-generating function of y is then given by

$$M(\theta : y) = \sum_{y=0}^{km} e^{\theta y} P(y) = \sum_{y=0}^{N} e^{\theta y} \sum_{x=0}^{x=k} C(k, x)p_1{}^x q_1{}^{k-x} C(mx, y)p_2{}^y q_2{}^{mx-y}$$

Interchanging the order of summation, and observing that $C(mx, y) = 0$ for $y > mx$, we have

$$M(\theta : y) = \sum_{x=0}^{x=k} C(k, x)p_1{}^x q_1{}^{k-x} \sum_{y=0}^{mx} C(mx, y)(p_2 e^{\theta})^y q_2{}^{mx-y}$$

Since the last summation is a binomial expansion, this gives

$$\sum_{x=0}^{x=k} C(k, x)p_1{}^x q_1{}^{k-x}(q_2 + p_2 e^{\theta})^{mx} = \sum_{x=0}^{x=k} C(k, x)[p_1(q_2 + p_2 e^{\theta})^m]^x q_1{}^{k-x}$$

and hence also

$$M(\theta : y) = [q_1 + p_1(q_2 + p_2 e^{\theta})^m]^k$$

Thus

$$E(y) = \left. \frac{dM(\theta : y)}{d\theta} \right|_{\theta=0} = kmp_1 p_2 = Np_1 p_2$$

and

$$\sigma_y^2 = v_2 - v_1{}^2 = \left. \frac{d^2 M(\theta : y)}{d\theta^2} \right|_{\theta=0} - k^2 m^2 p_1{}^2 p_2{}^2$$

$$= Np_1 p_2[mp_2(1 - p_1) + (1 - p_2)]$$

The mean value of the distribution of y is, therefore, $Np_1 p_2$ and is independent of the number of control units used.

The variance definitely depends on m and is smallest when $m = 1$ or, equivalently, when there are N control units. In this case, it is binomial where the variance is Npq with $p = p_1 p_2$ [$q = (1 - p_1 p_2)$]. The largest variance occurs when $m = N$, or there is only one control unit. If p_2 is large, the ratio of the largest to the smallest variance is about N, which is certainly important in design consideration. Higher moments can, of course, be obtained by computing higher derivatives, if more information about the shape of the distribution is desired in the form of more central moments.   ////

A major application of the moment-generating function lies in the convenient treatment of derived variates. If u is a continuous function of x, the moment-generating function of u can be found from the distribution of x without first deriving the distribution of u proper.   For example, if the distribution of x is given by $f(x)$ $(A \leq x \leq B)$, then

$$M(\theta : u) \equiv E(e^{\theta u}) = \int_A^B e^{\theta u} f(x)\, dx \qquad (8\text{-}21)$$

Once $M(\theta : u)$ is known, it is often possible to identify the distribution of u by an appeal to the uniqueness property.

EXAMPLE 8-12   Given $f(x) = 1$ $(0, 1)$ determine the distribution of $u = -2 \ln x$ from its moment-generating function.   Here $\theta u = \ln x^{-2\theta}$ and $e^{\theta u} = x^{-2\theta}$, whence

$$M(\theta : u) = \int_0^1 x^{-2\theta}\, dx = \frac{1}{1 - 2\theta}$$

Referring to Example 8-2 we recognize this as a special case of the moment-generating function of the gamma distribution with $\alpha = 0$, $\beta = 2$.   Hence u is distributed as chi-square with 2 degrees of freedom.   ////

From Eq. (8-21) we deduce that the effect of a linear transformation $u = a + bx$ $(b \neq 0)$ is to replace θ by $b\theta$ and multiply the resulting function by $e^{a\theta}$.   For, in the case considered,

$$M(\theta : a + bx) = \int_A^B e^{\theta(a + bx)} f(x)\, dx = e^{a\theta} \int_A^B e^{b\theta x} f(x)\, dx = e^{a\theta} M(b\theta : x) \qquad (8\text{-}22)$$

This same result holds in general.   An interesting special case of Eq. (8-22) is obtained by setting $a = -\mu$, $b = 1$, where $\mu = E(x)$.   Since u then becomes $x - \mu$, this transformation yields the generating function of the so-called *central* moments, i.e., moments taken about the mean as opposed to *raw* moments, which are the regular ones taken about the origin.   The result is

$$M(\theta : x - \mu) = e^{-\mu\theta} M(\theta : x) \qquad (8\text{-}23)$$

From the fact that $M(\theta : a + bx) = e^{a\theta} M(b\theta : x)$, it is readily shown that if $u = a + bx$ $(b \neq 0)$ and λ_r, λ_r' denote the respective rth cumulants of x and u, then

$$\lambda_1' = a + \lambda_1 \qquad \lambda_r' = b^r \lambda_r \qquad r > 1 \qquad (8\text{-}24)$$

On account of this simple behavior under a linear transformation of variables, the cumulants are also known as *semi-invariants*.

(*Suggested Problems 8-4 to 8-6*)

8-5   MULTIVARIATE MOMENT-GENERATING FUNCTIONS

A joint multivariate moment-generating function, $M(\theta_1, \theta_2, \ldots, \theta_n: x_1, x_2, \ldots, x_n)$ may be defined as the expected value of the exponential function $\exp(\theta_1 x_1 + \theta_2 x_2 + \cdots + \theta_n x_n)$. Joint functions of this type are treated in advanced works on mathematical statistics, but our exposition of the multivariate situation will be limited to the simple moment-generating function of a composite variate involving two or more random variables. If u is a continuous function of several variates $x_1, x_2, \ldots, x_n$, its moment-generating function $M(\theta : u)$, being the expected value of $e^{\theta u}$, can be found from the joint distribution of the x's as a special case of Eq. (7-22). Thus, if the admissible region is R_x,

$$M(\theta : u) = \iint \cdots \int_{Rx} e^{\theta u} \, dF(x_1, x_2, \ldots, x_n) \qquad (8\text{-}25)$$

In particular, if a joint density function exists over R_x, then

$$M(\theta : u) = \iint \cdots \int_{Rx} e^{\theta u} f(x_1, x_2, \ldots, x_n) \, dx_1 \, dx_2 \cdots dx_n \qquad (8\text{-}26)$$

EXAMPLE 8-13   By the method of moment-generating functions, find the distribution of $u = xy$ where

$$f(x, y) = xe^{-x(1+y)} \qquad (0, \infty \text{ each})$$

Here the moment-generating function is

$$M(\theta : u) = \int_0^\infty \int_0^\infty e^{\theta xy} xe^{-x-xy} \, dx \, dy = \int_0^\infty e^{-x}\left[\int_0^\infty xe^{-xy(1-\theta)} \, dy\right] dx$$

$$= \int_0^\infty \frac{e^{-x}}{1-\theta} \, dx = \frac{1}{1-\theta} \int_0^\infty e^{-x} \, dx = (1-\theta)^{-1}$$

This is a special case of the function $(1 - \beta\theta)^{-(\alpha+1)}$ with $\alpha = 0$, $\beta = 1$. Hence by Example 8-2 $f(u) = e^{-u}$ $(0, \infty)$ which agrees with previous results.   ////

By direct application of Eq. (8-25) we establish two theorems of considerable importance. The first concerns the sum of independent variates, and the second, the mean values derived from random samples of size n from the same infinite population.

Theorem 8-1   The moment-generating function of the sum of independent variates is equal to the product of their respective moment generating functions.

PROOF Let $u = x_1 + x_2 + \cdots + x_n$, where the x's are independent. If $F_i(x_i)$ represents the marginal-distribution function of x_i and R_i its range of definition, the product $e^{\theta u}\, dF(x_1, x_2, \ldots, x_n)$ factors, yielding

$$M(\theta : u) = \int_{R_1} e^{\theta x_1}\, dF_1(x_1) \int_{R_2} e^{\theta x_2}\, dF_2(x_2) \cdots \int_{R_n} e^{\theta x_n}\, dF_n(x_n)$$

$$= M(\theta : x_1)M(\theta : x_2) \cdots M(\theta : x_n) \qquad \qquad ////$$

Theorem 8-2 If $\bar{x}$ is the mean of a random sample of n observations from a population for which the moment-generating function is $M(\theta : x)$, the moment-generating function of $\bar{x}$ in repeated independent trials is

$$M(\theta : \bar{x}) = \left[M\left(\frac{\theta}{n} : x \right) \right]^n$$

PROOF Denoting the independent observations by $x_1, x_2, \ldots, x_n$ we can regard the x's as independent variates having the same distribution, whence, each has the same moment-generating function, $M(\theta : x)$. Putting $u = x_1 + x_2 + \cdots + x_n$, it follows from Theorem 8-1 that

$$M(\theta : u) = [M(\theta : x)]^n$$

Now $\bar{x} = u/n$ and from Eq. (8-22)

$$M(\theta : \bar{x}) = M\left(\frac{\theta}{n} : u \right) = \left[M\left(\frac{\theta}{n} : x \right) \right]^n \qquad \qquad ////$$

EXAMPLE 8-14 The length of time t required to accumulate a fixed number of counts c of radiational particles impinging upon the counting meter at an average rate r per unit time is distributed as follows: $f(t) = [r^c/\Gamma(c)]t^{c-1}e^{-rt}$ $(0, \infty; c \geq 1)$. If n independent tests are made under similar conditions and the average waiting time τ is computed, find the distribution of τ from its moment-generating function. Since t is a gamma variate with $\alpha = c - 1$, $\beta = 1/r$, its moment-generating function by Example 8-2 is $M(\theta : t) = (1 - \theta/r)^{-c}$. Therefore, by Theorem 8-2,

$$M(\theta : \tau) = (1 - \theta/nr)^{-nc}$$

This is the moment-gathering function of a gamma variate with $\alpha = nc - 1$, $\beta = 1/nr$. Therefore, the distribution of τ is

$$g(\tau) = \frac{(nr)^{nc}}{\Gamma(nc)}\, \tau^{nc-1}e^{-nr\tau} \qquad (0, \infty)$$

This result may be chcked by applying Theorem 6-2. $\qquad \qquad ////$

EXAMPLE 8-15 A linear combination is made of two independent random variables x_1 and x_2 which are distributed "binomially" with parameters (n_1, p_1) and (n_2, p_2), respectively. What are the conditions such that this linear combination is "binomial" also? What would be the conditions if x_1 and x_2 were Poisson-distributed with parameters λ_1 and λ_2, respectively?

Let the linear combination be $z = a_1 x_1 + a_2 x_2$. Since the MGF of a binomial variate is $M_x(\theta) = (q + pe^\theta)^n$, the MGF of z is

$$M_z(\theta) = (q_1 + p_1 e^{a_1 \theta})^{n_1} (q_2 + p_2 e^{a_2 \theta})^{n_2}$$

Thus the MGF will be that of a binomial variate only if $a_1 = a_2$ and $p_1 = p_2$. In that situation, we have

$$M_z(\theta) = (q_1 + p_1 e^{a_1 \theta})^{n_1 + n_2}$$

which is formally that of the binomial with parameters $(p_1, n_1 + n_2)$. However, in order that z have the proper range of a binomial variate, namely, $z = 0, 1, 2, 3,$ $\ldots, n_1 + n_2$, it is further necessary that $a_1 = 1$, whence $z = x_1 + x_2$.

In the second case where the random variables are "Poisson," the MGF of z is

$$M_z(\theta) = \exp[-\lambda_1]\exp[\lambda_1 e^{a_1 \theta}]\exp[-\lambda_2]\exp[\lambda_2 e^{a_2 \theta}]$$

since the MGF of a Poisson variate is $e^{-\lambda}e^{\lambda e^\theta}$. The function $M_z(\theta)$ will agree with the MGF of a Poisson variate only if $a_1 = a_2$ so that

$$M_z(\theta) = \exp[-(\lambda_1 + \lambda_2)]\exp[(\lambda_1 + \lambda_2)e^{a_1 \theta}]$$

where the parameter is now $\lambda_1 + \lambda_2$. Again, however, it is necessary that $a_1 = 1$ in order that z have the required range $0, 1, 2, \ldots$; thus, as before, $z = x_1 + x_2$.

////

EXAMPLE 8-16 The following linear combination is made of the three independent random variables x_1, x_2, and x_3 as

$$z = 2x_1 + \tfrac{1}{3}x_2 + 5x_3$$

where x_1 is Poisson-distributed with a mean of 3, x_2 is binomially distributed with parameters $(.3, 20)$, and x_3 has a probability density function which is a gamma distribution with parameters $(\alpha = 1.5, \beta = 7.1)$. What is the MGF of z? This is given by

$$M_z(\theta) = e^{-3}e^{3e^{2\theta}} \cdot (.7 + .3e^{\theta/3})^{20} \cdot \frac{1}{[1 - 5(7.1)\theta]^{2.5}}$$

This distribution is obviously complicated and consists of both a discrete and continuous part. Nevertheless, one could obtain the first few cumulants by expanding the $\ln[M_z(\theta)]$ in powers of θ. The characterizing parameters μ, σ, α_3, and α_4 would now indicate the approximate behavior of random samples taken from this population. ////

(*Suggested Problems 8-7 to 8-19*)

8-6 ASYMPTOTIC DISTRIBUTIONS

In Chap. 4 it was shown by specific examples that the normal distribution provides a useful approximation to several other distributions for sufficiently large values of their appropriate parameters. The normal distribution is said to be the *asymptotic* form of any distribution which tends toward it as a limit. In this section, we shall prove the central limit theorem under restricted conditions and establish the asymptotic forms of the binomial and multinomial distributions.

An elementary form of the central limit theorem asserts that the asymptotic distribution of the sample mean $\bar{x}$ of n independent observations of a variate x having finite mean and variance is normal as n tends to infinity. We shall prove this theorem under the restriction that the moment-generating function exists. This class includes all distributions with finite range and many with infinite range. For convenience, assume tentatively that x has zero mean and unit variance. Its cumulant function $K(\theta : x)$ can then be written as

$$K(\theta : x) = \frac{\theta^2}{2} + \sum_{r=3}^{\infty} c_r \theta^r$$

where $c_r = \lambda_r/r!$. Making the change of variable $y = \bar{x}\sqrt{n}$ in Theorem 8-2 we have

$$M(\theta : y) = \left[M\left(\frac{\theta}{\sqrt{n}} : x \right) \right]^n$$

whence, putting $k_r = c_r/(\sqrt{n})^{r-3}$ we obtain

$$K(\theta : y) = nK\left(\frac{\theta}{\sqrt{n}} : x \right) = \frac{\theta^2}{2} + \frac{1}{\sqrt{n}} \sum_{r=3}^{\infty} k_r \theta^r$$

Now the convergence of the series $c_r \theta^r$ for sufficiently small θ implies the convergence of the series $k_r \theta^r$. Hence, the sum $L(n) \equiv \sum_{3}^{\infty} k_r \theta^r$ is finite for all $n \geq 1$, and we may put

$$K(\theta : y) = \frac{\theta^2}{2} + \frac{1}{\sqrt{n}} L(n)$$

Therefore, in the limit as $n \to \infty$ we obtain

$$\lim_{n \to \infty} K(\theta : y) = \frac{\theta^2}{2}$$

showing that the asymptotic distribution of y is unit normal. Likewise, for any variate x with finite mean and variance μ, σ^2 the asymptotic distribution of the variate $y = [(\bar{x} - \mu)\sqrt{n}]/\sigma$ will be unit normal. Accordingly, for large n the distribution of $\bar{x}$ will be approximately normal with mean $= \mu$ and standard deviation $= \sigma/\sqrt{n}$.   ////

Consider, next, the asymptotic distribution of the binomial. The moment-generating function is $(q + pe^\theta)^n$ and that of the standardized variate $u = (x - \mu)/\sigma = (x - np)/\sqrt{npq}$ is given by

$$M(\theta : u) = e^{-np\theta/\sqrt{npq}}(q + pe^{\theta/\sqrt{npq}})^n$$

$$\equiv (qe^{-p\theta/\sqrt{npq}} + pe^{p\theta/\sqrt{npq}})^n$$

The cumulant function of u can be expressed as

$$K(\theta : u) = \frac{\ln(qe^{-p\theta/\sqrt{npq}} + pe^{q\theta/\sqrt{npq}})}{1/n}$$

a form to which L'Hospital's rule applies. Differentiating twice, we obtain

$$\lim_{n \to \infty} K(\theta : u) = \frac{\theta^2}{2}$$

showing that the asymptotic distribution of u is unit normal. Therefore, with large n, the distribution of the binomial variate x tends toward normality with $\mu = np$, $\sigma^2 = npq$.   ////

The labor entailed in calculating exact probabilities from the multinomial distribution for large n is so heavy that, in the absence of some feasible approximation, practical uses of this distribution would be nonexistent. Therefore, we shall consider a normal approximation and from it derive a variate which is distributed asymptotically as chi-square. The latter has many important applications, but we shall defer their study to Chap. 10.

Let us begin with the simplest multinomial situation—that of three categories. Substituting $x_3 \equiv n - x_1 - x_2$ we may write the joint distribution in bivariate form as

$$f(x_1, x_2) = \frac{n! \, p_1{}^{x_1} p_2{}^{x_2} p_3{}^{n-x_1-x_2}}{x_1! \, x_2! \, (n - x_1 - x_2)!}$$

where $p_1 + p_2 + p_3 = 1$; $x_1 = 0, 1, \ldots, n$; $x_2 = 0, 1, \ldots, n - x_1$. Since we have previously noticed that the marginal distribution of x_1 is binomial with parameters (n, p_1), let us revamp the joint distribution as

$$f(x_1, x_2) = f_1(x_1)\phi(x_2 \mid x_1)$$

where

$$f_1(x_1) = C(n, x_1)p_1{}^{x_1}q_1{}^{n-x_1} \qquad \phi(x_2 \mid x_1) = C(n - x_1, x_2)(p_2')^{x_2}(q_2')^{n-x_1-x_2}$$

in which

$$q_1 = 1 - p_1 = p_2 + p_3 \qquad p_2' = \frac{p_2}{q_1} \qquad q_2' = 1 - p_2' = \frac{p_3}{q_1}$$

The validity of this substitution may be appreciated by reviewing Example 2-6. Since the conditional distribution of x_2, given x_1, is binomial in form, we may write the conditional mean and variance by inspection. These are

$$\text{Mean } (x_2 \mid x_1) = (n - x_1)p_2'$$

$$\text{Variance } (x_2 \mid x_1) = (n - x_1)p_2' q_2' = \left(1 - \frac{x_1}{n}\right)np_2' q_2'$$

Now the expected value of the quantity $(1 - x_1/n)$ is equal to q_1 and the variance of it is $p_1 q_1/n$. Accordingly, with a relative error on the order of $1/\sqrt{n}$ (by Tchebysheff's inequality, large multiples of this are improbable), we may write

$$\text{Variance } (x_2 \mid x_1) \approx q_1 np_2' q_2' = \frac{np_2 p_3}{q_1}$$

Introducing the normal approximations with $\sigma_1{}^2 = np_1q_1$, $\sigma_2{}^2 = np_2p_3/q_1$, we then have

$$f(x_1, x_2) \approx \frac{1}{\sigma_1\sqrt{2\pi}} e^{-(x_1 - np_1)^2/2\sigma_1{}^2} \frac{1}{\sigma_2\sqrt{2\pi}} e^{-[x_2 - (n - x_1)p_2']^2/2\sigma_2{}^2}$$

If we now substitute $u = (x_1 - np_1)/\sigma_1$, $v = [x_2 - (n - x_1)p_2']/\sigma_2$ we find, to the same approximation, that u and v are independent unit normal variates. Consequently, their sum of squares $X^2 = u^2 + v^2$ is asymptotically distributed as chi-square with 2 degrees of freedom.

By straightforward, though fairly lengthy algebra, we may express this same X^2 in the following symmetrical form (note absence of q's):

$$X^2 = \frac{(x_1 - np_1)^2}{np_1} + \frac{(x_2 - np_2)^2}{np_2} + \frac{(x_3 - np_3)^2}{np_3} = \sum_{i=1}^{3} \frac{(x_i - E_i)^2}{E_i} \qquad (8\text{-}27)$$

where E_i is an abbreviation for $E(x_i) = np_i$ ($i = 1, 2, 3$). We shall omit the steps, but the reader needs only patience and moderate manipulative skill to fill them in. The outcome is that the function defined by Eq. (8-27) is distributed asymptotically as chi-square with 2 degrees of freedom. Now any multinomial probability function can be revamped into a continued product of binomial factors and the same line of argument applied. If there are k categories, we shall arrive at $k - 1$ variates which, asymptotically, are independent and unit normal. Their sum of squares can be expressed in the symmetrical form

$$X^2 = \sum_{i=1}^{k} \frac{(x_i - np_i)^2}{np_i} = \sum_{i=1}^{k} \frac{(x_i - E_i)^2}{E_i} \qquad (8\text{-}28)$$

and this variate is distributed asymptotically as chi-square with $k - 1$ degrees of freedom. ////

EXAMPLE 8-17   Show that the mean of n independent observations from the population

$$f(x) = 1/\beta e^{-x/\beta} \qquad 0 \le x < \infty$$

is asymptotically "normally" distributed. The MGF of $f(x)$ is obviously

$$M(\theta : x) = \int_0^\alpha e^{\theta x} \frac{1}{\beta} e^{-x/\beta} \, dx = \frac{1}{1 - \theta\beta}$$

Since

$$\left. \frac{dM(\theta : x)}{d\theta} \right|_{\theta = 0} = \beta \quad \text{and} \quad \left. \frac{d^2M(\theta : x)}{d\theta^2} \right|_{\theta = 0} = 2\beta^2$$

the mean of $f(x)$ is β and its variance β^2. Our interest thus centers on the behavior of

$$y = \left[\frac{\bar{x} - \beta}{\beta} \right] \sqrt{n} = \frac{\bar{x}\sqrt{n}}{\beta} - \sqrt{n}$$

as $n \to \infty$, where $\bar{x}$ is the mean of n observations.

The MGF of $\bar{x}$ is, by Theorem 8-2,

$$M(\theta : \bar{x}) = \left(\frac{1}{1 - \theta\beta/n} \right)^n$$

and thus the MGF of y by Eq. (8-22) is

$$M(\theta : y) = e^{-\sqrt{n}\theta} \left(\frac{1}{1 - \theta/\sqrt{n}} \right)^n$$

or

$$\ln[M(\theta : y)] = -\sqrt{n}\theta - n \ln\left(1 - \frac{\theta}{\sqrt{n}}\right)$$

$$= -\sqrt{n}\theta + n\left(\frac{\theta}{\sqrt{n}} + \frac{\theta^2}{2n} + \frac{\theta^3}{3n^{3/2}} + \cdots\right)$$

$$= \frac{\theta^2}{2} + \frac{\theta^3}{\sqrt{n}} + \cdots$$

as $n \to \infty$, $\ln M(\theta : y) \to \theta^2/2$ and so $y \to N(0, 1)$, i.e., a unit-normal variate.

$////$

EXAMPLE 8-18   We have previously derived a theorem which states that the moment-generating function of a linear combination of random variables is the product of the moment-generating functions (properly scaled) of each of the variables involved in the linear combination. Since the cumulant-generating function is the logarithm of the moment-generating function, the cumulants of the individual variates in the linear combination are additive (again properly scaled). If one wishes to compute the cumulants of the linear combination, this fact is a powerful tool in developing the moments of a complicated derived distribution as the following problem will illustrate.

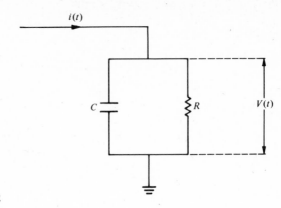

FIGURE 8-2

Let us consider the tank circuit of Fig. 8-2. A current $i(t)$ is flowing into the circuit and a vacuum-tube voltmeter measures the voltage V across the tank continuously, where V is given by

$$V(t) = \frac{1}{C} \int_0^t i(x) e^{-(t-x)/RC} \, dx \qquad (8\text{-}29)$$

and where the voltage at $t = 0$ is zero.

Let us now assume that the $i(t)$ consists of a series of discrete electric charges each of a magnitude of q coulombs which are entering the circuit completely at random as Poisson arrivals. The problem is to find the probability density of the voltage at time t after the tank circuit is connected to the source. This is a standard procedure for measuring the intensity of radioactive material.

Let us divide the time t into k equal intervals of time τ, giving $k + 1$ values of t as $t = 0, t = \tau, t = 2\tau, t = 3\tau, \ldots, t = k\tau$. Centered at these times consider $k + 1$ time intervals $r\tau \pm \tau/2$ ($r = 0, 1, 2, \ldots, k$), and let us assume that all the arrivals during one of these intervals occur instantaneously at the time designated by the middle of the interval. (See Fig. 8-3.) Thus n_0 is the number of

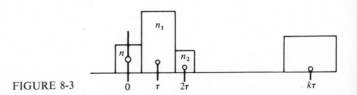

FIGURE 8-3

arrivals (Poisson) entering the tank circuit in the interval $0 - \tau/2$ to $0 + \tau/2$; we shall assume they all occur at $t = 0$. Similarly, n_1 is the number of arrivals in the interval $\tau - \tau/2$ to $\tau + \tau/2$ all assumed to occur at $t = \tau$, etc. If the average number of arriving charges q per unit time is a, then the n's are all Poisson-distributed with mean $\lambda = a\tau$.

The integral expression in Eq. (8-29) for the voltage V at time $t = k\tau$ can now be expressed as a summation as follows:

$$V(t) = \frac{1}{C} \int_0^t i(x) e^{-(t-x)/RC} \, dx$$

$$= \frac{q}{C} \left(n_0 \, e^{-k\tau/RC} + n_1 \, e^{-(k-1)\tau/RC} + \cdots + n_{k-1} \, e^{-\tau/RC} + n_k \right)$$

and if we let $\alpha = e^{-\tau/RC}$, then Eq. (8-29) becomes

$$V = \frac{q}{C} \left(n_0 \alpha^k + n_1 \alpha^{k-1} + \cdots + n_{k-1} \alpha + n_k \right) \tag{8-30}$$

The cumulant-generating function for each n_j is

$$K(\theta : n_j) = -\lambda + \lambda e^\theta$$

since the moment-generating function for the Poisson is $e^{-\lambda} e^{\lambda e^\theta}$, and the cumulant function for an arbitrary term in Eq. (8-30) is therefore

$$K\left(\theta : n_j \frac{q}{C} \alpha^{k-j} \right) = -\lambda + \lambda e^{\theta(q/C)\alpha^{k-j}} \qquad j = 0, 1, 2, \ldots, k$$

$$= -\lambda + \lambda \left[1 + \left(\frac{q}{C} \alpha^{k-j} \right) \frac{\theta}{1!} + \left(\frac{q^2}{C^2} \right)(\alpha^{k-j})^2 \frac{\theta^2}{2!} + \cdots \right]$$

Since V is the sum of $k + 1$ independent variables and the cumulants are additive, we have for the mth cumulant (the sum of the coefficients of $\theta^m/m!$)

$$K_m = \lambda \sum_{j=0}^{j=k} \left(\frac{q}{C} \alpha^{k-j} \right)^m = \frac{\lambda q^m}{C^m} (1 + \alpha^m + \alpha^{2m} + \cdots + \alpha^{km})$$

$$= \frac{a\tau q^m}{C^m} \frac{1 - (\alpha^m)^{k+1}}{1 - \alpha^m} = \frac{a\tau q^m}{C^m} \frac{1 - e^{-\tau m(k+1)/RC}}{1 - e^{-\tau m/RC}} \tag{8-31}$$

We shall now take the limit of this expression for K_m as $k \to \infty$ and $\tau \to 0$ in such a manner that $k\tau = t$. This is done by expanding the exponential in the denominator in a Taylor series in powers of $\tau m/RC$. This gives for the final cumulants of the distribution of voltage at time t

$$K_m = \frac{aRCq^m}{mC^m} (1 - e^{-mt/RC}) = \frac{aRq^m}{mC^{m-1}} (1 - e^{-mt/RC})$$

Thus

$$K_1 = \text{mean} = \mu = aRq(1 - e^{-t/RC})$$

$$K_2 = \text{variance} = \sigma^2 = \frac{aRq^2}{2C}(1 - e^{-2t/RC})$$

$$\text{Skewness} = \frac{\mu^3}{\sigma^3} = \frac{K_3}{(K_2)^{3/2}} \approx \frac{1}{\sqrt{aRC}} \qquad \text{for large } t$$

$$\text{Excess} = \frac{\mu^4}{\sigma_4} - 3 = \frac{K_4}{K_2^2} \approx \frac{1}{aRC} \qquad \text{for large } t$$

It is to be noticed that the effect of a large capacitance C retards the approach of the mean to its asymptotic value, although it keeps the variance small and makes the distribution more nearly gaussian with the higher K's approaching zero more rapidly. ////

(Suggested Problems 8-20 and 8-21)

PROBLEMS

8-1 Establish the following formulas expressing central moments in terms of raw moments. (a) $\mu_3 = \nu_3 - 3\mu\nu_2 + 2\mu^3$; (b) $\mu_4 = \nu_4 - 4\mu\nu_3 + 6\mu^2\nu_2 - 3\mu^4$.

8-2 Derive general expressions for the nth central moments of the following distributions: (a) the rectangular distribution $f(x) = 1$ (0, 1); (b) the normal distribution. *Ans.* (a) $[(\tfrac{1}{2})^{n+1} - (-\tfrac{1}{2})^{n+1}]/(n + 1)$.

8-3 Given that $\mu = np$, derive the following recursion formula for the central moments of the binomial distribution: $\mu_{k+1} = pq[nk\mu_{k-1} + (d\mu_k/dp)]$ where $k \geq 1$ and $\mu_0 \equiv 1$, $\mu_1 \equiv 0$.

8-4 A random variable can take on the values -2, -1, and 2, with probabilities $\tfrac{1}{8}$, $\tfrac{1}{4}$, $\tfrac{5}{8}$, respectively. Find the mean and standard deviation of the distribution, by direct calculation. Find the moment-generating function and use it to check the above results. *Ans.* $\mu = \tfrac{3}{4}$, $\sigma = \sqrt{43}/4$.

8-5 If $f(t) = [ab/(a - b)](e^{-bt} - e^{-at})$ ($t \geq 0$), obtain the moment-generating function of t. What is the variance of t? *Ans.* $\sigma^2 = (a^2 + b^2)/a^2b^2$.

8-6 Given that x is unit normal, find the moment-generating function of x^2 directly from the distribution of x. Then deduce at once the moment-generating function of chi-square, with n degrees of freedom. From this, infer the distribution.

8-7 What is the moment-generating function $z = 2x + 3y$, where the x and y are independent and $f(x) = 1$ (0, 1) and $g(y) = \tfrac{1}{2}$ (2, 4)?

8-8   Let x and y be independent random variables, and suppose x is normally distributed with mean μ and standard deviation σ, and y has density $e^{-y}, 0 \leq y < \infty$.
(a)   Find the mean of $z = ax + by$ (a, b constants).
(b)   Find the variance of $z = ax + by$.
(c)   Find the moment-generating function of $z = ax + by$.

8-9   A random variable has a moment-generating function

$$M(\theta) = \tfrac{1}{2}(1 + e^{\theta^2/2})$$

What is the variance of the mean of n independent observations from this population?   *Ans.* $1/(2n)$.

8-10   Four independent observations x_1, x_2, x_3, x_4 are taken on the discrete variate x, where $f(x) = \tfrac{1}{2}$ ($x = -1$, 1).   (a) State the mean and variance of $y = (x_1 + x_2 + x_3 + x_4)/2$.   (b) Check by obtaining the distribution of y and computing the mean and variance from it.   *Ans.*   (a) $\mu = 0$, $\sigma^2 = 1$.

8-11   The moment-generating function of the variate x is

$$M(\theta : x) = \cosh \theta = \tfrac{1}{2}(e^\theta + e^{-\theta})$$

Find the moment-generating function of $\bar{x}$ (the sample mean of n independent observations).   Also find the mean and variance of $\bar{x}$.   *Ans.* $\cosh^n(\theta/n)$.

8-12 $z = 2x + 3y$, where x is binomially distributed with parameters p and n, y is Poisson-distributed with parameter λ.
(a)   Find the first three cumulants of the distribution of z.
(b)   If $p \to 0$ and $n \to \infty$ in the binomial in such a way that $pn = \mu$, what happens to these cumulants?

8-13   The discrete variate u has two admissible values 0, 1 such that $f(0) = q, f(1) = p$, and, of course, $q + p = 1$.   (a) Find the mean and variance of u and thereby obtain the mean and variance of the variate $x = u_1 + u_2 + \cdots + u_n$, where $u_1, \ldots, u_n$ are independently distributed as u.   (b) Find the moment-generating function of u and thereby derive the moment-generating function of x.   (c) By expanding the latter and equating terms, prove that x is binomially distributed.

8-14   Prove that the rth cumulant of the sum of independent variates is equal to the sum of their respective rth cumulants.   Deduce at once that the sum of independent Poisson variates is Poisson-distributed.   Why cannot the same be said of any linear combination rather than just the sum?

8-15   Prove that the sum of two correlated Poisson variates does not have the Poisson distribution.   (*Hint:*   Show that the variance does not equal the mean.)

8-16   Given that $M(\theta : x + y) \equiv M(\theta : x)M(\theta : y)$, prove that x and y are independent.   [By subtraction, arrive at the equation $E([M(\theta : y) - M(\theta : y|x)]e^{\theta x}) = 0$ for all θ.   Hence $M(\theta : y|x) \equiv M(\theta : y)$, whereupon $\phi(y|x) = f(y)$.]

8-17   The admissible values of a certain discrete variate x are 0, 1, 2, $\ldots$, ∞ and all the cumulants of x have the same value μ.   From this information derive the distribution of x.

8-18   Show that $K(\theta : x) = \mu\theta + \ln M(\theta : x - \mu)$.   By the Maclaurin expansion, derive the coefficients of $\ln M(\theta : x - \mu)$ as far as θ^4.   Thus derive the expressions for the first four cumulants in terms of μ and the central moments.

8-19 A repairman notices over a long period of time that the number of machines lined up to be repaired each morning is distributed geometrically, i.e., the probability that there are n machines to be repaired is $(1-p)p^n$. The number of consequential defects per faulty machine is Poisson with parameter λ. Let $p(y)$ be the probability that among the machines lined up there is a total of y defects. Find the moment-generating function

$$M(\theta:y) = \sum_{y=0}^{\infty} p(y)e^{\theta y}$$

Also find the mean and variance of the number of consequential defects.

Ans. $M(\theta:y) = \dfrac{(1-p)\exp[\lambda(e^{\theta}-1)]}{1-p\exp[\lambda(e^{\theta}-1)]}$.

8-20 From the moment-generating function of the binomial distribution, determine the limiting form as $n \to \infty$ and $p \to 0$ in such a way that $np \to \mu$. Then expanding and equating terms, derive the Poisson distribution as an asymptotic form of the binomial.

8-21 Show that a Poisson variate is asymptotically normal as μ becomes indefinitely large.

9

MARKOV PROCESSES AND WAITING LINES

9-1 DEFINITION

Let us consider a probabilistic system which can assume any one of a finite number of states, N. These states are all distinct and identifiable and thus can be ordered by an index k, where $k = 1, 2, 3, \ldots, N$. At the start of certain prescribed intervals, the system may or may not change from its present state to one of the other $N - 1$ possible states. The start of these intervals at which the transitions take place need not be identifiable with equal intervals of time, but the concept of time is analogous to that used in our consideration of difference equations in previous chapters.

There exists a certain class of stochastic processes for which the transition probability p_{ij} for the system to change from state i to state j at the beginning of one of these intervals is constant and depends only upon i and j. In other words, the transition probability is independent of the past history of the systems or, equivalently, any past transitions.

Such a stochastic model is called a simple *Markov process*, and the observable results of a series of such trials from such a model is called a *Markov chain*. We shall now examine some of the mathematical properties of such a model.[1]

9-2 MATHEMATICAL PROPERTIES

Except for its initial state, apparently a Markov model or process can be completely described by what is called a stochastic matrix, where each element in the matrix p_{ij} is the probability that a system now in state i will transist to state j at the beginning of the next interval. Since it is possible for the system to remain in the same state from one interval to the next, p_{ii} is defined as the probability of this event. Thus the matrix becomes

$$\begin{pmatrix} p_{11} & p_{12} & \cdots & p_{1N} \\ p_{21} & p_{22} & \cdots & p_{2N} \\ \cdots\cdots\cdots\cdots\cdots\cdots \\ p_{N1} & \cdots & & p_{NN} \end{pmatrix} \tag{9-1}$$

All the p's in this square stochastic matrix (9-1) must have a value in the closed interval from 0 to 1, and the sum of the p's in each row must equal unity, since the system must be in some state during any particular interval. Thus

$$0 \le p_{ij} \le 1 \qquad i = 1, 2, \ldots, N, \; j = 1, 2, \ldots, N$$

and

$$\sum_{j=1}^{N} p_{ij} = 1 \qquad i = 1, 2, \ldots, N$$

If we define the probability that a Markov system is in state i during the nth period as $P_i(n)$, then

$$P_j(n + 1) = \sum_{i=1}^{N} p_{ij} P_i(n) \qquad j = 1, 2, \ldots, N \tag{9-2}$$

since the system can get to state j in the $(n + 1)$st interval only from one of the possible N states in the nth interval. It is evident, then, that if the initial

---

[1] It is of interest to note that if the conditional probability of i, given j, is also dependent upon the state of the system two intervals back, say l, as well as the previous state j such as $p(i|j, l)$, it is often possible to increase the number of states of the system by redefining new states involving combinations of the two past states of the system and the future one to reduce the increased state model to a simple Markov process. The same discussion applies to any number of states in the conditional dependency.

probabilities that the system is in state i at the start of the process, $P_i(0)$, are all known, then all the $P_j(1)$ can be computed from Eqs. (9-2) as

$$P_j(1) = \sum_{i=1}^{N} p_{ij} P_i(0) \qquad j = 1, 2, 3, \ldots, N$$

and recursively we could compute $P_j(2)$ from the probabilities $P_i(1)$, etc., up to and including $P_j(n)$, where n has any integral value. Thus all the probabilities $[(P_j(n)]$ that a system is in state j after n transitions are completely defined in terms of the initial probabilities $P_i(0)$.

There are two properties that a simple Markov process possesses in most cases:

1 The limiting value of $P_i(n)$ $(i = 1, 2, \ldots, N)$, as $n \to \infty$ is unique. These values are called the steady-state probabilities of the Markov process.
2 The limiting value of property 1 is independent of the state from which the system initially starts at $n = 0$.

In case both 1 and 2 are satisfied, the system is referred to as completely "ergodic." We shall discuss in detail later the conditions under which 1 and 2 are satisfied.

Condition 1 gives rise to the system of equations derived from Eqs. (9-2) $n \to \infty$ or

$$P_j(\infty) = \sum_{i=1}^{N} p_{ij} P_i(\infty) \quad j = 1, 2, \ldots, N \quad \text{where} \sum_{j=1}^{N} P_j(\infty) = 1 \qquad (9\text{-}3)$$

These conditions are equivalent to $N + 1$ equations, but since one of them is dependent on the others they usually may be solved uniquely for all the $P_j(\infty)$ $(j = 1, 2, \ldots, N)$.

EXAMPLE 9-1 Given the following stochastic matrix representing a Markov process, what are the steady-state probabilities?

$$\begin{pmatrix} .6 & .2 & .2 \\ .4 & .5 & .1 \\ .6 & .0 & .4 \end{pmatrix} \qquad (9\text{-}4)$$

Equations (9-3) for this system would be

$$P_1(\infty) = .6P_1(\infty) + .4P_2(\infty) + .6P_3(\infty)$$
$$P_2(\infty) = .2P_1(\infty) + .5P_2(\infty) + 0P_3(\infty)$$
$$P_3(\infty) = .2P_1(\infty) + .1P_2(\infty) + .4P_3(\infty)$$

where
$$P_1(\infty) + P_2(\infty) + P_3(\infty) = 1$$

The first two equations, together with the restriction $\sum_{i=1}^{3} P_i(\infty) = 1$, yield for the unique steady-state probabilities

$$P_1(\infty) = \tfrac{5}{9} \qquad P_2(\infty) = P_3(\infty) = \tfrac{2}{9}$$

The remaining equation is automatically satisfied.   ////

9-3   DIFFERENCE EQUATIONS

The N equations in (9-2) form a system of first-order difference equations in the $P_i(n)$ ($i = 1, 2, \ldots, N$).   These equations are linear, homogeneous, and with constant coefficients, so that a solution of the form $P_i(n) = A_i r^n$ may be assumed, as was demonstrated in Chap. 3.   If this solution form is substituted in (9-2), we develop the N equations as

$$A_1 p_{1j} + A_2 p_{2j} + \cdots + A_j(p_{jj} - r) + \cdots + A_N p_{Nj} = 0 \qquad j = 1, 2, \ldots, N \tag{9-5}$$

This system of linear, homogeneous equations in the A's [Eq. (9-5)] will have a solution other than $A_1 = A_2 = \cdots = A_N = 0$ only if the determinant of the coefficients of these A's is equal to zero, i.e.,

$$\begin{vmatrix} (p_{11} - r) & p_{21} & \cdots & p_{N1} \\ p_{12} & (p_{22} - r) & & p_{N2} \\ \cdots\cdots\cdots\cdots\cdots\cdots\cdots\cdots\cdots\cdots\cdots \\ p_{1N} & p_{2N} & \cdots & (p_{NN} - r) \end{vmatrix} = 0 \tag{9-6}$$

which, when evaluated, becomes an Nth-degree polynomial in r with N roots. Usually these N roots are distinct as $r_1, r_2, \ldots, r_k, \ldots, r_N$ and at present we shall assume this to be the case.

Since the solution is of the form $P_i(n) = A_i r^n$ ($i = 1, 2, \ldots, N$), there must be a different A_i for each root r_k which will be indicated by $A_i^{(k)}$ so that $P_i(n) = A_i^{(k)} r_k^n$ ($i, k = 1, 2, 3, \ldots, N$).   Therefore the general solution of the system of linear equations (9-2) must be

$$P_i(n) = \sum_{k=1}^{N} A_i^{(k)} r_k^n \qquad i = 1, 2, \ldots, N \tag{9-7}$$

The N^2 constants $A_i^{(k)}$ cannot all be arbitrary, since the system (9-2) can have only N arbitrary constants in the general solution.   In fact, when any one of the roots r_k is substituted for r in Eq. (9-5), this system is necessarily satisfied.   From the theory of algebra, it is known that there are $N - 1$ independent equations in this homogeneous system (9-5) provided the root r_k is distinct (i.e., not a multiple root).   Since we have assumed N distinct roots, there are $(N)(N - 1)$ independent linear equations connecting the A's. Thus the number of arbitrary constants $A_i^{(k)}$ must be $N^2 - N(N - 1) = N$, which is the correct number for the general solution of (9-2).

When the roots of (9-6) are not all distinct, there are no longer N^2 constants since some of these constants combine, nor are there usually $(N)(N - 1)$ independent relations connecting the A's. Sometimes it is possible in special situations to obtain a solution with N arbitrary constants in this instance, but in general the approach is to use a different assumed solution similar to that suggested in Example 3-7. It is beyond our present interest in Markov processes to pursue the subject of special cases, but the reader should be aware of their existence and the general approach to their solution.

If r is subtracted from the diagonal terms of (9-1) and the determinant of this matrix set equal to zero, we have

$$\begin{vmatrix} (p_{11} - r) & p_{12} & \cdots & p_{1N} \\ p_{21} & (p_{22} - r) & \cdots & p_{2N} \\ \cdots & \cdots & \cdots & \cdots \\ p_{N1} & & \cdots & (p_{NN} - r) \end{vmatrix} = 0 \qquad (9\text{-}8)$$

This determinant is just the transpose of (9-6) (rows and columns interchanged) and therefore has exactly the same N roots as (9-6). Equation (9-8) is often referred to as the secular equation of the square stochastic matrix and the roots of (9-8) as *characteristic values* or eigenvalues of the matrix (9-1).

It is clear that one of the characteristic values of (9-8) must be unity since, if $r = 1$ is substituted in the determinant, each row sums to zero because of the definition of a stochastic matrix. Also, the examination of the solution form of our system of difference equations given by (9-7) reveals that the other $N - 1$ roots cannot exceed unity in magnitude; otherwise the probabilities $P_i(n)$ would exceed unity in magnitude as n becomes large, and they could even become very large negatively with the proper choice of constants.

It can be shown by algebra that the roots of the secular equation of a stochastic matrix are such that $|r_k| \leq 1(k = 1, 2, \ldots, N)$. We shall show presently that the type of Markov process on which our interest centers has one and only one root equal to unity, and the remaining roots are less than 1 in magnitude.

Consider the system of difference equations for the Markov process defined by the stochastic matrix (9-4) as

$$P_1(n + 1) = .6P_1(n) + .4P_2(n) + .6P_3(n)$$
$$P_2(n + 1) = .2P_1(n) + .5P_2(n) \qquad (9\text{-}9)$$
$$P_3(n + 1) = .2P_1(n) + .1P_2(n) + .4P_3(n)$$

By substituting the assumed solution $P_i(n) = A_i r^n$ into Eqs. (9-9) we have the three equations

$$(.6 - r)A_1 + .4A_2 + .6A_3 = 0$$
$$.2A_1 + (.5 - r)A_2 = 0 \qquad (9\text{-}10)$$
$$.2A_1 + .1A_2 + (.4 - r)A_3 = 0$$

In order for (9-10) to have a solution for the A's other than $A_1 = A_2 = A_3 = 0$, it is necessary that the determinant of the coefficients equal zero as

$$\begin{vmatrix} (.6 - r) & .4 & .6 \\ .2 & (.5 - r) & 0 \\ .2 & .1 & (.4 - r) \end{vmatrix} = 0 \qquad (9\text{-}11)$$

or

$$r^3 - 1.5r^2 + .54r - .04 = (r - 1)(r - .4)(r - .1) = 0$$

which gives the three eigenvalues of the matrix as $r_1 = 1$, $r_2 = .4$, and $r_3 = .1$.

Thus the solution to this Markov process is

$$P_1(n) = A_1^{(1)} + A_1^{(2)}(.4)^n + A_1^{(3)}(.1)^n$$
$$P_2(n) = A_2^{(1)} + A_2^{(2)}(.4)^n + A_2^{(3)}(.1)^n \qquad (9\text{-}12)$$
$$P_3(n) = A_3^{(1)} + A_3^{(2)}(.4)^n + A_3^{(3)}(.1)^n$$

and if $r = 1$ is substituted into (9-10), then $A_1^{(1)} = \tfrac{5}{2}A_2^{(1)}$ and $A_3^{(1)} = A_2^{(1)}$. Observe that only two of the three equations are independent. If $r = .4$ is substituted in (9-10), then $A_1^{(2)} = -\tfrac{1}{2}A_2^{(2)}$ and $A_3^{(2)} = -\tfrac{1}{2}A_2^{(2)}$. Similarly the $r = .1$ gives $A_1^{(3)} = -2A_2^{(3)}$ and $A_3^{(3)} = A_2^{(3)}$. This reduces the values for the $P_i^{(n)}$ in (9-12) to

$$P_1(n) = \tfrac{5}{2}A_2^{(1)} - \tfrac{1}{2}A_2^{(2)}(.4)^n - 2A_2^{(3)}(.1)^n$$
$$P_2(n) = A_2^{(1)} + A_2^{(2)}(.4)^n + A_2^{(3)}(.1)^n \qquad (9\text{-}13)$$
$$P_3(n) = A_2^{(1)} - \tfrac{1}{2}A_2^{(2)}(.4)^n + A_2^{(3)}(.1)^n$$

Since $P_1(n) + P_2(n) + P_3(n) = 1$ for all n, $A_2^{(1)} = \tfrac{2}{9}$, which makes the steady-state probabilities (values as $n \to \infty$) correspond to those found in Example 9-1.

Finally, the remaining two arbitrary constants $A_2^{(2)}$ and $A_2^{(3)}$ are found from the initial conditions which must specify $P_1(0)$ and $P_2(0)$ [$P_3(0)$ is then uniquely determined since the sum of all three equals unity] or

$$P_1(0) = \tfrac{5}{9} - \tfrac{1}{2}A_2^{(2)} - 2A_2^{(3)}$$
$$P_2(0) = \tfrac{2}{9} + A_2^{(2)} + A_2^{(3)} \qquad (9\text{-}14)$$

For example, if the system started in state 1 so that $P_1(0) = 1$ and $P_2(0) = 0$, the solution of (9-14) for $A_2^{(2)}$ and $A_2^{(3)}$ would yield $A_2^{(2)} = 0$ and $A_2^{(3)} = -\frac{2}{9}$ so that (9-13) becomes

$$P_1(n) = \tfrac{5}{9} + \tfrac{4}{9}(\tfrac{1}{10})^n$$
$$P_2(n) = \tfrac{2}{9} - \tfrac{2}{9}(\tfrac{1}{10})^n \qquad (9\text{-}14a)$$
$$P_3(n) = \tfrac{2}{9} - \tfrac{2}{9}(\tfrac{1}{10})^n$$

whereas if $P_1(0) = 0$ and $P_2(0) = 0$ so that the system starts in state 3, making $P_3(0) = 1$, then (9-13) becomes

$$P_1(n) = \tfrac{5}{9} + \tfrac{1}{3}(\tfrac{4}{10})^n - \tfrac{8}{9}(\tfrac{1}{10})^n$$
$$P_2(n) = \tfrac{2}{9} - \tfrac{2}{3}(\tfrac{4}{10})^n + \tfrac{4}{9}(\tfrac{1}{10})^n \qquad (9\text{-}14b)$$
$$P_3(n) = \tfrac{2}{9} + \tfrac{1}{3}(\tfrac{4}{10})^n + \tfrac{4}{9}(\tfrac{1}{10})^n$$

It should be observed that one can approach the problem without involving any difference equations. We can find the eigenvalues of the original stochastic matrix (9-4), and these values would again be $r_1 = 1$, $r_2 = .4$, and $r_3 = .1$. This would immediately lead to the solution (9-12). The $A_1^{(1)}$, $A_2^{(1)}$, and $A_3^{(1)}$ are the steady-state probabilities, and their values were determined in Example 9-1. Since $P_1(n) + P_2(n) + P_3(n) = 1$ for all n, $A_3^{(2)} = -A_2^{(2)} - A_1^{(2)}$ and $A_3^{(3)} = -A_1^{(3)} - A_2^{(3)}$. This leaves only four constants to be determined: $A_1^{(2)}$, $A_1^{(3)}$, $A_2^{(2)}$, and $A_2^{(3)}$. If we set $n = 0$ in (9-12), the first two equations are

$$P_1(0) - P_1(\infty) = A_1^{(2)} + A_1^{(3)}$$
$$P_2(0) - P_2(\infty) = A_2^{(2)} + A_2^{(3)} \qquad (9\text{-}15a)$$

and if we set $n = 1$, these same two equations are

$$P_1(1) - P_1(\infty) = .4A_1^{(2)} + .1A_1^{(3)}$$
$$P_2(1) - P_2(\infty) = .4A_2^{(2)} + .1A_2^{(3)} \qquad (9\text{-}15b)$$

These four equations are sufficient to determine the four unknown A's. Of course, $P_1(1)$ and $P_2(1)$ can be computed from the initial probabilities at $n = 0$ from the relations

$$P_1(1) = p_{11}P_1(0) + p_{21}P_2(0) + p_{31}P_3(0)$$
$$P_2(1) = p_{12}P_1(0) + p_{22}P_2(0) + p_{32}P_3(0) \qquad (9\text{-}16)$$

9-4   ERGODIC PROCESS

It is now apparent that the condition that a Markov process be completely "ergodic" is that there be only one root equal to unity and that all other roots have a magnitude less than 1.   Since all but one root is numerically less than unity, the terms involving these roots in the solution will approach zero as $n \to \infty$ and the limit will be attained irrespective of the initial state.   Let us consider a few examples which illustrate the exceptions, and the rule.

EXAMPLE 9-2   Analyze the behavior of a Markov process described by the transition matrix

$$\begin{pmatrix} \frac{1}{3} & 0 & \frac{2}{3} \\ 0 & 1 & 0 \\ \frac{2}{3} & 0 & \frac{1}{3} \end{pmatrix}$$

We have the three equations for the steady-state probabilities as $P_1(\infty) = \frac{1}{3}P_1(\infty) + \frac{2}{3}P_3(\infty)$, $P_2(\infty) = P_2(\infty)$, $P_3(\infty) = \frac{2}{3}P_1(\infty) + \frac{1}{3}P_3(\infty)$,   and $P_1(\infty) + P_2(\infty) + P_3(\infty) = 1$.   These equations have the solution $P_1(\infty) = \alpha/2$, $P_2(\infty) = 1 - \alpha$, and $P_3(\infty) = \alpha/2$ for any α between 0 and 1.   Apparently the steady-state values are not unique.

If we find the eigenvalues of the stochastic matrix, they are the solutions of the equation

$$\begin{vmatrix} \frac{1}{3} - r & 0 & \frac{2}{3} \\ 0 & 1 - r & 0 \\ \frac{2}{3} & 0 & \frac{1}{3} - r \end{vmatrix} = 0$$

which yields the cubic equation

$$r^3 - \frac{5}{3}r^2 + \frac{1}{3}r + \frac{1}{3} = 0$$

and in turn has the roots 1, 1, and $-\frac{1}{3}$.   The solution for the probabilities of being in state i and n transitions would then be

$$P_i(n) = A_i + B_i n + C_i(-\tfrac{1}{3})^n \qquad i = 1, 2, 3$$

These equations cannot represent probabilities for all n unless $B_i = 0$ (all i). Even then, the process is not ergodic, because the steady-state values are not unique.   ////

EXAMPLE 9-3   Let us again consider a Markov process and its behavior where the process is characterized by the stochastic matrix

$$\begin{pmatrix} \tfrac{3}{5} & \tfrac{1}{5} & \tfrac{1}{5} \\ 0 & 0 & 1 \\ 0 & 1 & 0 \end{pmatrix}$$

This Markov process has steady-state probabilities which can be found in the usual manner, $P_1(\infty) = 0$, and $P_2(\infty) = P_3(\infty) = \tfrac{1}{2}$. The determinant for the eigenvalues would be

$$\begin{vmatrix} \tfrac{3}{5} - r & \tfrac{1}{5} & \tfrac{1}{5} \\ 0 & -r & 1 \\ 0 & 1 & -r \end{vmatrix} = 0$$

which gives the cubic equation

$$r^3 - \tfrac{3}{5}r^2 - r + \tfrac{3}{5} = 0$$

and thus the roots are $r = 1$, -1, and $\tfrac{3}{5}$. The probabilities for each state would have the form

$$P_i(n) = A_i + B_i \cos(\pi n) + C_i(\tfrac{3}{5})^n \qquad i = 1, 2, 3$$

Since $\cos(\pi n)$ does not converge to a limit, the process is not ergodic.   ////

EXAMPLE 9-4   A helicopter capable of carrying only one passenger besides the pilot connects an airport A with two industrial plants B_1 and B_2. According to the contract with the plants, the "copter" must keep to a 20-minute schedule, which is the time necessary to make any one of the three trips $A \rightleftarrows B_1$, $A \rightleftarrows B_2$, $B_1 \rightleftarrows B_2$ since the three helicopter ports are situated approximately at the vertices of an equilateral triangle.

The doctrine of operation is in the hands of the carrier except that, if the helicopter is at a particular port at the start of a 20-minute period and a passenger is waiting, he must be taken to his desired destination. If the "copter" waits at a particular port for the 20-minute cycle and a passenger arrives during this cycle, the takeoff time is still at the beginning of the next cycle. It is assumed that a second passenger arriving at any of the ports and seeing a passenger waiting will immediately seek another means of transportation. It is also assumed that any passenger will also seek immediately another means of transportation if he is waiting at a port and the "copter" fails to appear by the start of the next 20-minute period.

A survey indicates that the probabilities that a passenger will be waiting at the beginning of a cycle at the airport A, plant B_1, and plant B_2 are, respectively, $P_A = .80$, $P_{B_1} = .60$, and $P_{B_2} = .45$. It is also observed that the passengers waiting at A are about equally divided between destinations B_1 and B_2 whereas those waiting at either of the plants are twice as likely to want to go to A as to the other plant.

Many doctrines are available to the owners of the "copter" but only three are within the scope of the contract and also appear to be reasonable possibilities, namely:

1 Helicopter waits at any port until a passenger arrives. It then proceeds to the asked-for destination and waits there until another passenger arrives with a new destination.

2 "Copter" waits only at A or B_1 where the probabilities are highest for a new passenger. If it takes a load to B_2 and no one is waiting, it proceeds immediately to A empty.

3 Helicopter waits only at A and flies back there empty as soon as there is no passenger at the beginning of the next cycle when at either B_1 or B_2.

The decision which of these three policies to follow depends upon the objectives of the owners of the helicopter for what is termed "their measures of effectiveness." Here again there are obviously many possible ones, and we shall consider only two against which we shall determine the best strategy, namely:

1 It is desired to carry as many passengers as possible. By this showing, the owners hope to convince the source of capital that a larger operation would be feasible.

2 The owners wish to maximize the profit over a reasonably large number of cycles. The fare is $14 per person. The fixed charges per 20-minute period are $6, which includes the cost of the insurance, depreciation, repairs, etc., and the wages of the pilot. The variable cost per cycle for gas, oil, etc., when the "copter" runs between any of these ports, is $2.

Under doctrine 1 there are nine possible states for the system in any 20-minute period. They are:

1 Waiting at A
2 Going from A to B_1
3 Going from A to B_2

4 Waiting at B_1
5 Going from B_1 to A
6 Going from B_1 to B_2
7 Waiting at B_2
8 Going from B_2 to A
9 Going from B_2 to B_1

The stochastic matrix representing this simple Markov process is shown in Chart 1.

Chart 1

		$P_j(n+1)$								
		1	2	3	4	5	6	7	8	9
	1	.20	.40	.40	0	0	0	0	0	0
	2	0	0	0	.40	.40	.20	0	0	0
	3	0	0	0	0	0	0	.55	.30	.15
	4	0	0	0	.40	.40	.20	0	0	0
$P_i(n)$	5	.20	.40	.40	0	0	0	0	0	0
	6	0	0	0	0	0	0	.55	.30	.15
	7	0	0	0	0	0	0	.55	.30	.15
	8	.20	.40	.40	0	0	0	0	0	0
	9	0	0	0	.40	.40	.20	0	0	0

Denoting the steady-state probabilities $P_1(\infty)$ by P_1, $P_2(\infty)$ by P_2, etc., we have the following 10 equations which these probabilities must satisfy:

$P_1 = .20P_1 + .20P_5 + .20P_8$
$P_2 = .40P_1 + .40P_5 + .40P_8$ that is $P_2 = P_3 = 2P_1$
$P_3 = .40P_1 + .40P_5 + .40P_8$ $P_5 + P_8 = 4P_1$

$P_4 = .40P_2 + .40P_4 + .40P_9$
$P_5 = .40P_2 + .40P_4 + .40P_9$ that is $P_4 = P_5 = 2P_6$
$P_6 = .20P_2 + .20P_4 + .20P_9$ $2P_2 + 2P_9 = 3P_4$

$P_7 = .55P_3 + .55P_6 + .55P_7$
$P_8 = .30P_3 + .30P_6 + .30P_7$ that is $22P_9 = 11P_8 = 6P_7$
$P_9 = .15P_3 + .15P_6 + .15P_7$ $9P_7 = 11(P_3 + P_6)$

$P_1 + P_2 + \cdots + P_9 = 1$

Solving the nine independent equations, we arrive at the steady-state probabilities:

$$P_1 = P_6 = P_9 = \tfrac{3}{50} \qquad P_2 = P_3 = P_4 = P_5 = P_8 = \tfrac{6}{50} \qquad P_7 = \tfrac{11}{50}$$

Under doctrine 2, it is clear that state 7 is now eliminated. Thus, if a passenger is taken to B_2, the pilot immediately proceeds to A with probability .85 and to B_1 with probability .15. However, only $\tfrac{6}{17}$ of the time that he goes to A will he have a passenger. The trip to B_1 from B_2 will always have a pay load. The stochastic matrix is now only 8×8 with state 7 eliminated from both the rows and columns. Row 3 and row 6 would be altered as follows with the remaining rows unaltered:

Chart 2

		$P_j(n+1)$							
		1	2	3	4	5	6	8	9
$P_j(n)$	3	0	0	0	0	0	0	.85	.15
	6	0	0	0	0	0	0	.85	.15

The first six equations for the steady states are unaltered in (9-15) and the last three are now replaced by only two equations as

$$\begin{aligned} P_8 &= .85P_3 + .85P_6 \\ P_9 &= .15P_3 + .15P_6 \end{aligned} \qquad \text{that is} \qquad \begin{aligned} 3P_8 &= 17P_9 \\ 20P_9 &= 3(P_3 + P_6) \end{aligned}$$

together with

$$P_1 + P_2 + P_3 + P_4 + P_5 + P_6 + P_8 + P_9 = 1$$

The simultaneous solution of this set of linear equations yields

$$P_1 = \frac{57}{675} \qquad P_2 = P_3 = \frac{114}{675} \qquad P_4 = P_5 = \frac{92}{675}$$

$$P_6 = \frac{46}{675} \qquad P_8 = \frac{136}{675} \quad \text{and} \quad P_9 = \frac{24}{675}$$

If doctrine 3 is now considered, state 4 is also eliminated, leaving only seven states for the Markov process. Here again when the helicopter is in state 8 going from B_2 to A, $6/17$ of the time it has a passenger and when in state 5 going from B_1 to A, one-half the time it has a load.

The stochastic matrix for the process now becomes Chart 3:

Chart 3

		$P_j(n+1)$						
		1	2	3	5	6	8	9
	1	.20	.40	.40	0	0	0	0
	2	0	0	0	.80	.20	0	0
	3	0	0	0	0	0	.85	.15
$P_i(n)$	5	.20	.40	.40	0	0	0	0
	6	0	0	0	0	0	.85	.15
	8	.20	.40	.40	0	0	0	0
	9	0	0	0	.80	.20	0	0

The following eight equations determine the steady-state probabilities in this case:

$$\begin{aligned} P_1 &= .20P_1 + .20P_5 + .20P_8 \\ P_2 &= .40P_1 + .40P_5 + .40P_8 \\ P_3 &= .40P_1 + .40P_5 + .40P_8 \end{aligned} \quad \text{that is} \quad \begin{aligned} P_2 &= P_3 = 2P_1 \\ 4P_1 &= P_5 + P_8 \end{aligned}$$

$$\begin{aligned} P_5 &= .80P_2 + .80P_9 \\ P_6 &= .20P_2 + .20P_9 \end{aligned} \quad \text{that is} \quad \begin{aligned} P_5 &= 4P_6 \\ 5P_6 &= P_2 + P_9 \end{aligned}$$

$$\begin{aligned} P_8 &= .85P_3 + .85P_6 \\ P_9 &= .15P_3 + .15P_6 \end{aligned} \quad \text{that is} \quad \begin{aligned} 17P_9 &= 3P_8 \\ 20P_9 &= 3(P_3 + P_6) \end{aligned}$$

$$P_1 + P_2 + P_3 + P_5 + P_6 + P_8 + P_9 = 1$$

Hence

$$P_1 = \frac{97}{955} \qquad P_2 = P_3 = \frac{194}{955} \qquad P_5 = \frac{184}{955} \qquad P_6 = \frac{46}{955}$$

$$P_8 = \frac{204}{955} \qquad P_9 = \frac{36}{955}$$

Under the first measure of effectiveness, these three doctrines would be compared as to the expected value of the number of passengers carried per cycle in each case. Let us denote these by E_1, E_2, and E_3 as

$$E_1 = 0.P_1 + 1.P_2 + 1.P_3 + 0.P_4 + 1.P_5 + 1.P_6 + 0.P_7 + 1.P_8 + 1.P_9$$

$$= \frac{6}{50} + \frac{6}{50} + \frac{6}{50} + \frac{3}{50} + \frac{6}{50} + \frac{3}{50} = .600$$

$$E_2 = 0.P_1 + 1.P_2 + 1.P_3 + 0.P_4 + 1.P_5 + 1.P_6 + \frac{6}{17} \cdot P_8 + 1.P_9$$

$$= \frac{114}{675} + \frac{114}{675} + \frac{92}{675} + \frac{46}{675} + \frac{6}{17} \cdot \frac{136}{675} + \frac{24}{675} = \frac{146}{225} \approx .649$$

$$E_3 = 0.P_1 + 1.P_2 + 1.P_3 + \frac{1}{2} \cdot P_5 + 1.P_6 + \frac{6}{17} \cdot P_8 + 1.P_9$$

$$= \frac{194}{955} + \frac{194}{955} + \frac{1}{2}\left(\frac{184}{955}\right) + \frac{46}{955} + \frac{6}{17}\left(\frac{204}{955}\right) + \frac{36}{955}$$

$$= \frac{634}{955} \approx .664$$

Apparently, there is about a 10 percent improvement by not waiting at either B_1 or B_2. Thus doctrine 3 is the best if the number of passengers carried is the "measure of effectiveness."

Let us now consider the profit made per cycle with these three doctrines, which will be designated as F_1, F_2, and F_3. When the helicopter waits at a port, the cost is $6 for that 20-minute period. If it flies between ports with no load, the cost is $8 per cycle. The profit for a pay load is $14 - $8 = $6. Thus

$$F_1 = -6\left(\frac{3}{50}\right) + 6\left(\frac{6}{60}\right) + 6\left(\frac{6}{50}\right) - 6\left(\frac{6}{50}\right) + 6\left(\frac{6}{50}\right)$$

$$+ 6\left(\frac{3}{50}\right) - 6\left(\frac{11}{50}\right) + 6\left(\frac{6}{50}\right) + 6\left(\frac{3}{50}\right) = \$1.20$$

$$F_2 = -6\left(\frac{57}{675}\right) + 6\left(\frac{114}{675}\right) + 6\left(\frac{114}{675}\right) - 6\left(\frac{92}{675}\right) + 6\left(\frac{92}{675}\right)$$

$$+ 6\left(\frac{46}{675}\right) + \left[6\left(\frac{6}{17}\right) - 8\left(\frac{11}{17}\right)\right]\frac{136}{675} + 6\left(\frac{24}{675}\right)$$

$$F_2 = \$1.53$$

$$F_3 = -6\left(\frac{97}{955}\right) + 6\left(\frac{194}{955}\right) + 6\left(\frac{194}{955}\right) + \left(\frac{1}{2}\cdot 6 - \frac{1}{2}\cdot 8\right)\frac{184}{955}$$

$$+ 6\left(\frac{46}{955}\right) + \left[\left(\frac{6}{17}\right)6 - \left(\frac{11}{17}\right)\cdot 8\right]\frac{204}{955} + 6\left(\frac{36}{955}\right)$$

$$= \$1.50$$

Thus the second doctrine is the best for profit as a "measure of effectiveness." The value for F_2 and F_3 would be quite different if the variable cost were a larger percentage of the total. ////

9-5 MARKOV PROCESSES AND REWARDS

Associated with each change of state in a Markov process (or even with remaining in the same state) there is often a reward. This reward may be in the form of financial gain measured in dollars or in any other appropriate units. A negative reward in the first case would merely indicate a financial loss. Thus, associated with each transition probability p_{ij} there could exist a corresponding reward q_{ij} (either positive or negative) and therefore the q's could be represented as elements of a matrix corresponding to the p_{ij} in (9-1) as

$$Q = \begin{pmatrix} q_{11} & q_{12} & \cdots & q_{1N} \\ q_{21} & q_{22} & \cdots & q_{2N} \\ \cdots\cdots\cdots\cdots\cdots\cdots\cdots \\ q_{N1} & \cdots & & q_{NN} \end{pmatrix} \qquad (9\text{-}17)$$

where each element q_{ij} represents the gain or loss if the system goes from state i to state j. If one considers a specific Markov chain, it would be feasible to compute the total reward generated over that particular set of outcomes. It thus seems reasonable, since we are dealing with a probabilistic system, to consider the expected reward during the next n future transitions, assuming the system is in a particular state at the present time.

Therefore, let us define $Z_i(n)$ as the expected reward during the next n periods, assuming the system is now in state i.

The value attributable to the process with n transitions to go, if the system changes from i to j, would be

$$q_{ij} + Z_j(n - 1)$$

This merely states that an immediate reward is attributable to the process by going from i to j plus any additional amount which might be added because of the expected value of the rewards for the next $n - 1$ transitions starting from state j.

Thus the total expected reward starting from i would be

$$Z_i(n) = \sum_{j=1}^{N} p_{ij}[q_{ij} + Z_j(n-1)]$$

$$= \sum_{j=1}^{N} p_{ij} q_{ij} + \sum_{j=1}^{N} p_{ij} Z_j(n-1) \qquad (9\text{-}18)$$

Since $\sum_{j=1}^{N} p_{ij} = 1$, the first term on the right of (9-18) is just a weighted average of the q's in a particular row (the ith), which is designated now as S_i, or equivalently the expected reward for the ith row. Thus $Z_i(n)$ becomes

$$Z_i(n) = S_i + \sum_{j=1}^{N} p_{ij} Z_j(n-1) \qquad i = 1, 2, \ldots, N \qquad (9\text{-}19)$$

This system (9-19) is a set of N constant-coefficient difference equations of the first order in $Z_i(n)$ but nonhomogeneous. If the S_i are set equal to zero, the resulting homogeneous equations have the same solution as (9-2) since the matrix of the coefficients of (9-19) is just the transpose of the previous set. Since the S_i are constants, the particular solution for (9-19) must be of the form

$$Z_i(n) = C_i n + P_i(\infty) \qquad i = 1, 2, \ldots, N \qquad (9\text{-}20)$$

where the C_i are constants. The C_i are multiplied by n since a constant is already a solution of the homogeneous set. Since (9-19) is a set and not a single equation the constant solution of the homogeneous set must be added to the particular solution.

Substituting $Z_i(n)$ from (9-20) into (9-19) yields

$$nC_i + P_i(\infty) = S_i + \sum_{p=1}^{N} p_{ij}[C_j(n-1) + P_j(\infty)] \qquad i = 1, 2, \ldots, N$$

$$= S_i + n \sum_{j=1}^{N} p_{ij} C_j - \sum_{j=1}^{N} C_j p_{ij} + \sum_{j=1}^{N} p_{ij} P_j(\infty) \qquad (9\text{-}21)$$

Since (9-21) must be an identity for all n, the coefficients of n on both sides can be equated as

$$C_i = \sum_{j=1}^{N} p_{ij} C_j \qquad (9\text{-}22)$$

Equation (9-22) is obviously satisfied if all the C_i are the same since $\sum_{j=1}^{N} p_{ij} = 1$ and thus $C_i = C$.

Now, equating the constant terms on both sides of Eq. (9-21), we have the N equations

$$C = S_i + \sum_{j=1}^{N} p_{ij} P_j(\infty) - P_i(\infty) \qquad i = 1, 2, \ldots, N \qquad (9\text{-}23)$$

If both sides of (9-23) are multiplied by $P_i(\infty)$ and each term summed on i from 1 to N, there results

$$C \sum_{i=1}^{N} P_i(\infty) = \sum_{i=1}^{N} S_i P_i(\infty) + \sum_{i=1}^{N} P_i(\infty) \sum_{j=1}^{N} p_{ij} P_j(\infty) - \sum_{i=1}^{N} P_i^2(\infty)$$

By interchanging the order of summation in the term with the double sums, we see from Eq. (9-3) that

$$\sum_{j=1}^{N} P_j(\infty) \sum_{i=1}^{N} p_{ij} P_i(\infty) = \sum_{j=1}^{N} P_j(\infty) \cdot P_j(\infty) = \sum_{j=1}^{N} P_j^2(\infty)$$

so that the constant C equals the average of the S's weighted by the steady-state probabilities for an ergodic process or

$$C = \sum_{i=1}^{N} S_i P_i(\infty) = \bar{S} \qquad (9\text{-}24)$$

We can now interpret $\bar{S}$ as the average gain per transition for the ergodic Markov process, and the final solution of (9-19) becomes

$$Z_i(n) = A_i^{(1)} + A_i^{(2)} r_2{}^n + A_i^{(3)} r_3{}^n + \cdots + A_i^{(N)} r_N{}^n + \bar{S} n \qquad i = 1, 2, \ldots, N \qquad (9\text{-}25)$$

where the $A_i^{(k)}$ are constants and $r_2, r_3, \ldots, r_N$ the $N-1$ distinct roots less than unity.   If any of these $N-1$ roots are multiple, there is no problem and the form of the solution is changed to conform correctly in the usual manner.

The constants $A_i^{(k)}$ all depend upon the value assigned to the process at the end of the final transition period when $n = 0$.   It may be, for example, that a business which is to be sold when $n = 0$ is worth various amounts to a buyer, depending upon the state of the enterprise at the moment of sale.   There are only N arbitrary constants out of the N^2 in (9-25), and we have several choices in obtaining the necessary relationships between these constants so as to end up with the correct number which actually are arbitrary.

One reasonable way to obtain these relationships can easily be illustrated by examining a three-state system ($N = 3$).   Let $Z_1(0)$, $Z_2(0)$, and $Z_3(0)$ be the value of the process at $n = 0$.   By using Eq. (9-19) recursively, one can find the value of $Z_1(1)$, $Z_1(2)$, $Z_2(1)$, $Z_2(2)$, $Z_3(1)$, and $Z_3(2)$ all in terms of $Z_1(0)$,

$Z_2(0)$, $Z_3(0)$, and known parameters since they are uniquely determined.   Consider now Eqs. (9-25).   By fixing i (say $i = 1$) and allowing n to take on the value 0, 1, and 2 successively,

$$Z_1(0) = A_1^{(1)} + A_1^{(2)} + A_1^{(3)}$$

$$Z_1(1) = A_1^{(1)} + A_1^{(2)}r_2 + A_1^{(3)}r_3 + \bar{S}$$

$$Z_1(2) = A_1^{(1)} + A_1^{(2)}r_2^2 + A_1^{(3)}r_3^2 + 2\bar{S}$$

These three equations determine the $A_1^{(k)}$ in terms of the $Z_i(0)$ ($i = 1, 2, 3$), provided $r_2 \neq r_3$ and the system is ergodic.   Similar equations determine the $A_2^{(k)}$ and $A_3^{(k)}$ in terms of the same initial conditions.

EXAMPLE 9-5   Given the following probability transition matrix P defining a specific Markov process and an allied reward matrix Q as

$$P = \begin{pmatrix} .9 & .1 \\ .4 & .6 \end{pmatrix} \qquad Q = \begin{pmatrix} 10 & -40 \\ 5 & 30 \end{pmatrix}$$

What is the total gain for n transitions if the system is now in state 1? What would be the difference in value if the system started in state 2?   Assume in both cases that the value of the process is zero at the end.

The characteristic values of the determinant p satisfy the quadratic equation $r^2 - 1.5r + .5 = 0$ and so $r = 1$ or $.5$.   Also the steady-state probabilities for the Markov process satisfy the two equations $p_1 = .9p_1 + .4p_2$ and $p_1 + p_2 = 1$ which gives $p_1 = \frac{4}{5}$ and $p_2 = \frac{1}{5}$.   Therefore the solution for $Z_1(n)$ is

$$Z_1(n) = A_1^{(1)} + A_1^{(2)}(\tfrac{1}{2})^n + \bar{S}n$$

where $\bar{S} = \frac{4}{5}S_1 + \frac{1}{5}S_2$ and also where $S_1 = (.9)(10) + (.1)(-40) = 5$ and $S_2 = (.4)(5) + (.6)(30) = 20$.   Thus $\bar{S} = 8$, which is the gain per transition for the process, and the solution now becomes

$$Z_1(n) = A^{(1)} + A^{(2)}(\tfrac{1}{2})^n + 8n$$

As $Z_1(0) = Z_2(0) = 0$, $Z_1(1) = S_1$ and $Z_2(1) = S_2$, as may be seen from (9-19), so that (9-25) for $i = 1$ evaluates as

$$Z_1(0) = 0 = A_1^{(1)} + A_1^{(2)} \qquad \text{and} \qquad Z_1(1) = S_1 = A_1^{(1)} + A_1^{(2)}(\tfrac{1}{2}) + 8 = 5$$

Solving for $A_1^{(1)}$ and $A_1^{(2)}$, we have $A_1^{(1)} = -6$ and $A_1^{(2)} = 6$, which gives for the final value of $Z_1(n)$

$$Z_1(n) = -6 + 6(\tfrac{1}{2})^n + 8n$$

$A_2^{(1)}$ and $A_2^{(2)}$ may be evaluated in the same manner, giving for $Z_2(n)$

$$Z_2(n) = 40 - (40)(\tfrac{1}{2})^n + 8n$$

Since $Z_2(n) - Z_1(n) = 46$ for n reasonably large, it is worth 46 more monetary units to start in state 2 than in state 1.   One might be willing to pay an amount up to 46 units to get into state 2 if a reasonable number of transitions are to be expected in the future.   ////

9-6   STRATEGIES AND MARKOV PROCESSES

If a completely ergodic N-state system is specified, then for a reasonably large number of transitions n yet to go the expected reward is given from Eq. (9-23) by

$$Z_i(n) = A_i + \bar{S}n \qquad i = 1, 2, \ldots, N \qquad (9\text{-}26)$$

for initially starting in state i and where $\bar{S}$ is the gain per transition.

Although there may be some gain in starting in particular states because of differences in the A_i, nevertheless the significant parameter is $\bar{S}$ and attention is now focused on this constant.

In Sec. 9-4, the case was treated which involved a single transition matrix P and a reward matrix Q.   Sometimes it is possible with the use of additional effort (possibly spending money) to change the transition probabilities in P to new values P'.   In most cases this changes the reward matrix, because of the spending of part of the net gain, from Q to Q'.   It is possible that this new process consisting of P' and Q' has a greater net gain per transition, $\bar{S}'$, than the original $\bar{S}$.   However, there are other alternatives in that some rows in both the P and Q matrix could be combined with rows of the P' and Q' matrix to form square matrices P'' and Q'' which are a combination of both processes. Thus additional effort could be expended when the system is in certain states and no additional effort otherwise.   In general, there may be many possible different actions that can be taken when a system is in state i and each of these actions would produce different transition probabilities p_{ij} and corresponding rewards q_{ij}, where $j = 1, 2, \ldots, N$ in each case.   In fact, the number of different actions that may be taken could be different in number for each state i.   From all these possibilities, the problem is to pick the best strategy where "best" is defined as that Markov process and corresponding rewards which have the largest gain per transition.   This would mean determining a doctrine that would specify what action should be taken every time the system is in any one of its possible N states.

Ronald Howard[1] suggested the following technique for arriving at this best strategy. If Eq. (9-26) is substituted into Eqs. (9-19) there results

$$A_i + \bar{S}n = S_i + \sum_{j=1}^{N} p_{ij}[A_j + \bar{S}(n-1)] = S_i + \bar{S}(n-1) + \sum_{j=1}^{N} p_{ij} A_j$$

or

$$A_i + \bar{S} = S_i + \sum_{j=1}^{N} p_{ij} A_j \qquad i = 1, 2, \ldots, N \qquad (9\text{-}27)$$

If any given policy (good or bad) is decided upon, the transition matrix and the corresponding reward matrix are determined and thus the p_{ij} and S_i for each i. If these values for the p_{ij} and S_i are substituted in (9-27), there are generated N equations with $N + 1$ unknowns, namely, $A_1, A_2, \ldots, A_N$ and $\bar{S}$. With one more unknown than the number of equations, only the relative values of the A_i will be determined, but actually only the differences between the A_i are pertinent to any action. Therefore, if one of the A_i, say A_N, is set equal to zero, there are N nonhomogeneous equations with N unknowns which in general can be solved uniquely for $A_1, A_2, \ldots, A_{N-1}$ and $\bar{S}$.

Using this solution, the quantity

$$V_i = S_i^* + \sum_{j=1}^{N} p_{ij}^* A_j \qquad (9\text{-}28)$$

is examined for each i and the strategy that maximizes V_i is chosen in each case. This operation determines a new doctrine which has a transition matrix consisting of elements p_{ij}^* and corresponding reward matrix which produces the corresponding S_i^*. These values, p_{ij}^* and S_i^*, which maximized the V_i for each i, are now substituted in (9-27) and new A's and $\bar{S}$ computed. This process is then repeated recursively until the same value for $\bar{S}$ is obtained successively twice. It is easy to show that this technique always increases the value of $\bar{S}$ at each step until a maximum or best strategy is reached.[2]

EXAMPLE 9-6 Consider a two-state system each with three levels of effort E_1, E_2, E_3 which have different transition probabilities and corresponding rewards. What is the doctrine that maximizes the return per transition?

---

[1] Ronald Howard, "Dynamic Programming and Markov Processes," The M.I.T. Press and John Wiley & Sons, Inc., 1960.

[2] See Appendix at the end of this chapter for the proof.

		Transition State		Rewards State		Average Reward
		1	2	1	2	
State 1	E_1	.5	.5	40	−20	10
	E_2	.65	.35	35	−30	12.3
	E_3	.8	.2	25	−60	8
State 2	E_1	.6	.4	25	−40	−1
	E_2	.75	.25	15	−70	−6.2
	E_3	.9	.1	10	−120	−3

Let us start the process by choosing E_1 for both states since this has the maximum average reward. Then Eqs. (9-27) become

$$A_1 + \bar{S} = 10 + .5A_1 + .5A_2$$
$$A_2 + \bar{S} = -1 + .6A_1 + .4A_2$$

Setting $A_2 = 0$ and solving for A_1 and $\bar{S}$, we find $A_1 = 10$ and $\bar{S} = 5$. Thus the average return per transition is 5 with this particular strategy.

Using Eq. (9-28) and computing the value of V_i in each case, we have

$$\begin{aligned} &V(E_1) = 10 + (.5)(10) = 15 \\ \text{State 1} \quad &V(E_2) = 12.3 + (.65)(10) = 18.8 \\ &V(E_3) = 8 + (.8)(10) = 16 \end{aligned}$$

$$\begin{aligned} &V(E_1) = -1 + (.6)(10) = 5 \\ \text{State 2} \quad &V(E_2) = -6.2 + (.75)(10) = 1.3 \\ &V(E_3) = -3 + (.9)(10) = 6 \end{aligned}$$

Apparently the second choice in the iteration will be E_2 for state 1 and E_3 for state 2. Thus Eqs. (9-27) become

$$A_1 + \bar{S} = 12.3 + .65A_1 + .35A_2$$
$$A_2 + \bar{S} = -3 + .9A_1 + .1A_2$$

Again setting $A_2 = 0$, we now have $A_1 = 12.2$ and $\bar{S} = 8.0$. Thus we have gained a value of 3 per transition by this doctrine. Recomputing the V's we have

$$\begin{aligned} &V(E_1) = 10 + (.5)(12.2) = 16.1 \\ \text{State 1} \quad &V(E_2) = 12.3 + (.65)(12.2) = 20.2 \\ &V(E_3) = 8 + (.8)(12.2) = 17.7 \end{aligned}$$

$$\begin{aligned} &V(E_1) = -1 + (.6)(12.2) = 6.3 \\ \text{State 2} \quad &V(E_2) = -6.2 + (.75)(12.2) = 3.0 \\ &V(E_3) = -3 + (.9)(12.2) = 8.0 \end{aligned}$$

Thus we would pick E_2 for state 1 and E_3 for state 2.   This corresponds to the previous choice so now a maximum has been attained.   Therefore a doctrine has been determined and we should use a second level of effort for state 1 and maximum effort for state 2.   We would pay up to 12.2 more units of money to start off in state 1.   ////

9-7   CONTINUOUS MARKOV PROCESSES

With a few simple assumptions, it is possible to obtain continuous Markov processes as the limit of discrete processes as the length of the time interval between transition periods approaches zero.   Consider intervals which are time-measurable and each of length Δt.   Let p_{ij} be the probability that a system in state i at the beginning of Δt will change to state j during Δt so that at the beginning of the next interval of Δt the system will be in state j.   Then Eqs. (9-2) become

$$P_j(t + \Delta t) = \sum_{i=1}^{N} p_{ij} P_i(t) \qquad j = 1, 2, 3, \ldots, N \qquad (9\text{-}29)$$

If $P_j(t)$ is subtracted from both sides of (9-29) and these equations also divided by Δt, there results

$$\frac{P_j(t + \Delta t) - P_j(t)}{\Delta t} = \sum_{\substack{i=1 \\ i \neq j}}^{N} \frac{p_{ij} P_i(t)}{\Delta t} + P_j(t) \frac{p_{jj} - 1}{\Delta t} \qquad j = 1, 2, \ldots, N$$

In order to develop a continuous Markov process, the following three limits must exist:

$$\frac{P_j(t + \Delta t) - P_j(t)}{\Delta t}\bigg|_{\Delta t \to 0} = \frac{dP_j}{dt} \qquad \frac{p_{ij}}{\Delta t}\bigg|_{\Delta t \to 0} = a_{ij} \qquad \text{and} \qquad \frac{p_{jj} - 1}{\Delta t}\bigg|_{\Delta t \to 0} = a_{jj}$$

Furthermore, the existent a_{ij} and a_{jj} must be constant and thus independent of time to be equivalent to the Markov assumption in the previously discussed discrete case.   The reader should observe that

$$\frac{p_{ij}}{\Delta t} = a_{ij} + \varepsilon_{ij}$$

where $\varepsilon_{ij} \to 0$ as $\Delta t \to 0$ is equivalent to the statement that the probability that a system changes from state i to state j in dt is $a_{ij}\, dt$.   Obviously, the $a_{ij}\ (i \neq j)$ are all positive from the definition unless they are 0.   In Sec. 2-11 the probability of the occurrence of an event for a Poisson distribution in time dt was

$\lambda\, dt$, where λ was the average number of occurrences per unit time and was constant by hypothesis. Thus the two concepts are equivalent, and the a_{ij} are rates in the same sense as λ.

On the basis that the limits discussed do exist, Eqs. (9-29) yield

$$\frac{dP_j(t)}{dt} = \sum_{\substack{i=1 \\ i \neq j}}^{N} a_{ij} P_i(t) + a_{jj} P_j(t) = \sum_{i=1}^{N} a_{ij} P_i(t) \qquad j = 1, 2, \ldots, N \qquad (9\text{-}30)$$

There necessarily must exist some relationship between the a's since $\sum_{j=1}^{N} p_{ij} = 1$, and thus

$$\sum_{j=1}^{N} p_{ij} = (a_{i1} + \varepsilon_{i1})\,\Delta t + (a_{i2} + \varepsilon_{i2})\,\Delta t + \cdots + [(a_{ii} + \varepsilon_{ii})\,\Delta t + 1] + \cdots$$
$$+ (a_{iN} + \varepsilon_{iN})\,\Delta t = 1$$

so that

$$(a_{i1} + a_{i2} + \cdots + a_{iN})\,\Delta t + (\varepsilon_{i1} + \varepsilon_{i2} + \cdots + \varepsilon_{iN})\,\Delta t = 0$$

for $i = 1, 2, \ldots, N$.

Since t is an independent variable making Δt arbitrary and also since the second term in the above expression is an infinitesimal of higher order than Δt, it follows that

$$a_{i1} + a_{i2} + \cdots + a_{iN} = 0 \qquad (9\text{-}31)$$

as $\Delta t \to 0$ for $i = 1, 2, \ldots, N$.

As all the $a_{ij} \geq 0$ if $i \neq j$, then a_{jj} must be negative if (9-31) is true. In fact, the negative character of a_{jj} may be seen from its definition that $a_{jj} = (p_{jj} - 1)/\Delta t$ as $\Delta t \to 0$ since p_{jj} is approaching 1 through values less than 1.

One of the techniques for solving a system of k linear, homogeneous, first-order, differential equations with constant coefficients in the standard form, such as

$$\frac{dy_1}{dt} = a_{11} y_1 + a_{12} y_2 + \cdots + a_{1k} y_k$$

$$\frac{dy_2}{dt} = a_{21} y_1 + a_{22} y_2 + \cdots + a_{2k} y_k \qquad (9\text{-}32)$$

$$\cdots\cdots\cdots\cdots\cdots\cdots\cdots\cdots\cdots\cdots$$

$$\frac{dy_k}{dt} = a_{k1} y_1 + a_{k2} y_2 + \cdots + a_{kk} y_k$$

is to find the eigenvalues of the determinant of the coefficients as

$$
\begin{vmatrix}
a_{11} - r & a_{12} & \cdots & a_{1k} \\
a_{21} & a_{22} - r & \cdots & a_{2k} \\
\hdotsfor{4} \\
a_{k1} & a_{k2} & \cdots & a_{kk} - r
\end{vmatrix} = 0 \qquad (9\text{-}33)
$$

If (9-33) has k distinct roots, $r_1, r_2, \ldots, r_k$, then the solution of (9-32) would be

$$
y_i = A_{i1} e^{r_1 t} + A_{i2} e^{r_2 t} + \cdots + A_{ik} e^{r_k t} \qquad i = 1, 2, \ldots, k \qquad (9\text{-}34a)
$$

where the A_{ij} are constants.   Multiple roots are handled in the usual manner. For example, if r_1 is a triple root, then $e^{r_1 t}$, $t e^{r_1 t}$, and $t^2 e^{r_1 t}$ are all solutions of (9-32) so that

$$
y_i = (A_{i1} + A_{i2} t + A_{i3} t^2) e^{r_1 t} + A_{i4} e^{r_4 t} + \cdots + A_{ik} e^{r_k t} \qquad i = 1, 2, \ldots, k
$$
$$(9\text{-}34b)$$

Pairs of complex roots, such as $r_1 = \alpha + i\beta$ and $r_2 = \alpha - i\beta$, are usually written in the more convenient form

$$
A_{i1} e^{r_1 t} + A_{i2} e^{r_2 t} = e^{\alpha t}(a_{i1} \cos \beta t + a_{i2} \sin \beta t) \qquad (9\text{-}35)
$$

where a_{i1} and a_{i2} are constants also and expressible in terms of A_{i1} and A_{i2}.

Obviously, the k^2 constant coefficients in the solution of (9-32), as indicated, cannot all be independent since k first-order equations require only k arbitrary constants.   The relationship between the A_{ij} is necessary in order that the solution satisfy the equations.   If the appropriate solutions for the y_i are substituted in (9-32) and the coefficients of similar functions of t on both sides of the equations, i.e., coefficients of $e^{\alpha_1 t}$, $e^{\alpha_2 t}$, etc., and, possibly ($t e^{\alpha_1 t}$, $t^2 e^{\alpha_1 t}$, etc.), are set equal, there result k^2 equations relating the A_{ij}.   However, because each r_i is a root of the homogeneous system, one equation is dependent, so that actually there are only $k^2 - k$ equations, thus allowing all the coefficients A_{ij} to be expressed in terms of k of them, which k constitute the arbitrary constants of the system (9-32).   These k constants are those which are determined from the initial conditions imposed upon the system.

Returning to the system (9-30) for a continuous Markov process we observe that this system is in the standard form of (9-32).   The matrix of the coefficients, however, form what is known as a differential matrix, in which the sum of each row is equal to zero [see Eq. (9-31)].   The roots of the secular equation for the corresponding determinant have special properties which are very important in determining the behavior of a Markov process.

The discussion will be limited to a three-state process although the results are true for an N-state system. The secular equation for (9-30) with $N = 3$ would be as follows, if the appropriate restrictions are placed on the row totals:

$$\begin{vmatrix} -a_{12} - a_{13} - r & a_{12} & a_{13} \\ a_{21} & -a_{21} - a_{23} - r & a_{23} \\ a_{31} & a_{32} & -a_{31} - a_{32} - r \end{vmatrix} = 0 \qquad (9\text{-}36)$$

where $a_{12}, a_{13}, \ldots, a_{32}$ are all > 0.

If (9-36) is expanded, the following cubic equation for r is developed:

$$r[r^2 + (a_{12} + a_{21} + a_{13} + a_{31} + a_{32})r + a_{31}(a_{21} + a_{23})$$
$$+ a_{12}(a_{13} + a_{31} + a_{32}) + (a_{21} + a_{13})(a_{23} + a_{32})] = 0 \qquad (9\text{-}37)$$

Since the constants of the quadratic in r inside the brackets are all positive, the secular equation has no positive roots. If the roots are real, they are negative or zero, and if complex, as $\alpha + i\beta$, the α must be negative. It should also be noted that at least one root of this secular equation is equal to zero; this corresponds to the $r = 1$ in the discrete Markov process. The zero root is due to the fact that there is linear dependence in the system (9-30) since $p_1 + p_2 + \cdots + p_k = 1$ and, therefore, $dp_1/dt + dp_2/dt + \cdots + dp_k/dt = 0$.

Thus, in the case of a three-state continuous Markov process, (9-37) would indicate that, except in very exceptional situations, the solutions would be of the form

$$P_i(t) = A_i + B_i e^{-k_1^2 t} + C_i e^{-k_2^2 t} \qquad i = 1, 2, \text{ or } 3 \qquad (9\text{-}38)$$

or
$$P_i(t) = A_i + e^{-k_1^2 t}(B_i \cos \beta t + C_i \sin \beta t)$$

where, again, $i = 1, 2,$ or 3.

Actually, an N-state system would have solutions which would be made up of terms similar to these.

This property of the roots of the secular equation for the coefficients of (9-30) is thus very important inasmuch as, if we let $t \to \infty$, all the terms in the solution approach zero except the A_i ($i = 1, 2, \ldots, N$). Therefore, $P_i(\infty) = A_i$ in general. Multiple roots do not cause any difficulty here except in the exceptional case where more than one zero root exists, which then produces terms such as $B_i t$, $C_i t^2$, etc., which become infinite as $t \to \infty$.

The $P_i(\infty) = A_i$ when $i = 1, 2, \ldots, N$ (if they exist) are called the steady-state probabilities for the continuous Markov process, and they do not constitute a set of arbitrary constants. In fact, they are determined from the matrix of the coefficients of the set of differential equations in the same way as the transition matrix determined the steady-state probabilities in the discrete case.

These steady-state probabilities will occur when all the $dP_i(\infty)/dt = 0$ ($i = 1, 2,$ $..., N$); this restriction reduces (9-30) to the set of homogeneous equations

$$\sum_{i=1}^{N} a_{ij} P_i(\infty) = 0 \qquad j = 1, 2, ..., N \qquad (9\text{-}39)$$

This set (9-39), in conjunction with the fact that $\sum_{i=1}^{N} P_i(\infty) = 1$, is sufficient in general to determine the $P_i(\infty)$ ($i = 1, 2, ..., N$).

Fortunately, interest is often centered on the steady-state behavior of a Markov process so that the general solution is not necessary for the analysis, but only the solution of (9-39). Often, the steady-state probabilities can be found even with an infinite-state system since (9-39) can often be solved sequentially, as we shall see in the examples to follow. Obviously, the general solution of a Markov process with a large number of states becomes impractical, because of the difficulty of finding all the roots of the secular equation and the difficulty of solving for the relationship between the constants A_{ij}.

The reader should keep in mind in all this discussion and in the examples to follow that a Poisson process is always hypothesized and, therefore, the probability density of the time between occurrences is exponential.

EXAMPLE 9-7   A mouse is kept in a circular cage in the form of a ring which has three compartments, 1, 2, and 3, connected by passageways as indicated in Fig. 9-1. Over a long period of time, it is observed experimentally that he goes from compartment 1 to compartment 2 twice an hour and from 1 to 3 four times an hour. It is also observed that the mouse goes from 2 to 1 six times per hour, and from 2 to 3 four times per hour. Finally, he goes from 3 to 1 three times per hour, and from 3 to 2 once per hour. The mouse is observed in compartment 1 at $t = 0$. What is the probability that he will be in compartment 1 at the end of 15 minutes? The differential matrix portraying these values given by the data is

$$\begin{pmatrix} -6 & 2 & 4 \\ 6 & -10 & 4 \\ 3 & 1 & -4 \end{pmatrix}$$

and the differential equations for the continuous Markov process would therefore be (with P_1, P_2, and P_3 functions of time) from (9-30)

$$\frac{dP_1}{dt} = -6P_1 + 6P_2 + 3P_3 \qquad \frac{dP_2}{dt} = 2P_1 - 10P_2 + P_3$$

$$\frac{dP_3}{dt} = 4P_1 + 4P_2 - 4P_3$$

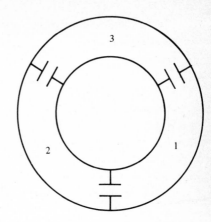

FIGURE 9-1

The steady-state probabilities satisfy the equations

$$-6P_1 + 6P_2 + 3P_3 = 0 \qquad 2P_1 - 10P_2 + P_3 = 0 \qquad 4P_1 + 4P_2 - 4P_3 = 0$$

and $P_1 + P_2 + P_3 = 1$ which gives as a solution

$$P_1(\infty) = \tfrac{3}{8} \qquad P_2(\infty) = \tfrac{1}{8} \qquad \text{and} \qquad P_3(\infty) = \tfrac{1}{2}$$

Thus characteristic roots of the coefficient matrix satisfy the equation

$$\begin{vmatrix} -6-r & 2 & 4 \\ 6 & -10-r & 4 \\ 3 & 1 & -4-r \end{vmatrix} = 0$$

whence $r = 0$, -8, and -12.

Therefore, the solution of the differential equations for the process will be of the form

$$P_1(t) = \tfrac{3}{8} + A_1 e^{-8t} + A_2 e^{-12t}$$
$$P_2(t) = \tfrac{1}{8} + B_1 e^{-8t} + B_2 e^{-12t}$$
$$P_3(t) = \tfrac{1}{2} + (-A_1 - B_1)e^{-8t} + (-A_2 - B_2)e^{-12t}$$

Notice that the coefficients of the third equation must be such that $P_1(t) + P_2(t) + P_3(t) = 1$ for all values of t.

If the solutions for $P_1(t)$, $P_2(t)$, and $P_3(t)$ are now substituted in any one of the three differential equations, here, for example, the first one, one obtains the relationship between the constants as

$$-8A_1 e^{-8t} - 12A_2 e^{-12t} = -6A_1 e^{-8t} - 6A_2 e^{-12t} + 6B_1 e^{-8t}$$
$$+ 6B_2 e^{-12t} + 3(-A_1 - B_1)e^{-8t} + 3(-A_2 - B_2)e^{-12t}$$

Hence, equating coefficients of e^{-8t} and e^{-12t}, the two following equations are developed

$$-8A_1 = -6A_1 + 6B_1 - 3A_1 - 3B_1$$
$$-12A_2 = -6A_2 + 6B_2 - 3A_2 - 3B_2$$

so that $B_1 = A_1/3$ and $B_2 = -A_2$.

The reader should observe that substituting in one equation is sufficient instead of two, which was suggested by the theory. The reason is, of course, that three constants have already been determined as steady-state values, and these values satisfy the system.

Thus, the solution for the differential equations for this particular Markov process is

$$P_1(t) = \tfrac{3}{8} + A_1 e^{-8t} + A_2 e^{-12t}$$

$$P_2(t) = \tfrac{1}{8} + \frac{A_1}{3} e^{-8t} - A_2 e^{-12t}$$

$$P_3(t) = \tfrac{1}{2} - \frac{4A_1}{3} e^{-8t}$$

The correct number of arbitrary constants are now present and must be determined from the initial conditions that $P_1(0) = 1$, $P_2(0) = 0$, and $P_3(0) = 0$ so that $1 = \tfrac{3}{8} + A_1 + A_2$ and $0 = \tfrac{1}{8} + A_1/3 - A_2$ whence $A_1 = \tfrac{3}{8}$ and $A_2 = \tfrac{2}{8}$.

The final solution satisfying the differential equations and the boundary conditions are

$$P_1(t) = \tfrac{3}{8} + (\tfrac{3}{8})e^{-8t} + (\tfrac{2}{8})e^{-12t}$$
$$P_2(t) = \tfrac{1}{8} + (\tfrac{1}{8})e^{-8t} - (\tfrac{2}{8})e^{-12t}$$
$$P_3(t) = \tfrac{1}{2} - (\tfrac{1}{2})e^{-8t}$$

and therefore the probability that the mouse is in 1 at $t = \tfrac{1}{4}$ would be

$$P_1(\tfrac{1}{4}) = \tfrac{3}{8} + (\tfrac{3}{8})e^{-2} + (\tfrac{2}{8})e^{-3} = (\tfrac{1}{8}) \frac{3e^3 + 3e + 2}{e^3} \approx 0.44$$

which is getting close to the steady-state value ≈ 0.38.   ////

9-8 ILLUSTRATIONS OF CONTINUOUS MARKOV PROCESSES

There are many situations where the assumption of a Markovian model is close enough to reality to produce a meaningful analysis of a given problem, and this solution often leads to a better understanding and improvement of the simulated operation.   Queuing theory and birth and death process are two subjects which have developed in some of their aspects through the use of the Markov model, and the interested reader is referred to the many books on these subjects.   However, a few illustrative examples point out the general approach to some problems in those categories and are within the scope of the present discussion.   In applying these general principles, the reader should keep in mind the necessity of always being able to identify each possible state, since it is usually the individual state of the system which can be assumed to be Poisson.   A lack of the reality of the Poisson assumption means that the probabilities are now dependent on the past history so that the solutions based on the ideas illustrated in the following examples would no longer be valid. Solutions can be found in the case of an infinite number of possible states provided the equations can be solved recursively, and a convergent set of probabilities result from the solution.   We shall also illustrate how it is possible, in some instances, to relax the assumption that the time between occurrences is exponential by redefining the problem, using additional states, and thus permitting the model to have wider applicability.

EXAMPLE 9-8   A hot-dog stand has room for only one customer to be served at a time, and additional customers queue up behind him and wait their turn.   Assuming the arrival rate is A customers per hour, the servicing rate is S customers per hour, and both arrivals and services are Poisson, what is the average length of the queue?

The Markov process has an infinite number of states with respective probabilities $P_0(t), P_1(t), P_2(t), \ldots, P_n(t), \ldots$, where $P_n(t)$ is the probability that there are n persons in the system at time t.   By the definition of such a process, the system can change only from its present state to an adjacent state in time dt. Thus the differential matrix is

State j

	0	1	2	3	4	5	6 $\cdots$
0	$-A$	A	0	0	0	0	0 $\cdots$
1	S	$(-S-A)$	A	0	0	0	0 $\cdots$
2	0	S	$(-S-A)$	A	0	0	0 $\cdots$
State i 3	0	0	S	$(-S-A)$	A	0	0 $\cdots$
4	0	0	0	S	$(-S-A)$	A	0 $\cdots$
5	0	0	0	0	S	$(-S-A)$	$A \cdots$

Hence the infinite set of differential equations describing the process as (9-30) would be

$$\frac{dP_0(t)}{dt} = -AP_0(t) + SP_1(t)$$

$$\frac{dP_1(t)}{dt} = A P_0(t) - (A + S)P_1(t) + SP_2(t)$$

. . . .

$$\frac{dP_n(t)}{dt} = AP_{n-1}(t) - (A + S)P_n(t) + SP_{n+1}(t)$$

. . . .

If only the steady-state probabilities are required, as in this problem, then all the $dP_n(t)/dt$ must equal zero for all n. If the $P_n(\infty)$ are denoted by P_n, the first equation yields $P_1 = (A/S)P_0$ and the second equation gives $P_2 = [(A + S)/S]P_1 - (A/S)P_0 = (A^2/S^2)P_0$. Solving the equation sequentially, we find $P_n = (A^n/S^n)P_0$.

Since the sum of the P_n to infinity must equal unity, then

$$\left(1 + \frac{A}{S} + \frac{A^2}{S^2} + \cdots + \frac{A^n}{S^n} + \cdots\right)P_0 = 1$$

which gives $P_0 = 1 - A/S$ so that in general

$$P_n = \left(1 - \frac{A}{S}\right)\left(\frac{A}{S}\right)^n \qquad \text{if} \quad \frac{A}{S} < 1$$

Thus the expected or average length of the queue would be

$$L = \sum_{n=0}^{\infty} nP = \left(1 - \frac{A}{S}\right) \sum_{n=0}^{\infty} n\left(\frac{A}{S}\right)^n = \frac{A/S}{1 - A/S}$$

which approaches infinity as $A/S \to 1$. ////

EXAMPLE 9-9 In the preceding example let us assume that the servicing facility of the hot-dog stand consists of two operations each one of which is Poisson in behavior with a rate of $2S$ per hour. Each customer's order must pass through both operations before the order of any other customer can be honored, but the total average time of service is the same as before since $1/2S + 1/2S = 1/S$. The effect of the two steps in the servicing facility is to change the density function of the time for a random order to be filled from $(1/S)e^{-t/s}$ to $(1/4S^2)te^{-t/2S}$ since the total time is now the sum of two exponential variates, each with density function $1/2S \, e^{-t/2S}$. This gamma density function may be much more realistic as a mathematical model for an order to be completed between t and $t + dt$. Assuming the arrival rate is A, but only four persons may be in line at once because of space limitations, what is the steady-state probability of 0, 1, 2, 3, or 4 persons in the queue? Assume that when there are four in the queue new arrivals merely go to another source. Associated with any number in the queue n, there are two possible states ($n1$) and ($n2$) where ($n1$) indicates that the order of the customer being served is in the first step and ($n2$) indicates that it is in the second step. Thus in this problem there are nine possible states and each one can be considered as behaving as a Poisson variate. It should be noted that when a customer's order is completed in the second phase the order of the next customer (if there is one) immediately starts in the first step. Thus the matrix is

State j

	0	11	12	21	22	31	32	41	42
0	$-A$	A	0	0	0	0	0	0	0
11	0	$(-2S-A)$	$2S$	A	0	0	0	0	0
12	$2S$	0	$(-2S-A)$	0	A	0	0	0	0
21	0	0	0	$(-2S-A)$	$2S$	A	0	0	0
State i 22	0	$2S$	0	0	$(-2S-A)$	0	A	0	0
31	0	0	0	0	0	$(-2S-A)$	$2S$	A	0
32	0	0	0	$2S$	0	0	$(-2S-A)$	0	A
41	0	0	0	0	0	0	0	$-2S$	$2S$
42	0	0	0	0	0	A	0	0	$-A$

Since interest is centered on the steady-state probabilities P_0, P_{11}, P_{12}, ..., P_{42}, we have the following equations connecting them, as (9-39):

$$-AP_0 + 2SP_{12} = 0 \qquad\qquad AP_0 - (2S + A)P_{11} + 2SP_{22} = 0$$

$$2SP_{11} - (2S + A)P_{12} = 0 \qquad\qquad AP_{11} - (2S + A)P_{21} + 2SP_{32} = 0$$

$$AP_{12} + 2SP_{21} - (2S + A)P_{22} = 0 \qquad\qquad AP_{21} - (2S + A)P_{31} + AP_{42} = 0$$

$$AP_{22} + 2SP_{31} + (2S + A)P_{32} = 0 \qquad\qquad AP_{31} - 2SP_{41} = 0$$

$$AP_{32} + 2SP_{41} - AP_{42} = 0$$

These equations may easily be solved for the steady-state probabilities and if $A/2S = K$, then $P_{12} = KP_0$, $P_{11} = (K + K^2)P_0$

$$P_{22} = (2K^2 + K^3)P_0 \qquad\qquad P_{21} = (K^2 + 3K^3 + K^4)P_0$$

$$P_{32} = (3K^3 + 4K^4 + K^5)P_0 \qquad\qquad P_{31} = (K^3 + 6K^4 + 5K^5 + K^6)P_0$$

$$P_{42} = (4K^3 + 10K^4 + 6K^5 + K^6)P_0 \qquad\qquad P_{41} = (K^4 + 6K^5 + 5K^6 + K^7)P_0$$

Since $P_0 + P_{11} + P_{12} + \cdots + P_{41} + P_{42} = 1$, this determines P_0 as

$$P_0 = \frac{1}{1 + 2K + 4K^2 + 12K^3 + 22K^4 + 18K^5 + 7K^6 + K^7}$$

The probability that the system is in state n is $P_{n1} + P_{n2} = P_n$. In this problem $A/2S$ is not restricted in value. For example, if $A/2S = 1$, then

$$P_0 = \tfrac{1}{67} \qquad P_{11} = \tfrac{2}{67} \qquad P_{12} = \tfrac{1}{67} \qquad P_{21} = \tfrac{5}{67} \qquad P_{22} = \tfrac{3}{67}$$

$$P_{31} = \tfrac{13}{67} \qquad P_{32} = \tfrac{8}{67} \qquad P_{41} = \tfrac{13}{67} \qquad P_{42} = \tfrac{21}{67}$$

It then follows that $P_0 = \tfrac{1}{67}$, $P_1 = \tfrac{3}{67}$, $P_2 = \tfrac{8}{67}$, $P_2 = \tfrac{21}{67}$, and $P_4 = \tfrac{34}{67}$.

////

EXAMPLE 9-10 An off-the-street parking lot has five spaces. There are two additional spaces, for live parking only, on the street where customers may wait for a vacancy. If these seven spots are filled, new arrivals must seek another place to park. The arrival rate is A per hour provided there is a vacancy in the lot's five spaces, but this drops to $A/2$ in case the five spaces are full and to $A/3$ in case there is a customer also waiting in a live parking space. After that, of course, the rate drops to zero when all seven spots are filled. The average time a car remains in the lot is $1/S$ while those waiting in the street never leave until a space is available. Assuming a Poisson process, what is the average number of spaces filled in the parking lot? The differential matrix for this system would be

		0	1	2	3	4	5	6	7
	0	$-A$	A	0	0	0	0	0	0
	1	S	$(-S-A)$	A	0	0	0	0	0
	2	0	$2S$	$(-2S-A)$	A	0	0	0	0
	3	0	0	$3S$	$(-3S-A)$	A	0	0	0
State i	4	0	0	0	$4S$	$(-4S-A)$	A	0	0
	5	0	0	0	0	$5S$	$(-5S-A/2)$	$A/2$	0
	6	0	0	0	0	0	$5S$	$(-5S-A/3)$	$A/3$
	7	0	0	0	0	0	0	$5S$	$-5S$

State j

Since, in this problem, interest is centered on the steady-state probability values only of $P_0, P_1, P_2, \ldots, P_7$, they must satisfy directly Eqs. (9-39) or

$$-AP_0 + SP_1 = 0 \qquad AP_0 - (S+A)P_1 + 2SP_2 = 0$$

$$AP_1 - (2S+A)P_2 + 3SP_3 = 0 \qquad AP_2 - (3S+A)P_3 + 4SP_4 = 0$$

$$AP_3 - (4S+A)P_4 + 5SP_5 = 0 \qquad AP_4 - \left(5S + \frac{A}{2}\right)P_5 + 5SP_6 = 0$$

$$\frac{A}{2}P_5 - \left(5S + \frac{A}{3}\right)P_6 + 5SP_7 = 0 \qquad \frac{A}{3}P_6 - 5SP_7 = 0$$

Thus

$$P_1 = \left(\frac{A}{S}\right)P_0 \qquad P_2 = \frac{(S+A)P_1 - AP_0}{2S} = \frac{1}{2!}\left(\frac{A}{S}\right)^2 P_0$$

$$P_3 = \frac{1}{3!}\left(\frac{A}{S}\right)^3 P_0 \qquad P_4 = \frac{1}{4!}\left(\frac{A}{S}\right)^4 P_0 \qquad P_5 = \frac{1}{5!}\left(\frac{A}{S}\right)^5 P_0$$

$$P_6 = \frac{1}{5S}\left[\left(5S + \frac{A}{2}\right)P_5 - AP_4\right]$$

$$= \frac{1}{5}\left[\frac{(5+A/2S)(A/S)^5}{5!} - \frac{(A/S)(A/S)^4}{4!}\right]P_0 = \frac{3}{5}\frac{(A/S)^6}{6!}P_0$$

and finally

$$P_7 = \left(\frac{7}{25}\right)\frac{(A/S)^7}{7!}P_0$$

Thus all the P's are uniquely determined and P_0 must equal

$$P_0 = \frac{1}{1 + A/S + (1/2!)(A/S)^2 + (1/3!)(A/S)^3 + (1/4!)(A/S)^4 + (1/5!)(A/S)^5}$$

$$+ \frac{3}{5}\frac{(A/S)^6}{6!} + \frac{7}{25}\frac{(A/S)^7}{7!}$$

The percentage of the time that the facility of five spaces will be full would be

$$P_5 + P_6 + P_7 = \frac{1}{5!}\left(\frac{A}{S}\right)^5\left[1 + \frac{1}{10}\left(\frac{A}{S}\right) + \frac{1}{150}\left(\frac{A}{S}\right)^2\right]P_0$$

The average or expected number of paying customers E would then be

$$E = \sum_{n=0}^{4} nP_n + 5(P_5 + P_6 + P_7)$$

$$= \frac{(A/S)[1 + A/S + (1/2!)(A/S)^2 + (1/3!)(A/S)^3 + (1/4!)(A/S)^4 + [1/2(5!)](A/S)^5 + [1/5(6!)](A/S)^6]}{1 + (A/S) + (1/2!)(A/S)^2 + (1/3!)(A/S)^3 + (1/4!)(A/S)^4 + (1/5!)(A/S)^5 + [3/5(6!)](A/S)^6 + [7/25(7!)](A/S)^7}$$

For small values of A/S, it is seen that the value of E is approximately equal to A/S since the numerator and denominator brackets are about equal.

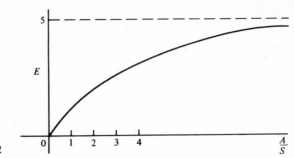

FIGURE 9-2

As A/S increases, the numerator becomes larger than the denominator and, for extremely large A/S, we have

$$E = \frac{\dfrac{1/5(A/S)^7}{6!}}{\dfrac{7/25(A/S)^7}{7!}} = \frac{25}{5} = 5$$

as would be expected. This suggests the graph of E versus A/S as in Fig. 9-2.

////

EXAMPLE 9-11 A particular car wash may be considered as consisting of a two-phase operation which each car must pass through sequentially. The average time spent in the first phase is $1/S_1$ hours, and the time in the second is $1/S_2$ hours, making the total average time for a car wash $1/S_1 + 1/S_2$. When a car is through the first phase, this facility is then available for a new car to enter and start its first phase. The arrival rate is A cars per hour, but there is no place to wait so that, unless the facility of phase 1 is open, the customer goes down the street to a competitor. What is the probability of their being 0, 1, or 2 cars in the car-wash facility and if $S_1 = S_2$, what is the average number of cars in the entire facility? Assume a Poisson process throughout.

In this problem, we have five possible states for the system in steady state: no cars in service, state 0 (P_0); one car in phase 1 and none in phase 2, state 10 (P_{10}); one car in phase 2 and none in phase 1, state 01 (P_{01}); two cars in the system and both being processed, state 11 (P_{11}); two cars in the system with work completed on the car in the first phase but it is blocked from moving, state $b1$ (P_{b1}). A blocked car obviously moves instantaneously when servicing in phase 2 is completed. The differential matrix is

State j

		0	10	01	11	$b1$
	0	$-A$	A	0	0	0
	10	0	$-S_1$	S_1	0	0
State i	01	S_2	0	$(-S_2 - A)$	A	0
	11	0	S_2	0	$(-S_1 - S_2)$	S_1
	$b1$	0	0	S_2	0	$-S_2$

which yields the steady-state equations

$$-AP_0 + S_2 P_{01} = 0 \qquad AP_0 - S_1 P_{10} + S_2 P_{11} = 0$$

$$S_1 P_{10} - (S_2 + A)P_{01} + S_2 P_{b1} = 0 \qquad AP_{01} - (S_1 + S_2)P_{11} = 0$$

$$S_1 P_{11} - S_2 P_{b1} = 0$$

Solving these equations sequentially, we have

$$P_{01} = \left(\frac{A}{S_2}\right)P_0 \qquad P_{11} = \frac{A^2 P_0}{S_2(S_1 + S_2)} \qquad P_{b1} = \frac{S_1 A^2 P_0}{S_2{}^2(S_1 + S_2)}$$

$$P_{10} = \frac{A(S_1 + S_2) + A^2}{S_1(S_1 + S_2)} P_0$$

Again, since the sum of the probabilities must equal unity, we have for P_0

$$P_0 = \frac{1}{1 + A(1/S_1 + 1/S_2) + A^2(S_1 S_2 + S_1{}^2 + S_2{}^2)/S_1 S_2{}^2(S_1 + S_2)}$$

When $S_1 = S_2 = S$, these probabilities become

$$P_{01} = \left(\frac{A}{S}\right)P_0 \qquad P_{11} = \frac{1}{2}\left(\frac{A}{S}\right)^2 P_0 \qquad P_{b1} = \frac{1}{2}\left(\frac{A}{S}\right)^2 P_0 \qquad P_{10} = \left[\frac{A}{S} + \frac{1}{2}\left(\frac{A}{S}\right)^2\right]P_0$$

$$P_0 = \frac{1}{1 + 2A/S + \tfrac{3}{2}(A/S)^2}$$

Therefore, the expected number of cars in service E would be

$$E = (P_{01} + P_{10}) + 2(P_{11} + P_{b1})$$

$$= \left[\frac{2A}{S} + \frac{1}{2}\left(\frac{A}{S}\right)^2\right]P_0 + 2\left(\frac{A}{S}\right)^2 P_0$$

$$= \frac{2A/S + \tfrac{5}{2}(A/S)^2}{1 + 2A/S + \tfrac{3}{2}(A/S)^2}$$

As $A/S \to \infty$ it is clear that the expected number approaches $\tfrac{5}{3}$, which shows the effect of the blocking since 2 would be the maximum.   ////

APPENDIX TO CHAPTER 9

9-9 CONVERGENCE OF THE MAXIMIZATION PROGRAM

Let S_i and p_{ij} be the reward and the transition probabilities chosen for each i to start the iteration process. Also, let S_i^* and p_{ij}^* be those values which maximize V_i for each i in Eq. (9-28). Then because of the way S_i^* and p_{ij}^* were chosen,

$$S_i^* + \sum_{j=1}^{N} p_{ij}^* A_j = S_i + \sum_{j=1}^{N} p_{ij} A_j + \epsilon_i^2 \qquad (9\text{-}40)$$

where $\epsilon_i^2 \geq 0$ and $i = 1, 2, \ldots, N$.

Since S_i and p_{ij} satisfy (9-27) with the appropriate A_j and $\bar{S}$ and S_i^* and p_{ij}^* also satisfy (9-27) with the appropriate A_i^* and $\bar{S}^*$, we have

$$\begin{aligned} A_i + \bar{S} &= S_i + \sum_{i=1}^{N} p_{ij} A_j \\ A_i^* + \bar{S}^* &= S_i^* + \sum_{i=1}^{N} p_{ij}^* A_i^* \end{aligned} \qquad (9\text{-}41)$$

where for a completely ergodic process $\bar{S}^* = \sum_{i=1}^{N} S_i^* p_i^*(\infty)$. The $p_i^*(\infty)$ are of course the steady-state probabilities for the Markov process defined by the transition probabilities p_{ij}^*.

If the first equation in (9-41) is subtracted from the second and if $S_i^* - S_i$ is replaced by its value from (9-40), there results

$$(A_i^* - A_i) + (\bar{S}^* - \bar{S}) = \epsilon_i^2 + \sum_{1}^{N} p_{ij}^*(A_j^* - A_j) \qquad (9\text{-}42)$$

for all $i = 1, 2, \ldots, N$.

Since Eqs. (9-42) are identical in form to (9-27) it is clear that the gain $\bar{S}^* - \bar{S}$ for the process, with A's replaced by $A_j^* - A_j$, must be

$$S^* - \bar{S} = \sum_{i=1}^{N} \epsilon_i^2 P_i^*(\infty) \qquad (9\text{-}43)$$

As both ϵ_i^2 and $P_i^*(\infty)$ are greater than or equal to zero, $\bar{S}^* - \bar{S} \geq 0$. The process thus gains on each iteration until $S^* = S$. Convergence to a "best" policy is usually very rapid with large gains realized during the first few iterations, and the value thus obtained after a few jousts is often sufficiently good for practical purposes. The value of the technique is realized when a 10-state process having 10 alternatives in each state has 10^{10} strategies, which would involve an extremely large computational effort if they were all to be tried one after the other in order to determine the "best" one.

PROBLEMS

9-1 A particular Markov process is described by the stochastic matrix P as

$$P = \begin{pmatrix} .6 & .4 \\ .3 & .7 \end{pmatrix}$$

The system is in state 1 at the beginning. What is the probability that it is in state 1 after three transitions? *Ans.* .444.

9-2 A Markov process is characterized by the stochastic matrix P as

$$P = \begin{pmatrix} \frac{2}{3} & \frac{1}{3} & 0 \\ \frac{1}{3} & \frac{1}{2} & \frac{1}{6} \\ 0 & \frac{1}{3} & \frac{2}{3} \end{pmatrix}$$

If the system starts in state 2, what is the probability that it will be in any one of the possible three states after n transitions? *Ans.* $P_1(n) = \frac{2}{5} - (\frac{2}{5})(\frac{1}{6})^n$.

9-3 Consider the behavior of the Markov process defined by

$$P = \begin{pmatrix} 0 & 1 & 0 \\ 0 & 0 & 1 \\ 1 & 0 & 0 \end{pmatrix}$$

by considering the solution of the difference equation.

9-4 Given the Markov process characterized by the stochastic matrix

$$\begin{bmatrix} .8 & .2 \\ .3 & .7 \end{bmatrix}$$

What is the difference in the probability that the system is in state 1 after three transitions, depending upon whether it started in state 1 or in state 2? *Ans.* $\frac{1}{8}$.

9-5 Assume a simple ergodic Markov process with only two states. We start the process in either one of the possible states by tossing an unbiased coin as a method of choice. After a large number of trials, it is determined experimentally that after one transition the system is in state 1 with probability $\frac{6}{20}$, and after two transitions it is in state 1 with probability $\frac{26}{100}$. What are the equations which determine the states as a function of any number of transitions n? *Ans.* $P_1(n) = \frac{1}{4} + (\frac{1}{4})(\frac{1}{5})^n$, $P_2(n) = \frac{3}{4} - (\frac{1}{4})(\frac{1}{5})^n$.

9-6 A man plans to speculate on the stock market by buying and selling one and only one share of stock of a certain company. He will retain his profit and not reinvest it. This stock oscillates between three possible values but remains the same price during an entire day since changes in price occur (if an occurrence takes place) just before the market opens each morning. The three prices of the stock are L (low) with price \$1, M (medium) with price \$2, and H (high) with price \$3.

The changes in states are given by the stochastic matrix

	L	M	H
L	.5	.3	.2
M	.2	.4	.4
H	.4	.4	.2

What would be his average gain per day, over a long period of time, if he bought low and sold high? Neglect the original money spent for the single share of stock which he now buys and sells.

Ans. $(2⁄7). Average cycle *LHL* takes 6.91 days.

9-7 A Markov process has four possible states. Within each state there are two possible sets of transition probabilities, depending upon which set of corresponding rewards one wishes to accept. What should be the 4×4 stochastic matrix if one wishes to maximize the process gain per transition over a long period of time? The transition probabilities and the corresponding rewards are given as

$$
\text{State 1} \quad
\begin{array}{c} \text{Transition} \\ \text{probabilities} \end{array}
\begin{bmatrix} .6 & .4 \\ .8 & .2 \\ .5 & .5 \\ .9 & .1 \end{bmatrix}
\qquad
\begin{array}{c} \text{Rewards} \end{array}
\begin{bmatrix} 4 & -2 \\ 3 & -3 \\ 5 & -1 \\ 2 & -6 \end{bmatrix}
$$

$$
\text{State 2} \quad
\begin{array}{c} \text{Transition} \\ \text{probabilities} \end{array}
\begin{bmatrix} .9 & .1 \\ .6 & .4 \\ .7 & .3 \\ .4 & .6 \end{bmatrix}
\qquad
\begin{array}{c} \text{Rewards} \end{array}
\begin{bmatrix} 1 & -6 \\ 3 & -4 \\ 2 & -3 \\ 8 & -5 \end{bmatrix}
$$

9-8 A mouse's cage consists of two compartments *A* and *B* connected by a passageway. It is observed that the mouse goes from *A* to *B* on an average of twice an hour. At $t = 0$, he is observed to be in *A*. Assuming a Markov process, what is the probability that he is in *A* 5 minutes later? *Ans.* .86.

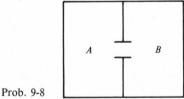

Prob. 9-8

9-9 A parking lot consists of five slots only. If a customer arrives and finds the lot full, he does not wait. The arrival of customers is Poisson with an average value of five per hour. The length of time a car stays is exponentially distributed with an average time of 15 minutes. Calculate the steady-state probabilities and the average number of cars in the lot. *Ans.* $P_0 = .287, E = 1.24$.

9-10 A gas station has a driveway that holds only three cars. No customer can wait out in the freeway. The arrival rate is different, depending upon how

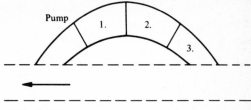

Prob. 9-10

many cars are lined up in the driveway; say λ_0, λ_1, λ_2 and $\lambda_3 = \lambda_4 \cdots =$ $\lambda_n = 0$. The service rate is μ per hour. Assuming a Poisson process, what are the steady-state probabilities and the average number of cars in the driveway?

9-11   A servicing facility consists of the sum of three operations, each Poisson with rate 3μ. Only one customer is in the facility at a time, and he progresses through the three operations consecutively. If there is room for only one customer to wait outside the service facility and additional ones leave, what is the steady-state probability for no customers, one customer, and two customers in the line if the arrival rate is λ?

9-12   A mouse is confined to a cage which has three compartments with connecting passageways (see Fig. 9-1). The rates at which the mouse goes between these compartments are observed to be as follows; the values are given as number of times per hour:

$1 \to 2$	2
$1 \to 3$	1
$2 \to 1$	3
$2 \to 3$	2
$3 \to 1$	3
$3 \to 2$	x

Assuming a continuous Markov process, is it possible to determine x so that the steady-state probability that the mouse is in compartment 2 is $\frac{1}{4}$? If so, what is this critical value?   *Ans.* $x = 1$.

9-13   The arrival rate of customers at a barber shop is three per hour if the barber is not busy cutting hair. If the barber is busy but no one is waiting, the arrival rate is two customers per hour. No customers come in if there is already someone waiting. If the barber cuts hair at the rate of four customers an hour, what is the probability that he will be busy?

9-14   A servicing facility consists of a two-stage affair. The object being served goes through the first unit and when it is finished passes on through the second one (a two-stage car wash, for example). Obviously, when a unit completes service in the first unit, it goes directly to the second unit unless that unit is filled. In that case the unit must wait until the second unit becomes vacant but will proceed directly to that second unit at the time of vacancy. This increases the number of states of the system, since we can have two units in service under two different conditions. In the first one both units are being serviced, and in the second one the first unit is merely sitting there waiting for a chance to proceed. Although the arrival rate is λ units per hour, unfortunately for the business no customer will wait so that they pass by the servicing facility if the first stage of service is occupied. It is suggested that a waiting place for one unit between the two service facilities should be installed so that a unit which has completed service in the first facility can move out of the way to wait for the second one to become vacant. This would allow a new arrival to come into the system if he showed up for service.

What is the expected gain in the number of cars in the system by installing this waiting station? Assume throughout that the process is Poisson in both servicing units and, of course, in the arrival. The time for service on the average is $\frac{1}{2}\mu$ for each of the servicing services.

10

SOME STATISTICAL USES OF PROBABILITY

10-1 INTRODUCTION

This chapter presents some of the tests and methods most commonly applied in statistical practice. Deliberate stress has been placed on reasoning as opposed to cut and dried rules of procedure. Much of the background developed in previous chapters is presupposed; here matters are brought to a head in specific applications to statistics. The chosen material represents an important part, but a small part, of this subject.

The chapter is organized around conventional uses of four major distributions: the normal, chi-square, t, and F. Many problems of hypothesis testing can be solved in terms of these four distributions, because a great many other distributions tend toward these as limiting forms or simply are sufficiently well approximated by them.

The normal distribution applies mainly to tests on large samples, unless the variate in question is normal to start with, in which case the sample size can be arbitrary. Some of the uses of chi-square assume fairly large samples, whereas others do not; as a rule it is the magnitude of an expected absolute frequency, rather than the sample size itself, which is critical in uses of chi-square.

With small samples, the t and F distributions fill the breach on certain tests to which otherwise the normal might be applied if the samples were larger; but they have broader uses as well.

The normal distribution is usually satisfactory for large-sample tests on percentages and means. Broadly speaking, hypotheses which characterize a distribution as a whole are tested by chi-square. For example, such hypotheses might define a distribution explicitly, assert that two or more variates have the same distribution without specifying its mathematical form, or state that certain attributes or variates are independent. Hypotheses which specify the mean value of one variate or the difference of the means of two variates are tested by t. Those which concern either the relative magnitudes of more than two means or the ratio of two variances are tested by F.

10-2 THE NORMAL TESTS

Use of the normal distribution as an approximation to the binomial distribution, in testing some hypothesized value of the probability of a given event, has been illustrated fully in Chap. 4. That topic will not be repeated in this chapter, but we shall take up two related topics which concern a pair of events. The first of these is the question whether two events have the same probability, or in sampling terms, whether two observed percentages are significantly different. This question was dealt with in Chap. 6 by employing the hypergeometric distribution. We shall see in Sec. 10-3 how the normal distribution can serve the same purpose as an approximation to the hypergeometric. The second topic is the test for association or independence of two events or attributes. This, also, is an application of the hypergeometric distribution and consequently falls within the scope of normal tests when the sample size is sufficiently large. How large a sample has to be is a question to which, at present, we can give only the empirical sort of answer stated in Sec. 10-3. Later on in that section (when we consider how to compensate for the continuity of the normal curve, as contrasted with the discrete increments of the hypergeometric probabilities) we shall reason as we did previously for the binomial (Chap. 4) but shall arrive at a correction term which differs in magnitude from that of the binomial because the interval of the variate concerned is of different width.

10-3 THE NORMAL TEST FOR THE DIFFERENCE BETWEEN TWO PERCENTAGES

When both a and N are large, the computation of $f(x)$ for the hypergeometric distribution becomes laborious. In that case, however, the hypergeometric distribution tends toward the binomial, and thence toward the normal. As n

becomes large, the asymptotic form of the distribution of x/n (the relative frequency of a given attribute A in random samples of size n) is normal with $\mu = p$, $\sigma^2 = pq/n$.

Suppose that in sample 1 of size N_1 there are x_1 observations exhibiting attribute A and in sample 2 of size N_2 there are x_2. Under the null hypothesis, the two samples are drawn from the same population. Assume that both N_1 and N_2 are large enough to make x_1/N_1 and x_2/N_2 approximately normal. Then by the additive property of the normal distribution, the difference

$$d = \frac{x_1}{N_1} - \frac{x_2}{N_2} \qquad (10\text{-}1)$$

is approximately normal with $\mu = 0$ and $\sigma^2 = pq(1/N_1 + N_2)$. Consequently the standard variate

$$z = \frac{d}{\sqrt{pq(1/N_1 + 1/N_2)}}$$

can be referred to unit normal tables. However, since p is ordinarily unknown, it is replaced by its sample estimate p', where

$$p' = \frac{x_1 + x_2}{N_1 + N_2} \qquad \text{with} \quad q' = 1 - p' \qquad (10\text{-}2)$$

and the test variate actually employed is

$$u = \frac{d}{\sqrt{p'q'(1/N_1 + 1/N_2)}} \qquad (10\text{-}3)$$

This, too, is treated as a unit normal variate, for

$$u \equiv z\sqrt{pq/p'q'}$$

and, in a probability sense, the ratio $pq/p'q'$ approaches unity as $N_1 + N_2$ becomes large. The accuracy of the normal approximation depends both on the sample size and the value of p. At present, rigorous proof of safe magnitudes is lacking, but it has been found empirically that good results are obtained provided that $np' > 5$ (that is, $N_1 p' > 5$ and $N_2 p' > 5$).

Trying the normal approximation on Example 6-27, we have

$$p' = \frac{8}{4,694} \approx .0017$$

and $np' = 4$, which falls a little short of the recommended limit. From Table 4-4, we find that the boundary of the critical region for the significance level .05 is -1.64. The numerator of u is

$$d = 0 - \frac{8}{2,347} = -.0034$$

and the denominator is

$$\sqrt{(.0017)(.9983)(.00085)} = .0012$$

whereupon

$$u = \frac{-.0034}{.0012} = -2.83$$

which lies well within the critical region. The normal probability $P(u \leq -2.83)$, corresponding to $x = 0$ in the stocking group, is .0023 as opposed to the value .004 given by the hypergeometric distribution. The decision, nevertheless, is the same, even though the apparent rarity of the observed event is exaggerated with the normal approximation.

EXAMPLE 10-1   The toy buyer of a department store observed that 8 out of 150 plastic dolls manufactured by company A were defective, whereas only 4 out of 130 manufactured by company B were defective. Is it reasonable to suppose that the two percentages represent a common level of quality?

Under the null hypothesis that both samples come from the same population, the difference

$$d = 8/150 - 4/130 = .053 - .031 = .022$$

represents a random draw from a distribution of possible differences for which the mean is zero and the variance is

$$\sigma^2 = pq(1/150 + 1/130) = pq(.0144)$$

where p is the proportion of defective articles in the common parent population and $q = 1 - p$. Sample estimates of p and q are

$$p' = \frac{8 + 4}{150 + 130} = 3/70 \qquad q' = 1 - p' = 67/70$$

whence the sample estimate of variance, say $\hat{\sigma}^2$, is

$$\hat{\sigma}^2 = p'q'(1/150 + 1/130) = (.0410)(.0144) = .000590$$

the corresponding square root being $\hat{\sigma} = .0243$.

Since the samples are fairly large and for both of them $np' > 5$, the test variate $u = d/\hat{\sigma}$ may be assumed to be unit normal in the sense that repeated independent trials would generate a series of values for which the true probability density function is adequately approximated by the unit normal. To decide whether an observed difference is consistent with the hypothesis that both samples represent a common population, we set up a two-sided critical region corresponding to the 5-percent level of significance. Table 4-4 shows that $2R(x) = .05$ when $x = 1.96$. Hence we shall accept the null hypothesis provided that

$$-1.96 < u < 1.96$$

but otherwise reject it.

Here the sample value of the test variate is

$$u = \frac{.022}{.024} = .92$$

which lies well within the acceptance region. Therefore, we accept the null hypothesis and conclude that the observed difference in proportions of defective articles is readily explainable on the basis of sampling fluctuations around the average.　　　////

EXAMPLE 10-2   Assuming equal sample sizes, how large must the sample size be in order that a given difference between two proportions will be judged significant?

Let N denote the sample size and p'_1, p'_2 (with $p'_1 > p'_2$) the observed proportions. Assuming N sufficiently large and $Np' > 5$, the difference will be judged significant at the .05 level if $u \geq 1.96$, whence $u^2 \geq 3.8416$. With equal sample sizes, we have

$$p' = \frac{p'_1 + p'_2}{2} \qquad q' = \frac{2 - p'_1 - p'_2}{2} \qquad d = p'_1 - p'_2$$

$$\hat{\sigma}^2 = p'q'\left(\frac{1}{N} + \frac{1}{N}\right) = \frac{(p'_1 + p'_2)(2 - p'_1 - p'_2)}{2N}$$

$$u^2 = \frac{d^2}{\hat{\sigma}^2} = \frac{2N(p'_1 - p'_2)^2}{(p'_1 + p'_2)(2 - p'_1 - p'_2)}$$

Therefore, to reach the 5-percent level of significance with a stated pair of proportions, the sample size must satisfy the following two conditions:

$$N \geq \frac{(1.9208)(p'_1 + p'_2)(2 - p'_1 - p'_2)}{(p'_1 - p'_2)^2}$$

$$N > \frac{10}{(p'_1 + p'_2)}$$

the first of which follows from the fact that $u^2 \geq 3.8416$ and the second from the fact that $Np' > 5$. The answer thus depends on both proportions and not merely on their difference. ////

As noted in Chap. 4 with reference to the normal approximation to the binomial distribution, a correction for continuity is a safeguard against an inflated estimate of significance. To arrive at a corresponding correction for the difference between relative frequencies, using the hypergeometric distribution, we recall that the difference d can be expressed in the form (Sec. 6-15)

$$d = \frac{N}{N_1 N_2} \left(x_1 - \frac{aN_1}{N} \right) \equiv \frac{N_1 + N_2}{N_1 N_2} (x_1 - N_1 p') \qquad (10\text{-}4)$$

Assume that our critical region is to be the right-hand tail of the distribution. Then in the binomial case we should reduce x_1 by one-half unit, and correspondingly, the value of d should be reduced by one-half times the coefficient of x_1, that is, $(N_1 + N_2)/2N_1 N_2$. (For it is to be noted that, although x_1 is binomially distributed, d is not.) Thus, the revised measures d' and u' are

$$d' = d - \frac{N_1 + N_2}{2N_1 N_2} \qquad u' = \frac{d'}{\hat{\sigma}} \qquad (10\text{-}5)$$

In the special case wherein $N_1 = N_2$ the correction reduces to $1/N_1$. For a left-hand region the correction is the same in magnitude but is added to d instead of being subtracted. Thus, in applying the normal approximation once again to Example 6-27, we put

$$d' = \frac{-8}{2,347} + \frac{1}{2,347} = \frac{-7}{2,347} = -.0030$$

whence

$$u' = \frac{-.0030}{.0012} = -2.50$$

and $P(u' \leq -2.50) = .0062$ which differs from .004 about as much as the former approximate value .0023 but in the opposite direction.

For a two-sided region these corrections should be applied to both sides. Or, what amounts to the same thing, the boundaries of the critical region should be pushed tailward in both directions and then the values of d and u taken as originally given.

10-4   THE NORMAL TEST FOR ASSOCIATION OR INDEPENDENCE

Two attributes A, B are independent in the probability sense when

$$P(A|B) = P(A)$$

Otherwise they are associated.   Therefore, although tests for association differ in motive from tests for independence, they are formally identical.   Furthermore, in the case of only two attributes, the test is the same as that for the difference between two percentages.   For if A and B are independent, the individuals exhibiting attribute B and those exhibiting the complementary one $\bar{B}$ may be regarded as making up independent samples as far as the relative frequencies of attribute A are concerned, because the observations themselves are random with respect to the two attributes, and by hypothesis, $P(A|B) = P(A) = P(A|\bar{B})$. Consequently, if the two conditional relative frequencies of A are significantly different, the two attributes A, B are significantly associated.   Thus the hypergeometric test of Sec. 6-15 applies, and under the conditions stated in Sec. 10-3, the normal test can be used as an approximation.   Ordinarily, a two-sided critical region would be appropriate.   When there are more than two attributes, the test for association or independence is based on chi-square.   The chi-square test is also applicable to two attributes, under the same conditions as the normal test, provided that the critical region is two-sided; for in that case, the chi-square test variate is identically the square of the normal test variate, assuming that continuity corrections are disregarded.

EXAMPLE 10-3   At the beginning of a semester the instructor announced that assigned problems were intended mainly for the students' own benefit as learning exercises and would not be counted heavily toward the course grade.   Unprepared for the wholesale hangings which his rope apparently invited, he decided to check up on the relationship, if any, between attention to homework in his course and the general average grade to date of the same students in their other courses, exclusive of his.

As an elementary procedure, he divided the class of 38 in half on the basis of homework performance (H = high, $\bar{H}$ = low) and, without reference to the first dichotomy, he divided the class in half also on the basis of general performance (G = high, $\bar{G}$ = low).   Assuming independence, the four joint categories HG, $\bar{H}G$, $H\bar{G}$, $\bar{H}\bar{G}$ are equally likely.   Letting x denote the number of individuals in one of these categories, say HG, the distribution of this symmetrical hypergeometric variate, under the null hypothesis, is

$$f(x) = \frac{C(19, x)C(19, 19 - x)}{C(38, 19)} \qquad x = 0, 1, \ldots, 19$$

Table 10-1 LOWER HALF OF THE HYPERGEOMETRIC DISTRIBUTION WITH
$a=19=N_1$, $N=38$

x	0	1	2	3	4	5	6	7	8	9
$f(x)$	.0000	.0000	.0000	.0000	.0004	.0038	.0208	.0718	.1616	.2414
$F(x)$	.0000	.0000	.0000	.0000	.0004	.0042	.0250	.0968	.2584	.4998

In this instance, because the parameters of the distribution are now known, we can establish the boundaries of the critical region without recourse to further data. From the lower half of the distribution, presented in Table 10-1, we find that for the .05 level of significance, the two parts of the critical region are $x \le 6$, $x \ge 13$.

Let us now look at the data (Table 10-2) which, being real, are a bit astonishing. Not only does the observed value of x (17) lie well within the critical region, but the total probability of the numerical class it represents (17, 18, 19, 2, 1, 0) is less than 1.7×10^{-6}.

As far as it goes, the evidence for association is arresting; still, we must be circumspect in its interpretation. First, consider the sample. Random samples would rarely give such an appearance of association if the attributes were, in fact, independent. Since this sample, however, was not drawn at random, we may question how well it typifies the academic population at large. We do not know and can only say that, although the results bear notice, they lack the essentials for valid inference. Suppose that the sample were not open to this very damaging objection, and let us grant that a high degree of association exists. Association per se does not imply a causal mechanism whereby one attribute can be acquired by cultivating the other; both may be outcomes of more basic conditions. So long as the population itself is not disturbed, the two attributes will continue hand in hand; but a deliberate change in one need not produce a corresponding change in the other. In particular, the instructor, in this example, has caught the students off guard and induced them to reveal their

Table 10-2 CONTINGENCY
TABLE FOR
EXAMPLE 10-3

Group	G	$\bar{G}$	Total
H	17	2	19
$\bar{H}$	2	17	19
Total	19	19	38

habits of work without inhibition. Perhaps, therefore, he has unintentionally measured their personal drive, which is reflected simultaneously in diligence and achievement. On the other hand, human behavior tends to support the belief that a favorable attitude of mind for success can be stimulated by activity characteristic of successful people. At any rate, the material for practice is presumably an aid to learning, and this is our assumption in furnishing exercises.

As an illustration of the normal approximation in this same example, we have $p = q = \frac{1}{2}$; $\sigma^2 = pq(\frac{2}{19}) = \frac{1}{38}$. To establish the inside boundary of the right half of the critical region we put $d\sqrt{38} = 1.9600$, $d = .3180$. The corrected value of d will then be $d' = d + \frac{1}{19} = .3706$ and for the corresponding limit on x we put

$$d' = \frac{2}{19}(x - 19\frac{1}{2}) \qquad x = (9.5)(1.3706) = 13.02$$

Taken literally, this limit would exlude 13, but since we know that the continuity adjustment overcorrects somewhat, we ignore the .02 and arrive at the same region as before, $x \leq 6$, $x \geq 13$. Needless to say, the labor is far less.

////

10-5 THE CHI-SQUARE TESTS

The distribution of chi-square was first derived by the mathematician-physicist, F. R. Helmert, in 1875 and was rediscovered independently by K. Pearson in 1900. Pearson was the first to realize the practical importance of chi-square. The principal uses of chi-square are (1) testing a hypothesis as to the mathematical form of a single distribution, (2) testing the hypothesis that two or more variates are identically distributed, and (3) testing the hypothesis that certain variates or attributes are independent. In applying these tests it is *not* assumed that the original populations are normal. The conventional process of eliminating unknown parameters, however, leads to a sum of positive quantities, the sampling behavior of which is closely approximated by that of the sum of squares of an appropriate number of independent unit normal variates. Hence, for practical purposes, the test variate may be regarded as chi-square.

In most applications of chi-square, we use a one-sided critical region of the form $\chi^2 \geq a$, where a is determined from the condition that $\int_a^\infty f(\chi^2)\, d\chi^2 = \alpha$ and α is the chosen level of significance. The lower extreme of the distribution is rarely of interest, because departures from the hypothesized form of distribution tend to increase the value of chi-square. Extremely small values of chi-square indicate abnormally close agreement of the sample with the hypothesized population but argue against the randomness of the sample rather than against the assumed type of population.

10-6   THE CHI-SQUARE TEST FOR GOODNESS OF FIT

The problem of testing the agreement of a sample frequency table with a hypothesized form of distribution was solved in the main by Karl Pearson in 1900,
although R. A. Fisher later contributed the important idea of "degrees of freedom," by which proper account is taken of parameters estimated from the same
data.   Let the total range of possibilities be divided into k mutually exclusive
and exhaustive categories.   Denote the observed frequencies by O_i ($i = 1$, 2,
..., k) and the expected frequencies (as determined by the hypothesis) by E_i.
The joint distribution of the frequencies in repeated independent trials is multinomial, and even if the use of this distribution were not prohibitively laborious
(which it is), we would still have the problem of deciding on the multiple regions
of acceptance and rejection.   A general theory of critical regions, based on
likelihood ratios, has been developed by J. Neyman and E. S. Pearson (the son
of Karl Pearson) but its application in this connection yields substantially the
same outcome as the earlier work of Karl Pearson.   Using the expected value
E_i as the norm for any category, it is reasonable to choose the quantity $(O_i - E_i)^2$
as a measure of departure from the norm.   However, the raw magnitudes of the
squared deviations would not be comparable from one category to another,
inasmuch as the scale of each is nearly proportional to E_i.   Therefore, a suitable
measure is $(O_i - E_i)^2/E_i$, and the measure of total discrepancy becomes

$$\hat{\chi}^2 = \sum_1^k \frac{(O_i - E_i)^2}{E_i} \qquad (10\text{-}6)$$

By expanding the quadratic and simplifying, we obtain the following equivalent
expression, which is sometimes more convenient for computational purposes:

$$\hat{\chi}^2 = \sum_1^k \frac{O_i^2}{E_i} - N \qquad (10\text{-}7)$$

where $N = \sum O_i = \sum E_i$.   We showed in Chap. 8 (Sec. 8-6) that this statistic is
distributed asymptotically as chi-square.   If the population parameters have
not been estimated from the sample, there are $k - 1$ degrees of freedom.   As a
rule, however, only the general form of the distribution is hypothesized, so that
in order to evaluate the expected frequencies, it is necessary to estimate the parameters from the data.   In that case, the number of degrees of freedom is di-
minished by the number of parameters estimated; if h parameters are estimated,
there are $k - 1 - h$ degrees of freedom.   The extent to which the true distribution of $\hat{\chi}^2$, under the given hypothesis, is approximated by that of chi-square is
affected primarily by the size of the *expected* frequencies (not the observed frequencies), and although the evidence on this score is partly empirical, it does

seem safe to apply the chi-square tables when none of the expected frequencies is less than five.

The logic of testing goodness of fit is the same as that of any other type of hypothesis test. We decide arbitrarily upon the value of α and thus determine the corresponding limit of the critical region. The rules of decision are then fixed; a sample value either does or does not fall within the acceptance region, and the issue is closed. Whether or not the majority of random samples would yield a better fit is beside the point; we first decide on what will be judged acceptable and then abide by the decision. On the other hand, when the sample value

Table 10-3 CRITICAL VALUES OF CHI-SQUARE FOR ONE-SIDED CRITICAL REGIONS AT STATED SIGNIFICANCE LEVELS

Degrees of freedom	Right-hand tail			Left-hand tail†		
	$\alpha=.10$	$\alpha=.05$	$\alpha=.01$	$\alpha=.10$	$\alpha=.05$	$\alpha=.01$
1	2.71	3.84	6.64	.02	.0039	.00016
2	4.61	5.99	9.20	.21	.10	.02
3	6.25	7.82	11.34	.58	.35	.12
4	7.78	9.49	13.28	1.06	.71	.30
5	9.24	11.07	15.09	1.61	1.15	.55
6	10.65	12.59	16.81	2.20	1.64	.87
7	12.02	14.07	18.48	2.83	2.17	1.24
8	13.36	15.51	20.09	3.49	2.73	1.65
9	14.68	16.92	21.67	4.17	3.33	2.09
10	15.99	18.31	23.21	4.87	3.94	2.56
11	17.28	19.68	24.73	5.58	4.58	3.05
12	18.55	21.03	26.22	6.30	5.23	3.57
13	19.81	22.36	27.69	7.04	5.89	4.11
14	21.06	23.69	29.14	7.79	6.57	4.66
15	22.31	25.00	30.58	8.55	7.26	5.23
16	23.54	26.30	32.00	9.31	7.96	5.81
17	24.77	27.59	33.41	10.09	8.67	6.41
18	25.99	28.87	34.81	10.87	9.39	7.02
19	27.20	30.14	36.19	11.65	10.12	7.63
20	28.41	31.41	37.57	12.44	10.85	8.26
21	29.62	32.67	38.93	13.24	11.59	8.90
22	30.81	33.92	40.29	14.04	12.34	9.54
23	32.01	35.17	41.64	14.85	13.09	10.20
24	33.20	36.42	42.98	15.66	13.85	10.86
25	34.38	37.65	44.31	16.47	14.61	11.52
26	35.56	38.89	45.64	17.29	15.38	12.20
27	36.74	40.11	46.96	18.11	16.15	12.88
28	37.92	41.34	48.28	18.94	16.93	13.57
29	39.09	42.56	49.59	19.77	17.71	14.26
30	40.26	43.77	50.89	20.60	18.49	14.95

† Left-hand tail used mainly for confidence limits on variance.

NOTE: For degrees of freedom $n > 30$ the variate $\sqrt{2\chi^2} - \sqrt{2n-1}$ is approximately unit normal.

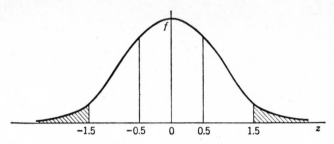

FIGURE 10-1

lies within the acceptance region, the result is often amplified by citing the further fact that there is (perhaps) a considerable margin of safety. In view of the fact that the test itself is only approximate, this practice is perfectly legitimate and even desirable; for it is relevant to note that the decision would still hold good even making liberal allowances for errors of approximation.

EXAMPLE 10-4 Using the data of Example 7-13, let us test whether the normal hypothesis is tenable as a description of the distribution of annual rainfall over the Susquehanna Valley.

In order to use the chi-square test for goodness of fit, the distribution of rainfall should be classified into mutually exclusive and exhaustive categories so chosen that random samples of N observations from the hypothetical population will, in the long run, yield an average of at least five observations in each category. Letting p_i represent the theoretical probability that an observation drawn at random will fall into category i, we must make this category broad enough so that $Np_i > 5$. Thus, with $N = 92$, the minimum value of p_i is $5/92$, or .0543.

To choose categories for a normal distribution, we begin with the tail area and work inward (Fig. 10-1). Representing a unit normal variate by z to avoid confusion with the notation x used for rainfall, we find from Table 4-4 that $R(z) = .0668$ when $z = 1.5$. True, the next higher value in the table, $z = 1.6$, yields $R(z) = .0548$, which would satisfy the minimum requirement, but to be on the safe side, let us decide upon $z = 1.5$. This means that the highest category will be $z > 1.5$ and the lowest category will be $z \leq -1.5$. Although there is no set rule for subdividing the intermediate range, equal intervals on the abscissa are most commonly employed; however, a division into equal areas is sometimes preferred. Here, to make this illustration easy to

Table 10-4 THEORETICAL PROBABILITIES
AND EXPECTED NUMBERS

i	Category	p_i	$E_i = Np_i$
1	$z \leq -1.5$	.07	6.4
2	$-1.5 < z \leq -.5$	.24	22.1
3	$-.5 < z \leq .5$	.38	35.0
4	$.5 < z \leq 1.5$	.24	22.1
5	$z > 1.5$	.07	6.4
Total		1.00	92.0

follow computationally, let us divide the range $-1.5 < z \leq 1.5$ into three equal intervals of z. We thus arrive at the expected numbers shown in Table 10-4.

Since the parameters μ and σ of the normal distribution have been estimated from the data, and we have decided to use five categories, $\hat{\chi}^2$ will be distributed as chi-square with $(5 - 1) - 2 = 2$ degrees of freedom. Using a one-sided critical region and the 5 percent level of significance, the acceptance region is $\hat{\chi}^2 < 5.99$.

In this example, the variate z is defined as $(x - \bar{x})/s$, where $\bar{x} = 39.492$, $s = 3.657$. The corresponding values of x and z are shown in Table 10-5. Hence, in terms of actual rainfall, the categories are (1) $x \leq 34.00$; (2) $34.01 < x \leq 37.66$; (3) $37.67 < x \leq 41.32$; (4) $41.33 < x \leq 44.97$; (5) $x \geq 44.98$. The computation of chi-square is exhibited in Table 10-6, the actual numbers of observations in the categories being found from Table 7-3.

Table 10-5 CORRESPONDING VALUES
OF x AND z

z	-1.5	-0.5	0.5	1.5
x	34.007	37.663	41.321	44.977

Table 10-6 COMPUTATION OF CHI-SQUARE

i	O_i	E_i	$O_i - E_i$	$(O_i - E_i)^2$	$(O_i - E_i)^2/E_i$
1	5	6.4	-1.4	1.96	.306
2	20	22.1	-2.1	4.41	.200
3	43	35.0	8.0	64.00	1.829
4	18	22.1	-4.1	16.81	.761
5	6	6.4	$-.4$	.16	.025
Total	92	92.0	.0	$\cdots$	$\hat{\chi}^2 = 3.121$

We accept the hypothesis and conclude that the distribution of annual rainfall is approximately normal, for the obtained value of $\hat{\chi}^2$ falls well within the acceptance region. Noticing that $O - E$ is positive for category 3 and negative for the other four categories, one is tempted to say that the normal curve must be too low in the immediate vicinity of the mean and too high elsewhere. This feature, however, should be discounted as an accidental property of the sample itself, at least as far as the present statistical evidence is concerned; for this pattern is merely one of the many whereby a random sample can yield a run of the mill value of $\hat{\chi}^2$. Obviously the distribution of rainfall could not be exactly normal, because the lower limit is zero and the upper limit is certainly not infinity. Nevertheless, the normal distribution seems to be a satisfactory approximation within the operating range of the variate. Few theoretical models are all-inclusive.   ////

10-7   THE CHI-SQUARE TEST FOR HOMOGENEITY

The problem is to decide whether s independent samples can be regarded as belonging to a common population. Divide the total range of possibilities into k mutually exclusive, exhaustive categories $(i = 1, 2, \ldots, k)$, each having the same definition with respect to all samples. Let N_j denote the total number of observations in sample j $(j = 1, 2, \ldots, s)$, E_{ij} the expected number of observations in category i for a sample of size N_j, and O_{ij} the corresponding observed number. Then the test variate

$$\hat{\chi}^2 \equiv \sum_j \sum_i \frac{(O_{ij} - E_{ij})^2}{E_{ij}} \qquad (10\text{-}8)$$

is distributed approximately as chi-square, with degrees of freedom governed by the number of estimates made from the data. As in testing goodness of fit, we may rewrite the above in the following equivalent form:

$$\hat{\chi}^2 \equiv \sum_j \sum_i \frac{O_{ij}^2}{E_{ij}} - N \qquad (10\text{-}9)$$

where $N \equiv \sum_j N_j$.

By hypothesis, the probability that a random observation will fall in category i has the same value, say p_i, for all samples, and of course, $\sum_i p_i = 1$. If the individual probabilities are known in advance, the expected number of observations in category i for sample j can be evaluated, for $E_{ij} = N_j p_i$. In that event, since the samples are assumed independent, we may express $\hat{\chi}^2$ of the above equations as the sum of s independent components:

$$\hat{\chi}^2 = \sum_{j=1}^{s} \hat{\chi}_j^2 \qquad \hat{\chi}_j^2 \equiv \sum_{i=1}^{k} \frac{(O_{ij} - N_j p_i)^2}{N_j p_i} \qquad (10\text{-}10)$$

Here the effect of the test is the same as though each sample were tested separately for goodness of fit and the respective chi-squares added together. (Note analogy with pooling evidence from independent samples.) Each χ_j^2 has approximately the chi-square distribution with $k - 1$ degrees of freedom, and so the pooled result χ^2 is distributed approximately as chi-square with $s(k - 1)$ degrees of freedom. Commonly, however, the individual probabilities are not known in advance, and it is necessary to substitute sample estimates, $p_j' \equiv \sum_j O_{ij}/N$. To reckon degrees of freedom, therefore, we must diminish the quantity $s(k - 1)$ by the number of parameters estimated. Effectively, there are only $k - 1$ parameters, inasmuch as $\sum p_i = 1 = \sum p_j'$. Consequently, the number of degrees of freedom in the corresponding χ^2 is given by the following equation:

$$\text{Degrees of freedom} = s(k - 1) - (k - 1) = (s - 1)(k - 1) \quad (10\text{-}11)$$

Hence, in the typical situation, χ^2 is given by

$$\hat{\chi}^2 = \sum_j \sum_i \frac{(O_{ij} - N_j p_i')}{N_j p_i'} \quad (10\text{-}12)$$

and is distributed approximately as chi-square with $(s - 1)(k - 1)$ degrees of freedom. An alternative computation formula is

$$\hat{\chi}^2 = \sum_j \sum_i \frac{O_{ij}^2}{N_j p_i'} - N \quad (10\text{-}13)$$

EXAMPLE 10-5   Table 10-7 summarizes the distributions of first-term ratings of four classes of M.I.T. freshmen, exclusive of war veterans from 1948–1951. Test for homogeneity.

Table 10-7   OBSERVED FREQUENCIES

Category	I	II	III	IV	Composite
0–1.00	12	8	13	5	38
1.01–1.50	23	10	27	26	86
1.51–2.00	39	32	49	29	149
2.01–2.50	58	39	48	38	183
2.51–3.00	101	89	106	66	362
3.01–3.50	112	136	135	110	493
3.51–4.00	92	88	123	72	375
4.01–4.50	51	79	63	61	254
4.51–5.00	25	35	37	28	125
Total	513	516	601	435	2,065

The probability p_i that a random observation will fall into category i ($i = 1, 2, \ldots, 9$) being unknown, we shall substitute the sample estimate $p_i' = \sum_j O_{ij}/N$, where $N = 2,065$ is the size of the composite sample made up of the four classes combined. Accordingly, with 9 categories and 4 samples there will be $(9 - 1)(4 - 1) = 24$ degrees of freedom. Using the 5-percent level of significance, the critical region will be $\chi^2 \geq 36.42$, whence the acceptance region will be $\chi^2 < 36.42$.

The expected number of observations (expected frequency) in category i for sample j is estimated from the equation $E_{ij} \approx N_j p_i'$. For example, $9.4 = (513)(38/2,065)$. These estimates are shown in Table 10-8, and the other computations leading to χ^2 are exhibited in Tables 10-9 and 10-10.

Table 10-8   ESTIMATES OF EXPECTED FREQUENCIES

Category	I	II	III	IV	Composite
0–1.00	9.4	9.5	11.1	8.0	38.0
1.01–1.50	21.4	21.5	25.0	18.1	86.0
1.51–2.00	37.1	37.2	43.4	31.4	149.1
2.01–2.50	45.5	45.7	53.3	38.5	183.0
2.51–3.00	89.9	90.5	105.4	76.3	362.1
3.01–3.50	122.5	123.2	143.5	103.9	493.1
3.51–4.00	93.2	93.7	109.1	79.0	375.0
4.01–4.50	63.1	63.5	73.9	53.5	254.0
4.51–5.00	31.1	31.2	36.4	26.3	125.0
Total	513.2	516.0	601.1	435.0	2,065.3

Table 10-9   VALUES OF $O - E$

Category	I	II	III	IV	Total
0–1.00	2.6	−1.5	1.9	−3.0	.0
1.01–1.50	1.6	−11.5	2.0	7.9	.0
1.51–2.00	1.9	−5.2	5.6	−2.4	−.1
2.01–2.50	12.5	−6.7	−5.3	−.5	.0
2.51–3.00	11.1	−1.5	.6	−10.3	−.1
3.01–3.50	−10.5	12.8	−8.5	6.1	−.1
3.51–4.00	−1.2	−5.7	13.9	−7.0	.0
4.01–4.50	−12.1	15.5	−10.9	7.5	.0
4.51–5.00	−6.1	3.8	.6	1.7	0
Total	−.2	.0	−.1	.0	−.3

Table 10-10   VALUES OF $(O - E)^2/E$

Category	I	II	III	IV	Total
0–1.00	.719	.237	.325	1.125	2.406
1.01–1.50	.120	6.120	.160	3.448	9.848
1.51–2.00	.097	.727	.723	.183	1.730
2.01–2.50	3.434	.982	.527	.006	4.949
2.51–3.00	1.371	.025	.003	1.390	2.789
3.01–3.50	.900	1.330	.503	.358	3.091
3.51–4.00	.015	.347	1.771	.620	2.753
4.01–4.50	2.320	3.783	1.608	1.051	8.762
4.51–5.00	1.196	.463	.010	.110	1.779
Total	10.172	14.014	5.630	8.291	38.107

Since this value 38.107 of $\hat{\chi}^2$ falls in the *critical* region, we *reject* the hypothesis of homogeneity and conclude that the first-term ratings of these freshman classes at M.I.T. do not behave as random samples from a common population.

////

10-8   THE CHI-SQUARE TEST FOR ASSOCIATION OR INDEPENDENCE

Association might be the object of investigation in one instance and independence in another, but in any event, the null hypothesis is that the variates or attributes are independent.   This implies that their joint distribution is the product of the corresponding marginal distributions.   The logic of the test applies equally well to quantitative and qualitative material or combinations of both, and there is no definite[1] restriction on the number of dimensions or characteristics that can be analyzed simultaneously; but, for simplicity, we shall consider merely two variates or attributes A, B.   Since these need not be commensurable, the categories into which each is subdivided are not assumed to be similarly defined.

Divide A into k mutually exclusive, exhaustive categories $A_1, A_2, \ldots, A_k$ and B into s mutually exclusive, exhaustive categories $B_1, B_2, \ldots, B_s$.   If α_i denotes the probability that a random observation will fall in category A_i and if β_j denotes the probability that a random observation will fall in category B_j (where $\sum_i \alpha_i = 1 = \sum_j \beta_j$), then, by hypothesis, the probability of observing

---

[1] Practical limitations are imposed by the size of the sample needed in order that the distribution be approximated by chi-square.

A_i and B_j simultaneously is the product $\alpha_i \beta_j$. The joint distribution is indicated schematically in Table 10-11. In successive random samples of N observations the joint distribution of the observed numbers $O_{11}, \ldots, O_{ij}, \ldots, O_{ks}$ in categories $A_1 B_1, \ldots, A_i B_j, \ldots, A_k B_s$ will be multinomial of the order ks, and with a fixed value of N, there will be $ks - 1$ independent quantities since $\sum O_{ij} = N$. The problem thus becomes one of testing the goodness of fit of the assumed bivariate distribution of A, B. The expected value of O_{ij} is simply $N\alpha_i \beta_j$ and so the statistic

$$\chi^2 \equiv \sum_i \sum_j \frac{(O_{ij} - N\alpha_i \beta_j)^2}{N\alpha_i \beta_j} \qquad (10\text{-}14)$$

is distributed approximately as chi-square with $ks - 1$ degrees of freedom.

When the individual probabilities are unknown, sample estimates are substituted in the usual way. The estimates a_i, b_j of α_i, β_j are

$$a_i = \sum_{j=1}^{s} \frac{O_{ij}}{N} \qquad b_j = \sum_{i=1}^{k} \frac{O_{ij}}{N} \qquad (10\text{-}15)$$

There are only $(k - 1) + (s - 1)$ independent parameters because $\sum \alpha_i = 1 = \sum \beta_j$. Hence

Degrees of freedom $= (ks - 1) - (k - 1) - (s - 1) = (k - 1)(s - 1)$ (10-16)

and the resulting test variate

$$\chi^2 \equiv \sum_i \sum_j \frac{(O_{ij} - Na_i b_j)^2}{Na_i b_j} \qquad (10\text{-}17)$$

is distributed approximately as chi-square with $(k - 1)(s - 1)$ degrees of freedom. For computational purposes it is more convenient to define

$$N_j \equiv Nb_j \equiv \sum_{i=1}^{k} O_{ij} \qquad Na_i b_j \equiv N_j a_i \qquad (10\text{-}18)$$

Table 10-11   JOINT DISTRIBUTION

Class	B_1	B_1	$\cdots$	B_s	Total
A_1	$\alpha_1 \beta_1$	$\alpha_1 \beta_2$	$\cdots$	$\alpha_1 \beta_s$	α_1
A_2	$\alpha_2 \beta_1$	$\alpha_2 \beta_2$	$\cdots$	$\alpha_2 \beta_s$	α_2
$\cdots$	$\cdots$	$\cdots$	$\cdots$	$\cdots$	$\cdots$
A_k	$\alpha_k \beta_1$	$\alpha_k \beta_2$	$\cdots$	$\alpha_k \beta_s$	α_k
Total	β_1	β_2	$\cdots$	β_s	1

Then, by expanding the quadratic in Eq. (10-17) and simplifying, we obtain

$$\hat{\chi}^2 = \sum_i \sum_j \frac{O_{ij}^2}{N_j a_i} - N \qquad (10\text{-}19)$$

a form clearly equivalent to that obtained for testing homogeneity. Thus, several uses of chi-square, while logically distinct, are mathematically identical.

EXAMPLE 10-6 A chronological listing of annual amounts of rainfall over the Susquehanna Valley for a series of 92 years (1860–1951) is presented in Table 10-12. Test for significant association between the amounts observed in consecutive years.

Table 10-12 CHRONOLOGICAL LISTING OF ANNUAL AMOUNTS OF RAINFALL OVER THE SUSQUEHANNA VALLEY

Year	Amount	Class†	Year	Amount	Class†	Year	Amount	Class†
1860	39.74	C	1891	41.96	D	1921	38.86	B
1861	38.91	B	1892	38.51	B	1922	35.76	A
1862	36.63	A	1893	39.13	B	1923	36.35	A
1863	44.82	D	1894	40.14	C	1924	39.99	C
1864	38.89	B	1895	31.21	A	1925	37.79	B
1865	41.45	C	1896	38.07	B	1926	40.60	C
1866	38.08	B	1897	37.79	B	1927	45.06	D
1867	42.26	D	1898 c	41.72	D	1928	40.08	C
1868	40.40	C	1899 b	37.13	A	1929	42.85	D
1869	42.09	D	1900	34.34	A	1930	28.04	A
1870	40.41	C	1901	42.03	D	1931	35.43	A
1871	37.72	B	1902	44.56	D	1932	39.08	B
1872 b	37.15	A	1903	43.44	D	1933	43.38	D
1873	44.76	D	1904	36.86	A	1934	37.96	B
1874	37.00	A	1905	41.18	C	1935	38.81	B
1875	36.70	A	1906	40.14	C	1936	40.67	C
1876	40.75	C	1907	40.57	C	1937	44.46	D
1877	36.65	A	1908	36.62	A	1938	37.77	B
1878	40.78	C	1909	34.25	A	1939	34.59	A
1879	32.11	A	1910	37.33	B	1940	41.23	C
1880	33.02	A	1911	40.15	C	1941	31.63	A
1881	36.39	A	1912 c	41.85	D	1942	46.51	D
1882	35.11	A	1913	40.54	C	1943	38.04	B
1883	38.15	B	1914	38.18	B	1944	38.43	B
1884	40.23	C	1915	43.17	D	1945	48.61	D
1885	36.62	A	1916	40.48	C	1946 c	39.55	B
1886	39.82	C	1917	38.65	B	1947	40.62	C
1887	37.57	B	1918 c	39.56	B	1948	44.34	D
1888 c	41.80	D	1919	41.12	C	1949	36.68	A
1889	48.22	D	1920	40.06	C	1950	45.89	D
1890	47.31	D				1951 c	41.90	D

† $A = 37.15$ or less; $B = 37.16$ to 39.56; $C = 39.57$ to 41.45; $D = 41.46$ and over.

Here the direct antithesis of the working hypothesis that consecutive annual amounts are associated is that they are independent. A more interesting and somewhat stronger null hypothesis would be that all observations (consecutive or not) are random draws from a common population. With a total of 92 observations we cannot test the latter hypothesis exhaustively, but we shall incorporate it as far as we can into the null hypothesis we actually do test. Considering successive pairs of observations (1860, 1861; 1862, 1863; etc.), we shall test the null hypothesis that both members of any pair are random draws from the same population and that this population is common to all pairs. Even if this hypothesis proves tenable, it will still be conceivable that other significant patterns not reflected in successive pairs exist in the data.

Let us try to divide the rainfall amounts into four exhaustive, equally likely categories A, B, C, D. Since there are 92 observations in all, each category should contain 23 cases. Referring to Table 7-3, we find that 23 of the observed amounts are less than or equal to 37.15 in., 46 are less than or equal to 39.56 in., and 69 are less than or equal to 41.45 in. Therefore, if we define category A as $x \le 37.15$, category B as $37.15 < x \le 39.56$, category C as $39.56 < x \le 41.45$, and category D as $x > 41.45$, we shall have four exhaustive, mutually exclusive categories, each containing 23 cases. Table 10-12 indicates the classification of each year of record; the lowercase letters occurring in eight places will be explained presently.

Under the null hypothesis the time sequence is unimportant; therefore, we are interested in 10 combinations of categories: AA, AB, AC, AD, BB, BC, BD, CC, CD, DD. No distinction being made as to order of occurrence, the probability for each of the six mixed combinations (AB, AC, etc.) is $\frac{1}{8}$, and for each of the four duplicate combinations (AA, BB, etc.) the probability is $\frac{1}{16}$. Since there are 46 pairs of observations, the expected frequency of any duplicate combination is only $\frac{46}{16} = 2.875$. Therefore, we combine these into two composite classes $(AA + BB)$, $(CC + DD)$, thus doubling the expected frequency in order to obtain a satisfactory approximation of the $\hat{\chi}^2$ distribution by chi-square. Accordingly, we have eight categories, and the expected frequency of each is $\frac{46}{8} = 5.75$. Since E_i is constant, we may express the formula for $\hat{\chi}^2$ in the following simplified form, wherein the categories are numbered serially 1 through 8:

$$\hat{\chi}^2 = \frac{8}{46} \sum O_i^2 - 46$$

In determining class limits, we have estimated three fractiles from the data, namely, those corresponding to $F = .25$, .50, .75 (where F denotes cumulative probability). Hence there will be $8 - 1 - 3 = 4$ degrees of freedom, and the acceptance region for the .05 level of significance is $\chi^2 < 9.49$. The observed frequencies are shown in Table 10-13.

Table 10-13

Category	$(AA+BB)$	AB	AC	AD	BC	BD	CD	$(CC+DD)$	Total
Frequency	8	2	6	5	10	5	5	5	46
Square	64	4	36	25	100	25	25	25	304

The calculations are as follows:

$$\hat{\chi}^2 = \frac{8(304)}{46} - 46 = 52.87 - 46 = 6.87$$

Since this value falls well within the acceptance region, we accept the hypothesis that the paired values are random draws from a common population which is the same for all the years tested. Thus, if any association whatever exists, it is not demonstrated in consecutive pairs of years.

Let us now examine a somewhat more specific hypothesis. Having shown that the distribution is approximately normal, let us determine the class limits from our normal fit, thereby gaining 1 degree of freedom. For the unit normal distribution the quartiles are $-.6745$, 0, $.6745$, and for a normal distribution with $\mu = 39.492$, $\sigma = 3.657$, the quartiles are 37.025, 39.492, 41.959. The four classes are thus defined as follows:

A: $x \le 37.025$
B: $37.025 < x \le 39.429$
C: $39.492 < x \le 41.959$
D: $x > 41.959$

It turns out that 84 of the years fall into the same categories as before, and the remaining eight differ by one category, as indicated by lower-case letters. Since we have fitted but two constants $\bar{x}$ and s to the data, we now have 5 degrees of freedom and the acceptance region is $\hat{\chi}^2 < 11.07$. The revised frequencies are shown in Table 10-14, and the calculation of $\hat{\chi}^2$ follows the table.

Table 10-14

Category	$(AA+BB)$	AB	AC	AD	BC	BD	CD	$(CC+DD)$	Total
Frequency	8	2	6	3	9	6	6	6	46
Square	64	4	36	9	81	36	36	36	302

$$\hat{\chi}^2 = \frac{8(302)}{46} - 46 = 52.52 - 46 = 6.52$$

This time we obtain very nearly the same value of $\hat{\chi}^2$ as before (actually, a trifle smaller) and yet have a broader acceptance region. Therefore, we accept the more specific hypothesis that the pairs are all drawn at random from the same normal population, with $\mu = 39.492$, $\sigma = 3.675$. ////

10-9 THE DISTRIBUTIONS OF $\bar{x}$ AND s^2 IN A NORMAL POPULATION

The threefold object of this section is to show that $\bar{x}$ and s^2, computed from the same random sample of size n from a *normal* population, are *independently* distributed in repeated independent trials, to derive the distribution of s^2/σ^2, and to relate these results to t and F.

In repeated independent trials the joint distribution of the observations $x_1, x_2, \ldots, x_n$ is

$$f(x_1, x_2, \ldots, x_n) = K_0\, e^{-Q_0/2} \qquad (-\infty, \infty \text{ all variates})$$

where

$$K_0 = (\sigma\sqrt{2\pi})^{-n} \qquad Q_0 = \frac{\sum_1^n (x_i - \mu)^2}{\sigma^2}$$

Making the change of variables

$$\bar{x} = \frac{x_1 + x_2 + \cdots + x_n}{n} \qquad u_i = x_i - \bar{x} \qquad i = 1, 2, \ldots, n-1$$

we have

$$x_i = \bar{x} + u_i \quad \text{for } i = 1, 2, \ldots, n-1 \quad \text{and} \quad x_n = \bar{x} - \sum_1^{n-1} u_i$$

The equation for x_n is obtained from the fact that $\sum_1^n (x_i - \bar{x}) = 0$, whence

$$x_n - \bar{x} = -\sum_1^{n-1} (x_i - \bar{x}) = -\sum_1^{n-1} u_i$$

The Jacobian, which we simplify by adding the first $n-1$ rows to the last row, is

$$J = \begin{vmatrix} 1 & 1 & 0 & 0 & \cdots & 0 & 0 \\ 1 & 0 & 1 & 0 & \cdots & 0 & 0 \\ \multicolumn{7}{c}{\cdots\cdots\cdots\cdots\cdots\cdots\cdots} \\ 1 & 0 & 0 & 0 & \cdots & 0 & 1 \\ 1 & -1 & -1 & -1 & \cdots & -1 & -1 \end{vmatrix} = \begin{vmatrix} 1 & 1 & 0 & 0 & \cdots & 0 & 0 \\ 1 & 0 & 1 & 0 & \cdots & 0 & 0 \\ \multicolumn{7}{c}{\cdots\cdots\cdots\cdots\cdots\cdots} \\ 1 & 0 & 0 & 0 & \cdots & 0 & 1 \\ n & 0 & 0 & 0 & \cdots & 0 & 0 \end{vmatrix}$$

and when this is expanded on the minors of its last row, the absolute value is found to be simply n. By virtue of the identity

$$\sum_1^n (x_i - \mu)^2 \equiv n(\bar{x} - \mu)^2 + \sum (x_i - \bar{x})^2$$

the quantity Q_0 may be expressed as a sum of two components:

$$Q_0 = \frac{n(\bar{x} - \mu)^2}{\sigma^2} + Q_1$$

where
$$Q_1 = \frac{\sum_1^n (x_i - \bar{x})^2}{\sigma^2} = \frac{\sum_1^{n-1} u_i^2 + \left(\sum_1^{n-1} u_i\right)^2}{\sigma^2}$$

Therefore, the joint distribution of the new variates is

$$g(\bar{x}, u_1, u_2, \ldots, u_{n-1}) = \left(\frac{\sqrt{n}}{\sigma\sqrt{2\pi}} e^{-n(\bar{x}-\mu)^2/2\sigma^2}\right)(K_1 e^{-Q_1/2}) \qquad (-\infty, \infty \text{ all})$$

where $K_1 = (\sigma\sqrt{2\pi})^{-(n-1)}\sqrt{n}$. Since the first factor of this expression is the marginal density function of $\bar{x}$, the second factor $K_1 e^{-Q_1/2}$ must necessarily be the joint density function of the u's; and from the fact that s^2 is a pure function of the latter, it follows that $\bar{x}$ and s^2 are independent. Although this strict independence of $\bar{x}$ and s^2 is a unique property of the normal population, the degree of dependence in many nonnormal populations is often slight enough to be negligible.

Because the definition of s^2 formally contains $\bar{x}$, readers are often puzzled by the assertion that s^2 and $\bar{x}$ are independent. It must be remembered that we are considering the behavior of two computed quantities in an infinite number of independent trials. Imagine all samples for which the mean $\bar{x}$ lies between any given values $\bar{x}'$ and $\bar{x}' + d\bar{x}$, and for this subset consider the conditional distribution of s^2. The quantity s^2 depends upon n terms of the type $x_i - \bar{x}$, and fixing the value of $\bar{x}$ still allows $n - 1$ of these deviations to assume any values whatever. Thus it is clear that s^2 is very far from determined by $\bar{x}$, and our proposition asserts that for a normal population the conditional distribution of s^2 is the same as its marginal distribution.

The distribution of s^2/σ^2 can now be ascertained from the moment-generating function. Let us put

$$S_0 = \frac{\sum (x_i - \mu)^2}{\sigma^2} \qquad S_1 = \frac{n(\bar{x} - \mu)^2}{\sigma^2} \qquad S_2 = \frac{\sum (x_i - \bar{x})^2}{\sigma^2}$$

Then, for any distribution whatever, we have the identity $S_0 \equiv S_1 + S_2$, but for the normal distribution S_1 and S_2 are independent and

$$M(\theta : S_0) = M(\theta : S_1)M(\theta : S_2)$$

whence
$$M(\theta : S_2) = \frac{M(\theta : S_0)}{M(\theta : S_1)}$$

Since $(x_i - \mu)/\sigma$ is unit normal, S_0 is distributed as chi-square with n degrees of freedom. Moreover, S_1 is distributed as chi-square with 1 degree of freedom. Now the moment-generating function of chi-square with, say, k degrees of freedom is given by $(1 - 2\theta)^{-k/2}$. Consequently,

$$M(\theta : S_2) = \frac{(1 - 2\theta)^{-n/2}}{(1 - 2\theta)^{-1/2}} = (1 - 2\theta)^{-(n-1)/2}$$

Therefore $S_2 = \sum (x_i - \bar{x})^2/\sigma^2$ is distributed as chi-square with $n - 1$ degrees of freedom, and s^2/σ^2 is distributed as the mean square with $n - 1$ degrees of freedom. By the same token, s/σ has the root-mean-square distribution with $n - 1$ degrees of freedom.

We are now in a position to construct test variates for several purposes. Recalling the definition of t (Chap. 6, Sec. 6-10), it follows at once that the quotient

$$\frac{(\bar{x} - \mu)\sqrt{n}}{s} \equiv \frac{(\bar{x} - \mu)\sqrt{n}/\sigma}{s/\sigma} \qquad (10\text{-}20)$$

is distributed as t with $n - 1$ degrees of freedom; for the variate $(\bar{x} - \mu)\sqrt{n}/\sigma$ is unit normal and the independent variate s/σ is distributed as the root-mean-square with $n - 1$ degrees of freedom. Moreover, if $\bar{x}_1$, s_1^2 and $\bar{x}_2$, s_2^2 are computed from two random samples of respectively n_1 and n_2 independent observations of the same normal variate x—that is, $\bar{x}_1 = \sum x_{1i}/n_1$, $s_1^2 = \sum (x_{1i} - \bar{x}_1)^2/(n_1 - 1)$ and $\bar{x}_2 = \sum x_{2j}/n_2$, $s_2^2 = \sum (x_{2j} - \bar{x}_2)^2/(n_2 - 1)$—then the quotient

$$\frac{s_1^2}{s_2^2} \equiv \frac{s_1^2/\sigma^2}{s_2^2/\sigma^2} \qquad (10\text{-}21)$$

will be distributed as F with $n_1 - 1$ and $n_2 - 1$ degrees of freedom. (Refer to Chap. 6, Sec. 6-11.) It is for this reason that F is called the "variance ratio." Let us now modify the previous conditions to the extent that the two samples are taken from normal populations with possibly different means μ_1, μ_2, but still with the same variance σ^2. Then, since the distribution of s^2 is independent of the population mean, the ratio s_1^2/s_2^2 will still be distributed as F with $n_1 - 1$ and $n_2 - 1$ degrees of freedom. Under the same conditions, the quantity

$$\frac{(n_1 - 1)s_1^2 + (n_2 - 1)s_2^2}{\sigma^2} \equiv \frac{\sum (x_{1i} - \bar{x}_1)^2 + \sum (x_{2j} - \bar{x}_2)^2}{\sigma^2} \tag{10-22}$$

will be distributed as chi-square with

$$(n_1 - 1) + (n_2 - 1) = n_1 + n_2 - 2 \qquad \text{degrees of freedom}$$

by the additive property of chi-square. Therefore, denoting the pooled estimator of variance as s^2, where

$$s^2 \equiv \frac{(n_1 - 1)s_1^2 + (n_2 - 1)s_2^2}{n_1 + n_2 - 2} \equiv \frac{\sum (x_{1i} - \bar{x}_1)^2 + \sum (x_{2j} - \bar{x}_2)^2}{n_1 + n_2 - 2} \tag{10-23}$$

it follows that the statistic s^2/σ^2 will have the mean-square distribution with $n_1 + n_2 - 2$ degrees of freedom and its square root s/σ will have the root-mean-square distribution with $n_1 + n_2 - 2$ degrees of freedom. The pooled estimator of variance s^2 will be independent of both sample means $\bar{x}_1$, $\bar{x}_2$, and likewise independent of any fixed function of them because s_1^2 and s_2^2 are independent of both. Consider, now, the difference $(\bar{x}_1 - \mu_1) - (\bar{x}_2 - \mu_2)$. Its variance is $\sigma^2(1/n_1 + 1/n_2)$, so that the quotient

$$\frac{(\bar{x}_1 - \mu_1) - (\bar{x}_2 - \mu_2)}{\sigma\sqrt{1/n_1 + 1/n_2}} \tag{10-24}$$

will be unit normal. When the latter expression is divided by s/σ, the unknown parameter σ cancels, with the result that the quotient

$$\frac{(\bar{x}_1 - \mu_1) - (\bar{x}_2 - \mu_2)}{s\sqrt{1/n_1 + 1/n_2}} \tag{10-25}$$

is distributed as t with $n_1 + n_2 - 2$ degrees of freedom. Much of the remainder of this chapter will be devoted to putting these results to work.

10-10 THE t TESTS

Prior to the work of W. S. Gosset in 1908, tests on sample means were made by replacing the unknown parameter σ^2 by a sample estimate and regarding the resulting variate as unit normal. Inexact as this approach was, it gave good results for large samples; but for small samples it gave (as is now known) an exaggerated impression of significance. Gosset (under the pseudonym "Student") cited several experimental situations wherein large samples could not be obtained and propounded the problem of determining the exact distribution of the test variate in samples from a normal population. In solving this problem, Gosset found that the exact distribution was free from unknown parameters—a fact of primary importance in the modern theory of hypothesis testing. Gosset's discovery was not appreciated until its implications were ably expounded and developed by R. A. Fisher, who, however, modified Gosset's test variate slightly to obtain the variate t as now used.   Fisher's own contributions to hypothesis testing as well as to the entire field of statistics are conceded to be the greatest of any single individual.

Although there are other applications of t, we shall confine our attention to its use in testing the significance of a single sample mean or the difference between two sample means, and its use in setting confidence limits for estimating the population mean.

The t tests are exact when the parent population is normal but approximate otherwise.   That is to say, the sampling behavior of the computed statistic is accurately described by the t distribution only in the case of a normal population. Empirical studies on nonnormal populations, however, have tended to show that the tests are not very sensitive to moderate departures from normality.   Similar remarks apply to F, except that F seems to be even less affected than t by non-normality.

10-11 THE t TEST FOR A SINGLE MEAN

The problem is to decide whether we can accept the hypothesis that the population mean equals some stated value μ_h.   The answer is found by obtaining a random sample of n observations $x_1, x_2, \ldots, x_n$, computing their average $\bar{x}$ and the unbiased estimate of variance s^2, and then computing the statistic t defined as

$$t = \frac{\bar{x} - \mu_h}{\sqrt{s^2/n}} \qquad (10\text{-}26)$$

Since there are $n - 1$ degrees of freedom associated with s^2, we refer to the t tables under $n - 1$ degrees of freedom.   Ordinarily we have equal interest in negative and positive deviations from μ_h, in which case we use a two-sided critical region such that $\int_{-\infty}^{-a} f(t)\, dt + \int_{a}^{\infty} f(t)\, dt = \alpha$.   On the other hand, if the working hypothesis is one-sided, as it would be if the object were to achieve a significant improvement over a standard performance, the practical implication of a deviation in the opposite direction is much the same as a confirmation of the null hypothesis proper, for the new technique will be abandoned.   The critical region will then be $t \geq a'$, where $\int_{a'}^{\infty} f(t)\, dt = \alpha$.   In either case, of course, the equation for t is the same, and for computational purposes it is convenient to write

$$\frac{s^2}{n} = \sum \frac{(x - \bar{x})^2}{n(n-1)}$$

whence

$$t = \frac{\bar{x} - \mu_h}{\sqrt{\sum (x - \bar{x})^2 / n(n-1)}} \qquad (10\text{-}27)$$

Table 10-15   CRITICAL MAGNITUDES OF t AT
STATED SIGNIFICANCE LEVELS

Degrees of freedom	One-sided region		Two-sided region	
	$\alpha = .05$	$\alpha = .01$	$\alpha = .05$	$\alpha = .01$
1	6.31	31.82	12.71	63.66
2	2.92	6.97	4.30	9.93
3	2.35	4.54	3.18	5.84
4	2.13	3.74	2.78	4.60
5	2.02	3.37	2.57	4.03
6	1.94	3.14	2.45	3.71
7	1.90	3.00	2.37	3.50
8	1.86	2.90	2.31	3.36
9	1.83	2.82	2.26	3.25
10	1.81	2.76	2.23	3.17
11	1.80	2.72	2.20	3.11
12	1.78	2.68	2.18	3.06
13	1.77	2.65	2.16	3.01
14	1.76	2.62	2.15	2.98
15	1.75	2.60	2.13	2.95
16	1.75	2.58	2.12	2.92
17	1.74	2.57	2.11	2.90
18	1.73	2.55	2.10	2.88
19	1.73	2.54	2.09	2.86
20	1.73	2.53	2.09	2.85
21	1.72	2.52	2.08	2.83
22	1.72	2.51	2.07	2.82
23	1.71	2.50	2.07	2.81
24	1.71	2.49	2.06	2.80
25	1.71	2.49	2.06	2.79
26	1.71	2.48	2.06	2.78
27	1.70	2.47	2.05	2.77
28	1.70	2.47	2.05	2.76
29	1.70	2.46	2.05	2.76
30	1.70	2.46	2.04	2.75
∞	1.64	2.33	1.96	2.58

Table 10-16

Person	A	B	C	D	E	F
Weight loss	10.1	6.3	7.4	11.4	11.9	5.7

EXAMPLE 10-7   A diet published in a newspaper is advertised as resulting in an average weight loss of 10 pounds over a period of 3 weeks. Six persons who tried the diet reported the decreases shown in Table 10-16. Is the mean of this sample consistent with the published average of 10 pounds?

With six observations there will be 5 degrees of freedom. Using a two-sided critical region, the acceptance region for the 5-percent level of significance of t with 5 degrees of freedom is $-2.57 < t < 2.57$. The sample value of t is computed as follows:

$$\sum x = 52.8 \qquad n = 6 \qquad \bar{x} = 8.8$$

$$\sum x^2 = 500.52 \qquad \sum (x - \bar{x})^2 = \sum x^2 - \bar{x} \sum x = 35.88$$

$$\frac{s^2}{n} = \frac{\sum (x - \bar{x})^2}{n(n - 1)} = \frac{35.88}{30} = 1.196$$

$$t = \frac{8.8 - 10.0}{\sqrt{1.196}} = \frac{-1.2}{1.094} = -1.1$$

Since the sample value of t lies inside of the critical limits, we accept the hypothesis $\mu = 10$ pounds. ////

10-12   THE t TEST FOR THE MEANS OF PAIRED OBSERVATIONS

When two processes are being compared, the technique of eliminating irrelevant sources of variation often calls for an experimental design in which paired observations are secured under controlled conditions. The two observational mates cannot be considered independent, but their difference is independent of a common mean value, and the mean difference can be tested as in a single sample.

Suppose we have n independent pairs of observations x_1, y_1; x_2, y_2; $\ldots$; x_n, y_n, and the hypothesis is that $\mu_x - \mu_y = \delta$. Put

$$d_i = x_i - y_i \qquad i = 1, 2, \ldots, n \qquad \bar{d} = \frac{\sum d_i}{n}$$

Compute t as follows:

$$t = \frac{\bar{d} - \delta}{\sqrt{\sum (d_i - \bar{d})^2/n(n-1)}} \qquad (10\text{-}28)$$

Under the hypothesis $\mu_x - \mu_y = \delta$ this will be distributed as t with $n-1$ degrees of freedom.   Usually a one-sided region is appropriate.

EXAMPLE 10-8   Let us apply the t test to the shellac problem introduced in Example 6-24.   In this problem, we have six paired measures of wear under alternative treatments; hence the t test for related samples will be based on 5 degrees of freedom.   The null hypothesis is that $\delta = 0$, whereas the working hypothesis is that $\delta > 0$, and this implies that we want a one-sided critical region. For 5 degrees of freedom the 5-percent level of t with a one-sided criterion is 2.02 and the acceptance region for the null hypothesis is $t < 2.02$.   The data and computations are as follows:

$$n = 6 \qquad \sum d = 18 \qquad \bar{d} = 3 \qquad \sum d^2 = 88$$

$$\sum (d - \bar{d})^2 = \sum d^2 - \bar{d} \sum d = 88 - 54 = 34$$

$$\frac{\sum (d - \bar{d})^2}{n(n-1)} = {}^{34}\!/_{30} = 1.133$$

The sample value of t is

$$t = \frac{3.0}{\sqrt{1.133}} = \frac{3}{1.064} = 2.82$$

Since this value lies outside the acceptance region, we reject the null hypothesis and conclude that the added ingredient improves the wearing quality of shellac. In this connection the t test is more sensitive than the magnitude test used in

Table 10-17

Bldg	Reg.	Fort.	Diff.
	Percentage damage		
1	5	2	3
2	11	7	4
3	4	5	-1
4	14	8	6
5	9	4	5
6	12	11	1

Chap. 6, for the probability of obtaining $t > 2.82$ is actually less than .02, as compared with the significance level of .047 reached with the magnitude test. The latter is sometimes advantageous, however, inasmuch as no appeal is made to the assumption of normality of sample means. ////

10-13 THE t TEST FOR THE DIFFERENCE OF THE MEANS OF TWO INDEPENDENT SAMPLES

When it is reasonable to assume common background conditions for all observations, the comparison of two means can be made with independent samples. This does not repudiate the logic of the previous section because uniform background conditions will not contribute to the variations among individual observations, and so we may legitimately regard the latter as independently distributed. With a fixed total number of observations, the precision is maximized by taking equal sample sizes.   However, if there are practical limitations on the size of one sample while the other can be made larger, the larger number of observations should be taken in the latter.   Hence, for generality, we allow for different sample sizes n_1, n_2.

Having independent samples $(x_1, x_2, \ldots, x_{n_1})$ and $(y_1, y_2, \ldots, y_{n_2})$ we wish to test the hypothesis that $\mu_x - \mu_y = \delta$.   The test of this hypothesis by t presupposes equal variances, that is, $\sigma_x^2 = \sigma_y^2 = \sigma^2$.   Denote the sample means as $\bar{x} = \sum x/n_1$, $\bar{y} = \sum y/n_2$.   An unbiased estimate of σ^2 is given by

$$s^2 = \frac{\sum (x - \bar{x})^2 + \sum (y - \bar{y})^2}{n_1 + n_2 - 2} \qquad (10\text{-}29)$$

and the test variate

$$t = \frac{(\bar{x} - \bar{y}) - \delta}{\sqrt{s^2(1/n_1 + 1/n_2)}} \qquad (10\text{-}30)$$

will be distributed as t with $n_1 + n_2 - 2$ degrees of freedom.   The critical region includes either one tail or two, depending upon the working hypothesis.

EXAMPLE 10-9   A factory has two machines both of which make $\frac{1}{8}$-inch steel washers. Thicknesses of independent samples of each machine were measured to the nearest thousandth of an inch, with the following results.   Are the mean thicknesses significantly different?

Machine 1:   .117, .130, .122, .139, .132
Machine 2:   .120, .127, .133, .115, .137, .126

Since the sample taken from machine 1 has five observations and that from machine 2 has six observations, the t test for independent samples will be based on $5 + 6 - 2 = 9$ degrees of freedom. Using a two-sided critical region, the acceptance region for t with 9 degrees of freedom is $-2.26 < t < 2.26$ at the 5-percent level of significance. The computations are as follows:

For machine 1 we have

$$n_1 = 5 \qquad \sum x = .640 \qquad \bar{x} = .128$$
$$\sum x^2 = .082218 \qquad \sum (x - \bar{x})^2 = .000298$$

For machine 2 we have

$$n_2 = 6 \qquad \sum y = .758 \qquad \bar{y} = .126$$
$$\sum y^2 = .096088 \qquad \sum (y - \bar{y})^2 = .000327$$

The combined estimate of variance is $s^2 = .000625/9 = .0000694$ whence

$$s^2 \left(\frac{1}{n_1} + \frac{1}{n_2} \right) = .0000254 \qquad \sqrt{s^2 \left(\frac{1}{n_1} + \frac{1}{n_2} \right)} = .00504$$

Here we hypothesize that $\mu_x = \mu_y$, that is, $\delta = 0$. The difference of the means is $\bar{x} = \bar{y} = .002$, and so

$$t = \frac{.002}{.005} = .4$$

Since this value of t falls inside the acceptance region, we accept the null hypothesis and conclude that the two machines are producing washers of equal average thickness. ////

Those who are ignorant of the vagaries of arithmetic sometimes laugh at the number of decimals carried in a statistical computation. The reader is warned not to confuse precision of measurement with precision of arithmetic; rounding errors can be devastating. As a case in point, we note that the correct value of $\sum (y - \bar{y})^2$ to three significant figures is .000327, this value being obtained from the formula

$$\sum (y - \bar{y})^2 = \frac{n \sum y^2 - (\sum y)^2}{n}$$

On the other hand, using the alternative formula $\sum (y - \bar{y})^2 = \sum y^2 - \bar{y} \sum y$ and retaining three decimals in $\bar{y}$, we get .000580. In order to obtain the correct result to three significant figures by the latter formula, it turns out that we must carry seven decimals in $\bar{y}$.

A common hypothesis is that $\delta = 0$, that is, $\mu_x = \mu_y$. In that case, readers often object to the estimate of σ^2 on the grounds that it would be more logical to carry out the full implications of the null hypothesis by employing a variance estimate based on a combined estimate of the mean. This point merits investigation. Suppose we put

$$m = \frac{\sum x + \sum y}{n_1 + n_2} = \frac{n_1 \bar{x} + n_2 \bar{y}}{n_1 + n_2} \qquad (10\text{-}31)$$

$$V = \sum (x - m)^2 + \sum (y - m)^2 \qquad S = \sum (x - \bar{x})^2 + \sum (y - \bar{y})^2 \qquad (10\text{-}32)$$

We then find identically

$$V \equiv S + n_1(\bar{x} - m)^2 + n_2(\bar{y} - m)^2$$

$$\equiv S + \frac{n_1 n_2 (\bar{x} - \bar{y})^2}{n_1 + n_2} \qquad (10\text{-}33)$$

Therefore, although V is independent of m, it is not independent of $\bar{x} - \bar{y}$ and could not be used for t. A perfectly usable parameter-free test variate could be derived from V, but it would not have the t distribution. As a matter of fact, if we put $v^2 = V/(n_1 + n_2 - 1)$ and define the statistic

$$w = \frac{\bar{x} - \bar{y}}{\sqrt{v^2(1/n_1 + 1/n_2)}} \qquad (10\text{-}34)$$

which is clearly analogous to Eq. (10-30) when $\delta = 0$, we find that the distribution of w is

$$f(w) = k\left(1 - \frac{w^2}{n_1 + n_2 - 1}\right)^{(n_1 + n_2 - 3)/2}$$

$$\left(-\sqrt{n_1 + n_2 - 1} \le w \le \sqrt{n_1 + n_2 - 1}\right) \qquad (10\text{-}35)$$

where

$$k = \frac{\Gamma[(n_1 + n_2)/2]}{\Gamma[(n_1 + n_2 - 1)/2]\sqrt{\pi(n_1 + n_2 - 1)}}$$

The test variate given by Eq. (10-30) is distributed as t only if $\sigma_x^2 = \sigma_y^2$; for if the variances are not equal, these parameters do not cancel. To gain qualitative insight into further effects of unequal variances, let us examine the correct formulation of t in this case. The variance of the quantity $(\bar{x} - \bar{y})$ $- (\mu_x - \mu_y)$ will be $\sigma_x^2/n_1 + \sigma_y^2/n_2$, and in order to obtain a denominator that is distributed as the root-mean square, we must replace s^2 by

$$\frac{(n_1 - 1)s_1^2/\sigma_x^2 + (n_2 - 1)s_2^2/\sigma_y^2}{n_1 + n_2 - 2}$$

Putting $\lambda = \sigma_x^2/\sigma_y^2$ and simplifying, we obtain

$$t = \frac{[(\bar{x} - \bar{y}) - (\mu_x - \mu_y)]\sqrt{n_1 + n_2 - 2}}{\sqrt{\sigma_x^2/n_1 + \sigma_y^2/n_2}\sqrt{(n_1 - 1)s_1^2/\sigma_x^2 + (n_2 - 1)s_2^2/\sigma_y^2}}$$

$$= \frac{[(\bar{x} - \bar{y}) - (\mu_x - \mu_y)]\sqrt{n_1 + n_2 - 2}}{\sqrt{\lambda/n_1 + 1/n_2}\sqrt{(n_1 - 1)s_1^2/\lambda + (n_2 - 1)s_2^2}} \tag{10-36}$$

If $\lambda = 1$, we have the same formula as before, Eq. (10-30). If $\lambda \neq 1$, one factor of the denominator is smaller than before and the other is larger, so that the resultant effect may go either way, depending upon the particular n's and s^2's. To be on the safe side in applying Eq. (10-30), therefore, we should first test for equality of variances. This test is described in Sec. 10-16.

Several approximate tests have been proposed in case $\sigma_x^2 \neq \sigma_y^2$. Because of its comparative simplicity, we have chosen to present the method of Cochran and Cox.[1] Put

$$t' = \frac{(\bar{x} - \bar{y}) - \delta}{\sqrt{s_1^2/n_1 + s_2^2/n_2}} \tag{10-37}$$

For a two-sided critical region the acceptance region is $-a < t' < a$, where a is a certain weighted function [Eq. (10-38)] of tabular values. Let a_1 be the right-hand limit of the two-sided critical region for t with $n_1 - 1$ degrees of freedom and a_2 the corresponding limit for t with $n_2 - 1$ degrees of freedom. Also put $w_1 = s_1^2/n_1$, $w_2 = s_2^2/n_2$. Then

$$a = \frac{w_1 a_1 + w_2 a_2}{w_1 + w_2} \tag{10-38}$$

For a one-sided critical region we have an analogous definition of the acceptance region. Thus, if b_1, b_2 are the respective limits with $n_1 - 1$ and $n_2 - 1$ degrees of freedom, then the limit b appropriate to t' is

$$b = \frac{w_1 b_1 + w_2 b_2}{w_1 + w_2} \tag{10-39}$$

By way of heuristic justification of the Cochran-Cox test, let us examine the case of equal sample sizes $n_1 = n_2 = n$. The two limits a_1, a_2 are then identical, and a (being the same) becomes the ordinary limit for $n - 1$ degrees of freedom. Moreover, the quantity $s_1^2/n_1 + s_2^2/n_2$ reduces to $(s_1^2 + s_2^2)/n$, whence

$$t' = \frac{(\bar{x} - \bar{y}) - \delta}{\sqrt{(s_1^2 + s_2^2)/n}} \tag{10-40}$$

[1] W. G. Cochran and G. M. Cox, "Experimental Designs," pp. 91–93, John Wiley & Sons, Inc., New York, 1950.

Since the samples are independent, there is no logical basis for pairing.[1] Nevertheless, we can form the differences $d_i = x_i - y_i$ $(i = 1, 2, \ldots, n)$ between the two sets of observations in whatever order they happen to occur. By the assumption of independence, all of the d's (in repeated sampling) have the same distribution; their expected value is $\mu_x - \mu_y$, and their sample mean value is $\bar{d} \equiv \bar{x} - \bar{y}$. Hence, a statistic t'' having the same formal definition as Eq. (10-28) could be employed here and (under the usual assumption that the parent populations are normal) would be rigorously distributed as t with $n - 1$ degrees of freedom. As to t', we note that since $s_1^2 + s_2^2$ is an unbiased estimate of $\sigma_x^2 + \sigma_y^2$, the true variance of the d's, an unbiased estimate of the variance of $\bar{d}$ is $(s_1^2 + s_2^2)/n$. True, the quantity $(s_1^2 + s_2^2)/(\sigma_x^2 + \sigma_y^2)$ is not distributed as the mean square, and as a consequence, t' is not distributed as t. But of all the quantities $\sum (d_i - \bar{d})^2/(n - 1)$ that could be formed by distinct rearrangements of the original data, the mean value is $s_1^2 + s_2^2$. Hence we should expect the distribution of t' to fall off more rapidly than that of t'' itself, and we infer that the stated limits of acceptance are on the conservative side, that is, broader than the true limits. If $n_1 \neq n_2$, the behavior of t' would follow some middle way between what would happen if both sample sizes were equal to n_1 and if both were equal to n_2. The weights of the critical limits are proportional to the estimated variances of the respective means; although somewhat arbitrary, this scheme is altogether reasonable.

10-14 RESPECTIVE CONFIDENCE LIMITS FOR μ AND σ^2

In random samples of n observations from a normal population, we have seen that the statistic $(\bar{x} - \mu)\sqrt{n/s^2}$ is distributed as t with $n - 1$ degrees of freedom, and $(n - 1)s^2/\sigma^2$ is distributed as chi-square with $n - 1$ degrees of freedom. From these statistics we may obtain separate pairs of confidence limits for μ and σ^2; although separate limits suffice for most purposes, they are not the same as simultaneous limits, for which a different approach is needed.

Adopting the confidence probability $1 - \alpha$, let $a < b$ denote any pair of values for which $\int_a^b f(t)\, dt = 1 - \alpha$, where $t = (\bar{x} - \mu)\sqrt{n/s^2}$; that is $P(a \leq t \leq b) = 1 - \alpha$. Substituting for t and simplifying, we find that the inequality $a \leq t \leq b$ implies that $\left(sa/\sqrt{n} - \bar{x}\right) \leq -\mu \leq \left(sb/\sqrt{n} - \bar{x}\right)$ whence

$$\bar{x} - \frac{sb}{\sqrt{n}} \leq \mu \leq \bar{x} - \frac{sa}{\sqrt{n}} \qquad (10\text{-}41)$$

[1] The argument we are about to give shows that pairing produces a valid test variate, but the reader is warned that it has less precision than the standard t test for independent samples when equal variances can be assumed, and no correlation between members of a pair.

When we assert that the latter inequality represents confidence limits for μ corresponding to a confidence probability $1 - \alpha$, we mean that in repeated independent trials the *variable interval* $\bar{x} - sb/\sqrt{n}$ to $\bar{x} - sa/\sqrt{n}$ will include the true value of μ a certain percentage of the time, and in the long run, the relative frequency with which such intervals include μ will be $1 - \alpha$. This statement must not be construed as implying that μ itself is distributed or subject to variation. (The latter idea belongs to Bayesian analysis.)

Since the requirement that $P(a \le t \le b) = 1 - \alpha$ does not suffice to determine the limits a, b, further criteria are in order. A criterion with intuitive appeal is that the confidence interval be as short as possible. Accordingly, representing the interval by D, where

$$D = \left(\bar{x} - \frac{sa}{\sqrt{n}}\right) - \left(\bar{x} - \frac{sb}{\sqrt{n}}\right) \equiv \frac{s}{\sqrt{n}}(b - a) \qquad (10\text{-}42)$$

we minimize D subject to the requirement that $\int_a^b f(t)dt = 1 - \alpha$. The latter requirement yields

$$f(b)\, db - f(a)\, da = 0 \qquad (10\text{-}43)$$

The unimodal form of the distribution permits us to minimize D by setting $dD = 0$, whence $da = db$. Accordingly, $f(a) = f(b)$, and since the distribution of t is symmetrical about zero, it follows that $a = -b$. Therefore, the upper limit b is determined from the relation $\int_b^\infty f(t)\, dt = \alpha/2$, and the confidence limits are $\bar{x} - sb/\sqrt{n}$ to $\bar{x} + sb/\sqrt{n}$; that is,

$$\bar{x} - \frac{sb}{\sqrt{n}} \le \mu \le \bar{x} + \frac{sb}{\sqrt{n}} \qquad (10\text{-}44)$$

If the distribution had been J-shaped, the minimization of D would have required an examination of the sign of the derivative.

At this point the reader might feel that he has been put through an involved process only to reach an obvious conclusion. We are about to discover that the conclusion is not so obvious when the distribution is skewed, and we shall need the method we have just applied.

With regard to σ^2, the direct probability statement is

$$P\left[a \le \frac{(n - 1)s^2}{\sigma^2} \le b\right] = 1 - \alpha$$

and by transforming the bracketed inequality we obtain, as confidence limits,

$$\frac{(n - 1)s^2}{b} \le \sigma^2 \le \frac{(n - 1)s^2}{a} \qquad (10\text{-}45)$$

Again representing the confidence interval by D, where

$$D = \frac{(n-1)s^2}{a} - \frac{(n-1)s^2}{b} \equiv (n-1)s^2\left(\frac{1}{a} - \frac{1}{b}\right) \qquad (10\text{-}46)$$

we minimize D subject to the requirement that $\int_a^b g(\chi^2)\, d\chi^2 = 1 - \alpha$. The intermediate equations are

$$g(a)\, da = g(b)\, db \qquad \frac{da}{a^2} = \frac{db}{b^2} \qquad (10\text{-}47)$$

and the result is

$$a^2 g(a) = b^2 g(b) \qquad (10\text{-}48)$$

Unfortunately, convenient tables for the determination of these limits are not widely available, so that for practical purposes other limits are often substituted. For small to moderate numbers of degrees of freedom, it turns out that nearly optimum results can be obtained by putting $b = \infty$. That is, we use only the left-hand tail of chi-square, with the result that $D = (n-1)s^2/a$, and the corresponding confidence limits are

$$0 \le \sigma^2 \le \frac{(n-1)s^2}{a} \qquad (10\text{-}49)$$

EXAMPLE 10-10 Suppose that a random sample of 11 observations from a normal population yields $\sum x = 37.4$, $\sum x^2 = 251.73$. Then

$$\bar{x} = 3.4 \qquad \sum(x - \bar{x})^2 = \sum x^2 - \bar{x}\sum x = 251.73 - 127.16 = 124.57$$

$$s^2 = \frac{\sum(x-\bar{x})^2}{10} = 12.457 \qquad s = 3.53 \qquad \frac{s}{\sqrt{11}} = 1.06$$

With a confidence probability of .95, the limits for t with 10 degrees of freedom are $t = \pm 2.23$, from which we obtain $(2.23)s/\sqrt{11} = 2.36$. Therefore the .95 confidence limits for μ are 3.4 ± 2.36, whence

$$1.04 \le \mu \le 5.76$$

With regard to σ^2, we note that for 10 degrees of freedom the distribution of χ^2 is

$$g(\chi^2) = \frac{1}{2^5 \Gamma(5)}(\chi^2)^4 e^{-\chi^2/2} \qquad (0, \infty)$$

Whence Eq. (10-48) becomes

$$a^6 e^{-a/2} = b^6 e^{-b/2}$$

or $$\ln \frac{b}{a} = \frac{b-a}{12}$$

Solving for a and b by successive approximations, we obtain

$$a = 3.8845 \qquad b = 27.2638$$

which corresponds to splitting the specified area .05 between the extremes of chi-square in such a way that the portion in the lower tail has an area of .0476 and the portion in the upper tail has an area of only .0024. These values of a and b yield

$$\frac{(n-1)s^2}{a} = 32.0685 \qquad \frac{(n-1)s^2}{b} = 4.5691 \qquad D = 27.4994$$

Therefore, the .95 confidence limits for σ^2 are

$$4.57 \le \sigma^2 \le 32.07$$

Had we simply applied Eq. (10-49) with $a = 3.94$, our confidence interval would have been $0 \le \sigma^2 \le 31.62$, which compares favorably with the shortest interval. ////

10-15 THE F TESTS

Nearly all the problems to which the F distribution applies were originally solved by R. A. Fisher, who constructed an appropriate parameter-free variate, derived its distribution, provided the necessary tables, and in all respects made it available as a research tool of extraordinary versatility. The variate introduced by Fisher was actually $(\frac{1}{2})\ln F$, but F itself is now commonly used because it is easier to compute.

In view of its definition as the ratio of two mean squares, the most direct use of the variate F lies in testing whether two independent sample values of s^2 may be considered representative of a common variance. Although not obvious from the definition, the F ratio is also appropriate for testing the significance of the difference between two sample means, in this capacity serving as an alternative to the t test, or, what is much more important, testing the significance of the differences between *several* sample *means*, thus going *beyond* the *scope* of the t *test*. This, the principal use of F, has a wide range of application, the elaboration of which comes under the heading of *analysis* of *variance*.

At this point we call the reader's attention to a widespread misuse of the t test by research workers with scanty training in statistics.   A succession of t tests on the means of several samples is not a substitute for the F test applied simultaneously to the entire data.   The separate t's are not independently distributed; therefore the evaluation of their combined significance must take proper account of their joint distribution.   Fortunately, however, this end is accomplished automatically by using the F test in the first place.

Standard tables of the F distribution give limits for one-sided critical regions.   Most tables give critical values for 5 percent and 1 percent levels of significance, but limits for intermediate levels can be approximated by linear interpolation.   When a two-sided critical region is appropriate, the *lower* critical limit c can be found from tabulated values by virtue of the relation

$$\int_0^c f(F)\, dF = \int_{1/c}^{\infty} g\left(\frac{1}{F}\right) d\left(\frac{1}{F}\right) \qquad (10\text{-}50)$$

the second member of which is simply a transformation of the first.   But from the definition of F with a and b degrees of freedom, it follows immediately that the reciprocal variate $1/F$ is analogously distributed with b and a degrees of freedom.   Hence, by taking the reciprocal of the upper critical limit $1/c$ of the F with interchanged degrees of freedom, we obtain the lower limit c for the F in question.

10-16   TESTS FOR EQUALITY OF VARIANCES

The technical term " homoscedasticity " is used to denote " equality of variances." The appropriate test for homoscedasticity depends on whether there are two or more samples involved.   The F test applies if there are two samples, but not otherwise.   Several tests, based on suitable ratios or their logarithms, have been devised for more than two samples.   The one we have selected yields a variate which is distributed approximately as chi-square.

Before applying any significance test, one should decide upon an appropriate critical region.   Let us examine this issue with reference to supporting the t test by a check on homoscedasticity.   Since very small values of the variance ratio and very large values are both unfavorable to the hypothesis of homoscedasticity, it is plausible that the critical region should be two-sided.   Further arguments, which we are about to present, lead to the conclusion that the two portions of the critical region should have equal areas.

We reason to equal subdivision of area as follows: Defining the ratios $F = s_x^2/s_y^2$, $F' = s_y^2/s_x^2$, it is purely arbitrary whether we choose to compute F

Table 10.6.8 CRITICAL VALUES OF F WITH n_1 AND n_2 DEGREES OF FREEDOM. ONE-SIDED CRITICAL REGION (RIGHT TAIL) WITH $\alpha=.05, .01$ (UPPER HALF, $\alpha=.05$; LOWER HALF, $\alpha=.01$)

n_1

n_2	1	2	3	4	5	6	7	8	9	10	15	25	50	100	∞
1	161	200	216	225	230	234	237	239	241	242	246	249	252	253	254
2	18.51	19.00	19.16	19.25	19.30	19.33	19.36	19.37	19.38	19.39	19.43	19.45	19.47	19.49	19.50
3	10.13	9.55	9.28	9.12	9.01	8.94	8.88	8.84	8.81	8.78	8.70	8.64	8.58	8.56	8.53
4	7.71	6.94	6.59	6.39	6.26	6.16	6.09	6.04	6.00	5.96	5.85	5.77	5.70	5.66	5.63
5	6.61	5.79	5.41	5.19	5.05	4.95	4.88	4.82	4.78	4.74	4.62	4.53	4.44	4.40	4.36
6	5.99	5.14	4.76	4.53	4.39	4.28	4.21	4.15	4.10	4.06	3.94	3.84	3.75	3.71	3.67
7	5.59	4.74	4.35	4.12	3.97	3.87	3.79	3.73	3.68	3.63	3.50	3.41	3.32	3.28	3.23
8	5.32	4.46	4.07	3.84	3.69	3.58	3.50	3.44	3.39	3.34	3.21	3.11	3.03	2.98	2.93
9	5.12	4.26	3.86	3.63	3.48	3.37	3.29	3.23	3.18	3.13	3.00	2.89	2.80	2.76	2.71
10	4.96	4.10	3.71	3.48	3.33	3.22	3.14	3.07	3.02	2.97	2.84	2.73	2.64	2.59	2.54
15	4.54	3.68	3.29	3.06	2.90	2.79	2.70	2.64	2.59	2.55	2.41	2.28	2.18	2.12	2.07
25	4.24	3.38	2.99	2.76	2.60	2.49	2.41	2.34	2.28	2.24	2.08	1.95	1.84	1.77	1.71
50	4.03	3.18	2.79	2.56	2.40	2.29	2.20	2.13	2.07	2.02	1.87	1.73	1.60	1.52	1.44
100	3.94	3.09	2.70	2.46	2.30	2.19	2.10	2.03	1.97	1.92	1.77	1.62	1.48	1.39	1.28
500	3.86	3.02	2.62	2.39	2.23	2.12	2.03	1.96	1.90	1.85	1.69	1.53	1.38	1.28	1.12
1,000	3.85	3.00	2.61	2.38	2.22	2.10	2.02	1.95	1.89	1.84	1.67	1.52	1.36	1.26	1.08
∞	3.84	2.99	2.60	2.37	2.21	2.09	2.01	1.94	1.88	1.83	1.66	1.51	1.35	1.24	1.00
1	4,052	4,999	5,403	5,625	5,764	5,859	5,928	5,981	6,022	6,056	6,158	6,238	6,302	6,334	6,366
2	98.49	99.01	99.17	99.25	99.30	99.33	99.34	99.36	99.38	99.40	99.43	99.46	99.48	99.49	99.50
3	34.12	30.81	29.46	28.71	28.24	27.91	27.67	27.49	27.34	27.23	26.87	26.58	26.30	26.23	26.12
4	21.20	18.00	16.69	15.98	15.52	15.21	14.98	14.80	14.66	14.54	14.19	13.91	13.69	13.57	13.46
5	16.26	13.27	12.06	11.39	10.97	10.67	10.45	10.27	10.15	10.05	9.72	9.45	9.24	9.13	9.02
6	13.74	10.92	9.78	9.15	8.75	8.47	8.26	8.10	7.98	7.87	7.56	7.30	7.09	6.99	6.88
7	12.25	9.55	8.45	7.85	7.46	7.19	7.00	6.84	6.71	6.62	6.31	6.05	5.85	5.75	5.65
8	11.26	8.65	7.59	7.01	6.63	6.37	6.19	6.03	5.91	5.82	5.52	5.27	5.06	4.96	4.86
9	10.56	8.02	6.99	6.42	6.06	5.80	5.62	5.47	5.35	5.26	4.96	4.72	4.51	4.41	4.31
10	10.04	7.56	6.55	5.99	5.64	5.39	5.21	5.06	4.95	4.85	4.56	4.32	4.12	4.01	3.91
15	8.68	6.36	5.42	4.89	4.56	4.32	4.14	4.00	3.89	3.80	3.52	3.27	3.07	2.97	2.87
25	7.77	5.57	4.68	4.18	3.86	3.63	3.46	3.32	3.21	3.13	2.85	2.61	2.40	2.29	2.17
50	7.17	5.06	4.20	3.72	3.41	3.18	3.02	2.88	2.78	2.70	2.42	2.17	1.94	1.82	1.68
100	6.90	4.82	3.98	3.51	3.20	2.99	2.82	2.69	2.59	2.51	2.22	1.96	1.73	1.59	1.43
500	6.69	4.65	3.82	3.36	3.06	2.84	2.68	2.55	2.45	2.36	2.07	1.81	1.56	1.41	1.18
1,000	6.66	4.62	3.80	3.34	3.04	2.82	2.66	2.53	2.43	2.34	2.05	1.79	1.54	1.38	1.11
∞	6.64	4.60	3.78	3.32	3.02	2.80	2.64	2.51	2.41	2.32	2.03	1.77	1.52	1.36	1.00

NOTE: Missing values may be filled in by linear interpolation with respect to reciprocals of degrees of freedom. Accuracy improved by interpolating on log F instead of F itself.

or F'.   Consequently, if B is that value such that $P(F \geq B) = \varepsilon$, and also B' is that value such that $P(F' \geq B') = \varepsilon$, then there is neither more nor less logical justification for attaching significance to the event $F \geq B$ than to the event $F' \geq B'$, for both events are equally rare and both are damaging to the hypothesis that $\sigma_x^2 = \sigma_y^2$.   But $F \leq 1/B'$ whenever $F' \geq B'$.   Therefore, the logical implications of large or small ratios are essentially equivalent with reference to this particular hypothesis, and for its test of significance the critical region should be divided equally between the upper and lower tails of the F distribution.

Tables of F at the levels .005 and .025 have been constructed for use with two-sided critical regions corresponding to the conventional significance levels, $\alpha = .01$ and $\alpha = .05$, respectively.   Without deprecating the construction of such tables, we think that in the present context it is desirable to enlarge the critical region.   When the object of the test for equal variances is to complement the t test on the difference of two means, one cannot afford to be quite so liberal as usual with regard to accepting the hypothesis of homoscedasticity, because an erroneous assumption here might vitiate the test on the means.   Therefore, in making the F test, it is advisable to increase the probability of rejecting the null hypothesis that the variances are equal by choosing a somewhat lower level of significance.   For instance, one might take α as .10 instead of .05.   Accordingly, if B and B' represent the upper .05 levels of F and F', respectively, then the acceptance region of F at the .10 level of significance is given by $1/B' < F < B$.

EXAMPLE 10-11   Using the data of Example 10-9, test the hypothesis that $\sigma_x^2 = \sigma_y^2$.   There will be 4 and 5 degrees of freedom if we take F as the ratio s_x^2/s_y^2, and so F' will have 5 and 4 degrees of freedom.   Choosing $\alpha = .10$, the two critical values are $B = 5.19$ and $B' = 6.26$, corresponding respectively to the upper .05 levels of F and F'.   Thus, the acceptance region $1/B' < F < B$ becomes $.160 < F = 5.19$.   The computations are as follows:

$$s_x^2 = \frac{.000298}{4} = .0000745 \qquad s_y^2 = \frac{.000327}{5} = .0000654 \qquad F = \frac{.0000745}{.0000654} = 1.14$$

Since this value of F falls within the acceptance region, we accept the hypothesis of equal variances.   Thus, in the t test of Example 10-9 we are justified in assuming that $\sigma_x^2 = \sigma_y^2$.   ////

With regard to more than two variances an exact test for equality is lacking, but approximate tests have been devised by Neyman and Pearson, Bartlett, and others.   The test we present here was proposed by M. S. Bartlett,[1] who showed that the distribution is approximately that of chi-square.

Given k independent, unbiased variance estimates $v_1, v_2, \ldots, v_k$ based, respectively, on $a_1, a_2, \ldots, a_k$ degrees of freedom.   To test the hypothesis that the v's are all estimates of a common variance σ^2, put $A = \sum_1^k a_i$, $\bar{v} = \sum a_i v_i / A$, and compute the statistic Z defined thus:

$$Z = \frac{A \ln \bar{v} - \sum a_i \ln v_i}{C} \qquad (10\text{-}51)$$

where

$$C = 1 + \frac{\sum (1/a_i) - 1/A}{3(k - 1)} \qquad (10\text{-}52)$$

The statistic Z is distributed approximately as chi-square with $k - 1$ degrees of freedom.   The nature of the test is such that any differences among the v's tend to increase Z; therefore, the appropriate critical region is the right-hand tail of the chi-square distribution.   Since $C > 1$, it is often possible to save arithmetic by noticing whether the numerator of Z would fall short of the critical limit before dividing by C.

In the special case of uniform sample size $a_i = a$ (all i), the quantity C [Eq. (10-52)] reduces to

$$C = \frac{1 + k(3a + 1)}{3ak} \qquad (10\text{-}53)$$

and Z assumes the simple form

$$Z = \frac{3a^2 k(k \ln \bar{v} - \sum \ln v_i)}{1 + k(3a + 1)} \qquad (10\text{-}54)$$

The purpose of this test is usually to justify the assumption of homoscedasticity, which is involved in significance tests for the differences among several means.   In this connection, it is well to choose .10 as the level of significance.

EXAMPLE 10-12   In a study on ultraviolet irradiation of sterols to produce a form of vitamin D, Bunker, Harris, and Mosher[2] obtained the experimental

[1] M. S. Bartlett, Properties of Sufficiency and Statistical Tests, *Proc. Roy. Soc. (London)* ser. A, vol. 160, p. 273, 1937.

[2] J. W. Bunker, R. S. Harris, and L. M. Mosher, Relative Efficiency of Active Wave Lengths of Ultraviolet in Activation of 7-Dehydrocholesterol, *J. Am. Chem. Soc.*, vol. 62, pp. 508–511, 1940.

data which are summarized in Table 10-19. The measure of effectiveness x was a quantitative rating of healing response when the test animals, rachitic rats, were given doses of sterol (7-dehydrocholesterol) which had been irradiated at specified wavelengths. Later on, we shall test the mean healing rates for significant differences, but at present we shall test for equality of variances.

Since there are seven wavelength groups, $k = 7$ and chi-square has 6 degrees of freedom. At the .10 level, the acceptance region will be $Z < 10.65$. The computations are as follows:

$$A = \sum a_i = 301 \qquad \bar{v} = \frac{\sum a_i v_i}{A} \equiv \frac{\sum_i \sum_j (x_{ij} - \bar{x}_i)^2}{A} = \frac{114.3546}{301} = .3799$$

$$\ln \bar{v} = -0.9678 \qquad A \ln \bar{v} - \sum a_i \ln v_i = -291.3078 + 297.9490 = 6.6412$$

The fact that the latter value falls within the acceptance region as it stands means that we do not need to compute the denominator C in order to accept the hypothesis of equal variances. However, to complete the illustration, we do compute C as follows:

$$3(k - 1) = 18 \qquad \sum \frac{1}{a_i} = .1646 \qquad \frac{1}{A} = .0033$$

$$\frac{\sum (1/a_i) - 1/A}{3(k - 1)} = \frac{.1613}{18} = .0090$$

$$C = 1.0090 \qquad Z = \frac{6.6412}{1.0090} = 6.582$$

As anticipated, this value of Z falls within the acceptance region, and so the hypothesis of equal variances is sustained. ////

Table 10-19

Wave-length, Å	Sample size n_i	$\sum_j x_{ij}$	$\sum_j x_{ij}^2$	$\bar{x}_i$	$\sum_j(x_{ij} - \bar{x}_i)^2$	v_i	$\ln v_i$	a_i
2,483	40	64.0	114.50	1.6000	12.1000	.3103	−1.1702	39
2,537	40	65.5	121.75	1.6375	14.4938	.3716	−0.9899	39
2,652	40	66.0	119.00	1.6500	10.1000	.2590	−1.3509	39
2,804	50	83.0	154.00	1.6600	16.2200	.3310	−1.1056	49
2,894	49	86.0	173.00	1.7551	22.0612	.4596	−0.7774	48
2,967	49	97.0	211.50	1.9796	19.4796	.4058	−0.9019	48
3,025	40	62.0	116.00	1.5500	19.9000	.5103	−0.6728	39
Sum	308	523.5	1,009.75	·······	114.3546	······	·········	301

10-17 THE F TEST FOR EQUALITY OF MORE THAN TWO MEANS

Given m independent samples, the question as to whether their means $\bar{x}_1$, $\bar{x}_2$, ..., $\bar{x}_m$ are significantly different may become the deciding issue in a research problem. A major object of experimental science is to find relationships between variables or attributes. The mere existence of a significant association can be tested by chi-square, but contingency methods require very large samples if it is important to investigate the tail ends of the distributions. Usually large samples are unavailable, and in any event, it is often sufficient to determine the answer to a more specific question—namely, whether the mean value of a variate x assumes significantly different levels depending upon the particular level (or category) assumed by another variate (or attribute) A. This question can be answered by the *analysis of variance*, which is due in the main to R. A. Fisher. Variance analysis has been developed extensively in the theory of experimental design, but we shall consider only the simplest case.

Suppose A is divided into m discrete levels or categories $a_1, a_2, \ldots, a_m$. For each class of A take independent observations of x, letting $n_1, n_2, \ldots, n_m$ be the respective numbers of such observations and $N = \sum n_i$. Let x_{ij} be any observation of x when $A = a_i$, and let

$$\bar{x}_i = \frac{1}{n_i} \sum_j x_{ij}$$

be the computed mean value of x for this class. The problem is then to decide whether the several mean values $\bar{x}_1$, $\bar{x}_2$, ..., $\bar{x}_m$ are significantly different.

The actual differences among the means can be measured by the weighted sum of squares of their deviations from the pooled sample mean $\bar{x}$, where $\bar{x} \equiv (1/N) \sum_i \sum_j x_{ij} \equiv (1/N) \sum_i n_i \bar{x}_i$. This measure of variability is

$$S_1 = \sum_i n_i (\bar{x}_i - \bar{x})^2 \tag{10-55}$$

The question now hinges on whether S_1 falls within acceptable limits as fixed by the null hypothesis. If the samples are all drawn from a common normal population, it can be shown by the methods employed in Sec. 10-9 that $\bar{x}$ and S_1 are independent and that S_1/σ^2 is distributed as chi-square with $m - 1$ degrees of freedom. However, as we shall presently see, if the populations from which the samples are drawn have unequal means, then S_1 will tend to be greater than the value expected when the population means are equal.

Our test for equality of means presupposes equality of variances. To eliminate the parameter σ^2, we compute an unbiased estimate by combining

the sums of squares about the respective sample means.   Put $W_i = \sum_j (x_{ij} - \bar{x}_i)^2$
and

$$S_2 = \sum_i W_i = \sum_i \sum_j (x_{ij} - \bar{x}_i)^2 \qquad (10\text{-}56)$$

For any class the statistic W_i is independent of its own class mean $\bar{x}_i$ and certainly independent of the mean of any other class.   Hence, the statistic S_2 is independent of all the class means and, therefore, independent of S_1.   Now W_i/σ^2 is distributed as chi-square with $n_i - 1$ degrees of freedom, and by the additive property of chi-square (since the W's are mutually independent) it follows that S_2/σ^2 is distributed as chi-square with $N - m$ degrees of freedom [that is, $\sum_i (n_i - 1) = N - m$].   Therefore the variance ratio

$$\frac{S_1/(m-1)\sigma^2}{S_2/(N-m)\sigma^2} \equiv \frac{S_1/(m-1)}{S_2/(N-m)} \qquad (10\text{-}57)$$

will be distributed as F with $m - 1$ and $N - m$ degrees of freedom; and this statistic provides a test of the null hypothesis.

Now suppose that the samples are, in fact, drawn from populations with different means, $\mu_1, \mu_2, \ldots, \mu_m$.   Then the weighted sum of squares S_1 can be expressed identically as follows, where $\bar{\mu} = \sum_i n_i \mu_i / N$:

$$S_1 \equiv \sum_i n_i [(\bar{x}_i - \mu_i) - (\bar{x} - \bar{\mu}) + (\mu_i - \bar{\mu})]^2 \qquad (10\text{-}58)$$

the expected value of which is

$$E(S_1) = (m-1)\sigma^2 + \sum_i n_i(\mu_i - \bar{\mu})^2 \qquad (10\text{-}59)$$

Thus, any differences among the μ's will tend to increase S_1, so that the critical region will be the upper right-hand tail of F.

The reason for the name "analysis of variance" lies in the separation of the total sum of squares about the grand mean into additive components.   Thus, denoting the total sum of squares by S_0, where

$$S_0 = \sum_i \sum_j (x_{ij} - \bar{x})^2 \qquad (10\text{-}60)$$

we have identically

$$S_0 \equiv S_1 + S_2 \qquad (10\text{-}61)$$

The reader is invited to establish this identity on his own by putting $x_{ij} - \bar{x} \equiv (x_{ij} - \bar{x}_i) + (\bar{x}_i - \bar{x})$, squaring both sides, and noting that the cross-product term on the right sums to zero.

COMPUTATIONAL PROCEDURE   Although the foregoing definitions arise naturally out of the theoretical development, it is advantageous to work with algebraically simpler (although identical) terms for computational purposes.   We shall present the procedure itself first and later on prove its validity.   Define class totals $T_1, T_2, \ldots, T_m$ and the grand total T as follows:

$$T_i = \sum_{j=1}^{n_i} x_{ij} \qquad \text{whence } \bar{x}_i = \frac{T_i}{n_i}$$

$$T = \sum_i T_i \equiv \sum_i \sum_j x_{ij} \qquad \text{whence } \bar{x} = \frac{T}{N}$$

Compute these totals, the corresponding means, and the raw sum of squares:

1 $T_i, \bar{x}_i, T, \bar{x}.$

2 $\sum_i \sum_j x_{ij}^2 \equiv \sum_i \left(\sum_j x_{ij}^2 \right)$.   It is advisable to record subtotals of squares by classes to facilitate the test for uniform variability within classes.   Next compute the grand mean times the grand total as well as the sum of the products of class means and class totals:

3 $\bar{x}T, \sum_i \bar{x}_i T_i$.   In the special case of constant sample size, that is, $n_1 = n_2 \cdots = n_m = n$, the sum of products of class means and class totals reduces to

$$\sum_i \bar{x}_i T_i \equiv \frac{1}{n} \sum_i T_i^2 \qquad \text{(constant sample size only)}$$

This formula promotes arithmetical accuracy and saves work but must not be used unless all the n's are equal.

From the quantities thus far computed, derive the conventional terms of the analysis of variance table.   These are:

4   The "total sum of squares," $S_0 = \sum_i \sum_j x_{ij}^2 - \bar{x}T$.   This term equals $\sum_i \sum_j (x_{ij} - \bar{x})^2$, which is the total sum of squares of deviations from the grand mean.

5   The "sum of squares between classes," $S_1 = \sum_i \bar{x}_i T_i - \bar{x}T$, or where applicable, $S_1 = (1/n)\sum_i T_i^2 - \bar{x}T$.   This   term   equals $\sum_i n_i(\bar{x}_i - \bar{x})^2$, which is the weighted sum of squares of class means about the grand mean.

6   The "sum of squares within classes," $S_2 = S_0 - S_1$.   This term equals $\sum_i \sum_j (x_{ij} - \bar{x}_i)^2$, which is the pooled sum of squares of individual values about the means of their respective classes.

Table 10-20   ANALYSIS OF VARIANCE

Source	Degrees of freedom	Sum of squares	Mean square
Between classes	$m-1$	S_1	$S_1/(m-1)$
Within classes	$N-m$	S_2	$S_2/(N-m)$
Total	$N-1$	S_0	Not done

$$F = \frac{S_1/(m-1)}{S_2/(N-m)}$$

Compile the results in the form of an analysis-of-variance table as shown in Table 10–20. Look up F under $m-1$ and $N-m$ degrees of freedom. As a rule we are interested only in the one-sided alternative $S_1/(m-1) > S_2/(N-m)$ and so would not bother to compute F it if were obvious by inspection that the ratio is less than unity.

EXAMPLE 10-13   Using the healing response data of Example 10-12, Table 10-19, test the means for significant differences. Carrying four decimals in the means, but rounding the product sums to two decimals, we obtain the following results:

$$T = 523.5 \quad N = 308 \quad \bar{x} = 1.6997 \quad \bar{x}T = 889.79$$

$$\sum_i \sum_j x_{ij}^2 = 1009.75 \quad S_0 = 1009.75 - 889.79 = 119.96$$

$$\sum_i \bar{x}_i T_i = 895.40 \quad S_1 = 895.40 - 889.79 = 5.61$$

$$S_2 = S_0 - S_1 = 119.96 - 5.61 = 114.35$$

The analysis of variance is presented compactly in Table 10-21. There being 6 and 301 degrees of freedom, the critical region with $\alpha = .05$ is $F \geq 2.13$. From

Table 10-21   ANALYSIS OF VARIANCE OF HEALING RESPONSE DATA

Source	Degrees of freedom	Sum of squares	Mean square
Between wavelength classes	6	5.61	0.935
Within wavelength classes	301	114.35	0.380
Total	307	119.96	

Example 10-12 we already know that the assumption of homoscedasticity is justified.

$$F = \frac{.935}{.380} = 2.46$$

Since this value of F falls in the critical region, we reject the null hypothesis and conclude that activation at different wavelengths produces significantly different mean healing responses.   ////

JUSTIFICATION OF COMPUTATIONAL PROCEDURE   We first verify the computational formulas for S_0 and S_1 and then show that $S_0 \equiv S_1 + S_2$. Expanding the quadratic in the definition of S_0, identifying notation, and collecting terms, we have at once

$$S_0 \equiv \sum_i \sum_j (x_{ij} - \bar{x})^2 \equiv \sum_i \sum_j x_{ij}^2 - \bar{x}T$$

Next
$$S_1 \equiv \sum_i n_i(\bar{x}_i - \bar{x})^2 \equiv \sum_i (n_i \bar{x}_i^2 - 2\bar{x}n_i \bar{x}_i + n_i \bar{x}^2)$$

Now
$$n_i \bar{x}_i \equiv T_i \qquad n_i \bar{x}_i^2 \equiv \bar{x}_i T_i$$

Thus
$$S_1 \equiv \sum_i \bar{x}_i T_i - 2\bar{x} \sum_i T_i + \bar{x}^2 \sum_i n_i =$$
$$\sum_i \bar{x}_i T_i - 2\bar{x}T + \bar{x}^2 N = \sum_i \bar{x}_i T_i - \bar{x}T$$

as was to be proved.   Finally, to justify the formula for

$$S_2 \equiv \sum_i \sum_j (x_{ij} - \bar{x}_i)^2$$

we put
$$x_{ij} - \bar{x} \equiv (x_{ij} - \bar{x}_i) + (\bar{x}_i - \bar{x})$$

whereupon
$$S_0 \equiv \sum_i \sum_j (x_{ij} - \bar{x})^2 \equiv \sum_i \sum_j [(x_{ij} - \bar{x}_i) + (\bar{x}_i - x)]^2$$

When we expand the latter quadratic, the cross-product term vanishes, for

$$\sum_i \sum_j (\bar{x}_i - \bar{x})(x_{ij} - \bar{x}_i) \equiv \sum_i [(\bar{x}_i - \bar{x})\sum_j (x_{ij} - \bar{x}_i)] = 0$$

since $\sum_j (x_{ij} - \bar{x}_i) \equiv 0$.   Therefore

$$S_0 = \sum_i \sum_j (x_{ij} - \bar{x}_i)^2 + \sum_i \sum_j (\bar{x}_i - \bar{x})^2$$

But
$$\sum_i \sum_j (\bar{x}_i - \bar{x})^2 \equiv \sum_i n_i(\bar{x}_i - \bar{x})^2 \equiv S_1$$

Therefore $S_0 \equiv S_1 + S_2$, which completes the proof.

Analysis of variance forms the backbone of a large part of statistical methodology. Hand in hand with the development of the theory of experimental design, the subject of variance analysis has broadened and deepened to such an extent that it deserves specialized study. We do not pursue this subject further here, because an adequate treatment of its theoretical and practical aspects requires an entire book, whereas a sketchy treatment gives the misleading impression of pat answers.

PROBLEMS

10-1 According to theory the probability of a certain trait is .25. Taking .05 as the level of significance and using the normal approximation corrected for continuity, establish the limits of a two-sided critical region to which the absolute frequency of the trait in a sample of 192 random observations can be referred directly.

$$\textit{Ans. } x < 36, \, x > 60.$$

10-2 The variate x is binomially distributed with parameters p, N. (*a*) Assuming the normal limit of the binomial distribution, explain why the sample relative frequency ($p' = x/N$) of occurrences may be regarded as normally distributed with $\mu = p$, $\sigma^2 = pq/N$. (*b*) Derive appropriate corrections for continuity.

Ans. (*a*) Theorem 6-1. (*b*) If $p' \geq p + \frac{1}{2}N$, use $p' - p - \frac{1}{2}N$; if $p' \leq p - \frac{1}{2}N$, use $p - (p' + \frac{1}{2}N)$.

10-3 Given a binomial distribution with $p = \frac{1}{2}$, how large a sample must be taken in order that the probability will be at least .98 that the sample relative frequency p' will not deviate from $\frac{1}{2}$ by more than .05? Assume the normal approximation but disregard the correction for continuity. *Ans.* $N = 542$.

10-4 For a certain binomial variate x the true value of the parameter p is unknown, and it is desired to determine the minimum sample size N such that the probability will be at least as great as some specified value P that the sample relative frequency $p' = x/N$ will not deviate from p by more than some specified magnitude $h > 0$. (*a*) Assuming the normal approximation, but disregarding the correction for continuity, determine N as a function of p, h, and k, where k is that value of a unit normal variate such that $W(k) = P$. (*b*) If h is small compared with k^2pq, show that the correction for continuity would increase the required value of N by an amount approximately equal to $1/h$.

Ans. (*a*) $N = k^2pq/h^2$. (*b*) Since the larger value of N must be allowed for in considering positive versus negative deviations from p, the required value is the greater root of the equation $4h^2N^2 - 4(k^2pq + h)N + 1 = 0$ rather than the greater root of the alternative equation $4h^2N^2 - 4(k^2pq - h)N + 1 = 0$. Simplification of the radical by neglecting the term $-h^2$ yields the stated result.

10-5   Random samples of five observations each are taken from two populations in order to test the hypothesis that a certain attribute A has the same probability in the two populations. Assuming the populations so large that sampling depletions are negligible, use the normal approximation corrected for continuity to estimate the areas of appropriate two-sided regions associated with (a) relative frequencies of 0 and 1 respectively, and (b) relative frequencies of .2 and .8 respectively. (c) Compare with corresponding probabilities computed from the hypergeometric distribution in Prob. 6-35.

Ans. (a) .011, (b) .206, (c) .008 and .206.

10-6   First-term grades in mathematics and physics for a class of 728 freshmen showed the following joint frequencies of A's and non-A's where the first symbol of any pair designates the grade in mathematics. Consider two groups: AA, $\bar{A}A$ and $A\bar{A}$, $\bar{A}\bar{A}$.

Combination	AA	$A\bar{A}$	$\bar{A}A$	$\bar{A}\bar{A}$	Total
Frequency	53	57	27	591	728

(a) Choosing the .05 level of significance, determine the limits on the difference between the relative frequencies of AA and $A\bar{A}$ for a two-sided critical region based on the normal test for association, incorporating the continuity correction. (b) Is the observed degree of association significant? (c) What assumption is being made with regard to the sample? (d) If we were interested in whether or not the percentages of A's in mathematics and physics are significantly different, what would be wrong with applying the same form of normal test to this sample?

Ans. (a) $\pm.09$. (b) Yes. (c) Drawn at random from a hypothetical general population of this school's freshmen. (d) Condition of independence not satisfied.

10-7   The problem of testing the difference between two percentages pertaining to the same set of individuals was solved by McNemar.[1] Assume that a given individual may exhibit either, or both, or neither of two traits A and B, and suppose that in a sample of N individuals there are n_{11} instances of AB, n_{12} instances of $A\bar{B}$, n_{21} of $\bar{A}B$, and n_{22} of $\bar{A}\bar{B}$. Then, McNemar has shown that the appropriate test of the difference between the proportion of A's and the proportion of B's is given by $\hat{\chi}^2 = (n_{12} - n_{21})^2/(n_{12} + n_{21})$, a statistic which may be regarded as having the chi-square distribution with 1 degree of freedom. Provided that $n_{12} \neq n_{21}$, the continuity correction would be made by subtracting $\frac{1}{2}$ from the absolute value of the difference before taking the square; that is, $\hat{\chi}^2 = (|n_{12} - n_{21}| - \frac{1}{2})^2/(n_{12} + n_{21})$. (a) Using Table 10-3, apply the latter form of this test to the data of Prob. 10-6 to decide whether there is a significant difference between the percentage of A's in mathematics and the percentage of A's in physics. (b) If one preferred to use a unit normal test

[1] McNemar, *Psychometrika*, vol. 12, no. 2, p. 153, June, 1947.

variate instead of chi-square, how should he go about it? (c) Using this normal variate, estimate the significance measure of the given sample.

Ans. (a) $\hat{\chi}^2 = 10.3$, which is significant; hence the percentages are significantly different. (b) Compute $u = (|n_{12} - n_{21}| - \frac{1}{2})/\sqrt{n_{12} + n_{21}}$ and refer to $2R(u)$ in Table 4-4. (c) $2R(3.3) \approx .001$.

10-8   At a certain locality the probability of rain during any 24-hour period at a particular time of the year is .2. A random sample of 100 independent trials presented the following distribution of forecast and observed categories, where the subscripts f, o denote "forecast" and "observed" respectively.

Event	R_f, R_o	$R_f, \bar{R}_o$	$\bar{R}_f, R_o$	$\bar{R}_f, \bar{R}_o$	Total
Frequency	6	14	10	70	100

(a) Using the normal approximation corrected for continuity and setting the significance level at .05, test for significant association between forecast and and observed categories of rainfall. (b) Compute the mean score m based on the scoring scheme of Table 7-2. (c) Applying the central limit theorem, test for a significant degree of skill as compared with a random allocation of categories. (d) What accounts for the discrepancy between the conclusions of (a) and (c)?

Ans. (a) $u' = 1.57$, not significant. (b) $m = .175$. (c) $\sigma_m = .1$ $m/\sigma_m = 1.75$, which is significant at the .05 level. (d) The test for association takes the marginal totals for granted and employs a two-sided critical region. The test for skill leaves the marginal totals unspecified and uses a one-sided critical region, because only a positive degree of skill is relevant.

10-9   Contract specifications usually require the vendor to demonstrate a satisfactory level of quality by subjecting the articles offered for sale to certain prescribed tests. A typical specification reads: "The vendor shall establish, with a confidence coefficient of at least .90, that the fraction defective does not exceed .05." Since the manufacturer would ordinarily aim at a fraction defective appreciably lower than the maximum acceptable, in order to stand a good chance of passing the test, the numbers of defective items in random samples of constant size N may be assumed to be Poisson-distributed with $\mu = Np$, where p is the true fraction defective in the population from which the samples are drawn. Special tables giving confidence limits for the Poisson and binomial distributions are available, but good results can be obtained also by applying normalizing transformations.[1] The advantage of the transformation lies not only in the applicability of readily available tables of the normal distribution but also in the fact that the variance becomes virtually independent of the mean. If x is Poisson-distributed with parameter μ, the transformed variate $y = \sqrt{x + \frac{1}{2}}$ is approximately normal with mean equal to $\sqrt{\mu + \frac{1}{2}}$ and standard

[1] See M. S. Bartlett, The Use of Transformations, *Biometrics*, vol. 3, no. 1, p. 39, March, 1947, and J. H. Curtiss, On Transformations Used in Analysis of Variance, *Ann. Math. Stat.*, vol. 14, no. 2, p. 107, June, 1943.

deviation equal to $\frac{1}{2}$.   (a) Using this approximation, determine the maximum fraction defective in a random sample of 175 items for which it can be asserted with a confidence coefficient of .90 (i.e., the probability that statements of this sort will be correct is .90) that the fraction defective in the population does not exceed .05.   Correct for continuity by adding one-half unit to the absolute frequency (besides the $\frac{1}{2}$ already called for by the equation for y), but otherwise neglect the discrete nature of the data.   (b) Considering the fact that only discrete increments are possible, what would the actual upper limit of the fraction defective in a sample have to be?   (c) Estimate the probability that the vendor will meet the specified standards with a random sample of 175 items if the true fraction defective in the population sampled is .02.

　　　　Ans.   (a) .0272,   (b) 4/175,   (c) $F(.472) = .681$; the exact probability that x will not exceed 4 if $\mu = 3.5$ is .725.

10-10　The average weight of 100 regular-sized cigarettes is 4 ounces.   Hence, if .01 ounce is chosen as the unit, the weight of individual cigarettes will average out to four units.   Individual weights vary, of course, because the tobacco itself is not perfectly uniform, and even if it were, perfect precision could not be maintained in the mechanical processes of manufacture.   Therefore, production standards prescribe certain control limits within which the individual weights are allowed to vary and still be judged acceptable; and weight uniformity is checked intermittently by random sampling.   The general field of measuring the output and maintaining prescribed standards is called *quality control*. Assume that individual cigarette weights are normally distributed with $\mu = 4$, $\sigma = .15$.   (a) Determine 95 percent control limits for the mean weights of random samples of 5.   That is, find a constant C such that $P(\mu - C < \bar{x} < \mu + C)$ $= .95$, where $\bar{x}$ denotes the sample mean, and then compute $\mu - C$ and $\mu + C$. (b) Let $s^2 = \Sigma(x_i - \bar{x})^2/(n - 1)$.   If the value of $\bar{x}$ were satisfactory, would there be any objection to an unusually small value of s?   (c) Determine a one-sided control limit S for s such that $P(s < S) = .95$.   (d) A random sample of five observations gave the following results: $\Sigma x = 20.40$, $\Sigma x^2 = 83.3162$. Compute $\bar{x}$ and s and decide whether the sample indicates satisfactory control.

　　　　Ans.   (a) $3.87 < \bar{x} < 4.13$.   (b) No, the smaller the better.   (c) $s < .231$. (d) $\bar{x} = 4.08$, $s = .145$.   Both statistics lie in their respective acceptance regions; hence, satisfactory control is indicated.

10-11　Five cigarettes were picked at random from the production run of a given hour, with sufficient spacing to justify the assumption of independence, and the sampling was repeated hourly.   In samples from six consecutive hours, the individual weights in units of .01 ounce were as follows:

1.　4.15, 4.04, 4.25, 4.10, 3.86; $\Sigma x = 20.40$, $\Sigma x^2 = 83.3162$
2.　4.01, 3.63, 4.09, 3.89, 4.08; $\Sigma x = 19.70$, $\Sigma x^2 = 77.7636$
3.　4.19, 3.81, 3.84, 4.20, 4.16; $\Sigma x = 20.20$, $\Sigma x^2 = 81.7634$
4.　4.02, 3.91, 3.80, 3.85, 4.07; $\Sigma x = 19.65$, $\Sigma x^2 = 77.2759$
5.　3.79, 3.66, 3.82, 4.21, 3.97; $\Sigma x = 19.45$, $\Sigma x^2 = 75.8371$
6.　4.03, 3.98, 4.05, 3.55, 3.64; $\Sigma x = 19.25$, $\Sigma x^2 = 74.3359$

(a) Compute $\bar{x}$ and s for each sample, and plot a control chart. Use the upper half of the chart for $\bar{x}$ and the lower half for s, with the control limits established in Prob. 10-10 indicated as horizontal lines. (b) When is lack of control definitely indicated?

Ans. (a)

	Sample					
	1	2	3	4	5	6
$\bar{x}$	4.08	3.94	4.04	3.93	3.89	3.85
$\Sigma(x-\bar{x})^2$	.0842	.1456	.1554	.0514	.1766	.2234
s	.145	.191	.197	.113	.210	.236

(b) Lack of control indicated definitely in sample 6.

10-12 As a practical convenience in the cigarette industry, weights are customarily expressed in reciprocal units rather than in hundredths of an ounce. That is, the scale indicates the number of cigarettes, of a weight equal to the one on the scale, which would be needed in order to make up a total of 4 ounces. Assuming a normal distribution of actual weights, would reciprocals be normally distributed?

Ans. Not exactly; but since the standard deviation is so small compared with the mean, the normal approximation can be justified by expanding the reciprocal in powers of $x - \mu$ and neglecting terms beyond the first degree.

10-13 The reliability of an electronic component is defined as the probability of continuous, satisfactory operation over a stipulated interval of time T. If initial defects are weeded out by the manufacturer's inspection of finished pieces and failure due to wear-out is prevented by maintenance inspection on the part of the user, the remaining causes of failure may be regarded as random phenomena. Under these circumstances failures per unit time arising from a large number of similar components exposed to comparable conditions typically conform to the Poisson distribution. Hence, by Example 4-8, the time between failures will follow the exponential distribution, and by a property unique to the exponential distribution (Prob. 5-24), the distribution of service life is precisely the same as that of time between failures. Consequently, if the mean time between failures is equal to θ, then the distribution of service life t will be given by $f(t) = (1/\theta)e^{-t/\theta}$ $(0, \infty)$ and the reliability ρ will be given by $\rho = P(t \geq T) = e^{-T/\theta}$. An efficient test of reliability has been devised by Epstein and Sobel.[1] A convenient sample size n being chosen, n items are placed on test and the test is continued until a prescribed number k (determined in accordance with costs and available time) have failed. Denoting

[1] See B. Epstein, and M. Sobel, Life Testing, J. Am. Stat. Assoc., vol. 48, no. 263, p. 486, September, 1953.

the failure times by t_1, t_2, ..., t_k in ascending order, the maximum likelihood estimate of θ is given by

$$\hat{\theta} = \frac{t_1 + t_2 + \cdots + t_k + (n - k)t_k}{k} = \frac{t_1 + t_2 + \cdots + t_{k-1} + (n + 1 - k)t_k}{k}$$

and the corresponding estimate of reliability is $\hat{\rho} = e^{-T/\hat{\theta}}$. Confidence limits can be calculated from the critical levels of chi-square, for Epstein and Sobel have proved that the variate $2k\hat{\theta}/\theta$ is distributed as chi-square with $2k$ degrees of freedom. Although the sampling distribution of the estimate depends on k, the average length of time one has to wait for the occurrence of k failures decreases as the sample size n is increased.

Suppose that contract specifications call for 5 hours of continuous performance with a reliability of at least .95 demonstrated with a confidence coefficient of .90. The vendor decides to place seven units on test and wait until the first four have run to failure. The test data (hours to failure) are 23.5, 52.8, 72.0, 158.8.

(a) Estimate the mean life and the reliability. (b) Find the lower confidence limit R on reliability such that the confidence coefficient is .90 that $\rho \geq R$. (c) On the strength of this sample, does the lot pass the test?

Ans. (a) $\hat{\theta} = 195.9$ hr, $\hat{\rho} = .975$. (b) $8\hat{\theta}/13.36 = 117.3$; $5/117.3 = .0426$; $R = .958$. (c) Yes, but just barely.

10-14 An operations research group studying the M.I.T. libraries found that the number of books withdrawn by individual users during any one visit to the Science Library was geometrically distributed with $p = .4$; that is,

$$f(x) = .4(.6^x) \qquad x = 0, 1, \ldots$$

Test this hypothesis for goodness of fit, using the following data, which represent an independent random sample of 100 observations.

Value of x	0	1	2	3	≥ 4
Frequency	37	26	17	5	15

Ans. With 4 degrees of freedom the acceptance region is $\hat{\chi}^2 < 9.49$ at the .05 level of significance. The sample value of 2.68 falls within the acceptance region, and the hypothesis is sustained.

10-15 In terms of the sample estimates of the mean and standard deviation based on 310 independent observations, the frequency distribution of January temperatures (time: 1230 Z) at Bismarck, N.Dak., was found to be as summarized in the following table. On the hypothesis of a normal distribution, expected frequencies were computed to the nearest whole number. (a) Explain how the expected frequencies were obtained, and illustrate by calculating the expected frequencies in the intervals $> -.4$ to $\leq .4$ and $> .4$ to ≤ 1.2. (b) How many degrees of freedom are there? (c) Test the hypothesis that the temperature distribution is normal.

$(x-\bar{x})/s$	≤ -2.0	$> -2.0,$ ≤ -1.2	$> -1.2,$ $\leq -.4$	$> -.4,$ $\leq .4$	$> .4,$ ≤ 1.2	$> 1.2,$ ≤ 2.0	> 2.0
Observed frequency	7	34	67	87	78	32	5
Expected frequency	7	29	71	96	71	29	7

Ans. (a) $310W(.4) = 96$; $310[F(1.2) - F(.4)] = 71$. (b) There are $7 - 3 = 4$ degrees of freedom. (c) Acceptance region, $\hat{\chi}^2 < 9.49$; sample yields $\hat{\chi}^2 = 3.50$. Therefore the hypothesis of normal distribution is sustained.

10-16 The theoretical probabilities associated with the possible combinations of two attributes A and B are as follows:

Combination	A_1B_1	A_1B_2	A_1B_3	A_2B_1	A_2B_2	A_2B_3	A_3B_1	A_3B_2	A_3B_3
Probability	.20	.08	.04	.08	.20	.08	.04	.08	.20

The observed frequencies in a random sample of 100 were respectively 15, 12, 6, 11, 15, 10, 6, 10, 15. Is the theory tenable?

Ans. If nine classes are used, there will be 8 degrees of freedom, and the acceptance region at the 5 percent level of significance will be $\hat{\chi}^2 < 15.51$. If the two smallest classes are combined, there will be 7 degrees of freedom and the acceptance region at the 5 percent level of significance will be $\hat{\chi}^2 < 14.07$. It turns out that exactly the same sample value is obtained in either case, namely, $\hat{\chi}^2 = 9.875$. Since this value falls within the acceptance regions, the theory is tenable.

10-17 In a particular city, the sales of a certain commodity are dominated by four leading brands A, B, C, D; all other brands combined (represented by E) account for only 10 percent of the sales. Initially, the percentage distribution of market shares was as follows:

Brand	A	B	C	D	E	Sum
Share	25	25	20	20	10	100

After vigorous sales campaigns put on by the manufacturers of C and D, a random sample of 1,000 sales showed the following breakdown:

Brand	A	B	C	D	E	Sum
Sales	232	228	229	227	84	1000

Has the distribution undergone a statistically significant change?

Ans. With 4 degrees of freedom, the critical region is $\hat{\chi}^2 \geq 9.49$ at the 5 percent level of significance. The sample value 13.64 falls in the critical region and indicates a significant change in the distribution.

10-18   Annual sales of a certain type of sporting goods to individual retail dealers in a large eastern city have a markedly skewed distribution.   With economic data it is often found that the logarithm of the initial variate is approximately normal. In that case, the initial variate is said to have a log-normal distribution. Another transformation worth considering is the cube root, for if the initial variate has the gamma distribution, its cube root will be approximately normal. Both transformations were tried on the same sample of 150 observations, and in each case the mean and variance of the transformed variate were estimated from the sample.   Letting u represent the deviation from the mean in units of s, nine categories were defined thus: I($u \leq -1.75$), II($-1.75 < u \leq -1.25$), III($-1.25 < u \leq -.75$),   IV($-.75 < u \leq -.25$),   V($-.25 < u \leq .25$),   VI($.25 < u \leq .75$), VII($.75 < u \leq 1.25$), VIII($1.25 < u \leq 1.75$), IX($u > 1.75$).   The sample gave the following results:

| | Category | | | | | | | | | |
Comparison	I	II	III	IV	V	VI	VII	VIII	IX	Sum
Normal relative frequency	.04	.07	.12	.17	.20	.17	.12	.07	.04	1.00
Expected normal frequency	6.0	10.5	18.0	25.5	30.0	25.5	18.0	10.5	6.0	150
Observed frequency with logarithm	9	9	10	21	31	35	24	9	2	150
Observed frequency with cube root	4	11	20	27	29	27	16	7	9	150

Test each transformation for normality.

Ans.   With $9 - 3 = 6$ degrees of freedom, the acceptance region at the 5 percent level of significance is $\hat{\chi}^2 < 12.59$.   The log gives $\hat{\chi}^2 = 14.55$, which indicates a significant departure from normality.   The cube root gives $\hat{\chi}^2 = 4.01$, which lies well within the acceptance region and shows that the distribution is approximately normal.

10-19   Over the course of the year, the climatological " normal " temperature (normal in the sense of population average) from day to day exhibits, as one might expect, a pronounced seasonal cycle which is approximately sinusoidal.   The standard deviation has a seasonal cycle also, but the curve is flat compared with that of the " normal."   It turns out that, if the daily mean temperature is expressed as a deviation from normal and the difference is standardized by dividing by the standard deviation based on the climatological record for the whole month, the resulting standard score has very nearly the same (slightly non-gaussian) distribution in large samples, irrespective of season and location, within North America.   The time dependence among consecutive observations still remains, nevertheless, and shows up in a tendency (called " persistence ") for one day's score to deviate less from that of the previous day than from the population mean of zero. This fact, of course, is valuable from the viewpoint of weather forecasting, but it complicates a statistical analysis. The general effect of persistence is illustrated by the following example.

Standard scores of daily mean temperature for 3 years x, y, z at Station W during the month of September were classified into three categories: high, medium, low.

		Category		
Year	High	Medium	Low	Total
x	0	20	10	30
y	8	18	4	30
z	10	13	7	30

(a) Assuming independence, would these 3 years be considered a homogeneous group, or would the differences between them be judged significant? (b) Keeping the observed *relative* frequencies the same, determine the greatest number N (in place of 30) for which the samples would be regarded homogeneous.

Ans. (a) With 4 degrees of freedom, the acceptance region is $\hat{\chi}^2 < 9.49$ at the .05 level of significance. Hence the sample value 13.43 would be judged significant. (b) $N = 30(9.49)/13.43 = 21.2$; that is, $N = 21$.

10-20   The chi-square tests for association, independence, or homogeneity in the case of two samples with k categories each can be simplified by applying a formula due to Brandt and Snedecor. One way of writing this is as follows:

$$\hat{\chi}^2 = \frac{1}{pq} \sum r_i - \frac{a.}{q} \qquad k-1 \text{ degrees of freedom}$$

wherein a_i = observed frequency of category i in sample 1; b_i = observed frequency of category i in sample 2; $c_i = a_i + b_i$; $r_i = a_i^2/c_i$; $a. = \sum a_i$; $b. = \sum b_i$; $c. = \sum c_i$; $p = a./c.$, $q = b./c.$.

(a) Derive this formula by showing first that in category i, the combined contribution to chi-square can be expressed as

$$\frac{(a_i - pc_i)^2}{pc_i} + \frac{(b_i - qc_i)^2}{qc_i} = \frac{(a_i - pc_i)^2}{pqc_i}$$

Complete the proof by expanding the quadratic and simplifying the summation. (b) In case the sample sizes are equal ($p = \frac{1}{2}$), derive the still-simpler formula $\hat{\chi}^2 = \Sigma(d_i^2/c_i)$, where $d_i = a_i - b_i$; $c_i = a_i + b_i$.

10-21   Two discrete variates x, y have the same set of possible values, and random samples of each turned out as follows:

	Possible values					
	1	3	6	9	11	Total
x	5	13	64	13	5	100
y	9	2	78	2	9	100

Test the hypothesis that the distributions are the same.

Ans. At the .05 level of significance, the acceptance region with 4 degrees of freedom is $\hat{\chi}^2 < 9.49$.   Since the sample value 19.80 falls in the critical region, we conclude that the distributions are different.

10-22   In attempting to judge a certain forecaster's ability to forecast cloud coverage 24 hours in advance, the following information was compiled from a series of independent trials.

Category observed	Category forecast Cloudy	Scattered	Clear	Total
Cloudy	12	28	3	43
Scattered	8	55	24	87
Clear	1	8	36	45
Total	21	91	63	175

(*a*) Is there a significant degree of association between the forecast and observed categories of cloudiness?   (*b*) What would you say about the forecaster's ability?

Ans. (*a*) At the .05 level of significance, the acceptance region with 4 degrees of freedom is $\hat{\chi}^2 < 9.49$.   The sample value 62.61 falls in the critical region, thereby indicating a significant degree of association between the forecast and observed categories of cloudiness.   (*b*) An analysis of marginal totals by formula (*b*) of Prob. 10-20 reveals significant differences.   At the .05 level of significance, the acceptance region with 2 degrees of freedom is $\hat{\chi}^2 < 5.99$, whereas the sample value is 10.65.   Hence the sample excess of forecasts of "clear" and deficit of forecasts of "cloudy" must be judged significant departures from observation.   Since errors of two categories are comparatively rare, we may surmise that when in doubt the forecaster tends to call "cloudy," "scattered" and "scattered," "clear."

10-23   The mean of sample 6, Prob. 10-11, was found to differ significantly from the population value $\mu = 4$ when the standard deviation was known to be .15. If the standard deviation were unknown, would lack of control be detected with this sample at the same level of significance?

Ans. At the .05 level of significance, the acceptance region for *t*, corresponding to a two-sided critical region, is $-2.78 < t < 2.78$, with 4 degrees of freedom.   Since the sample value $t = -1.42$ is not significant, lack of control would not be detected with this sample if the standard deviation were unknown.

10-24   Seven students were sent to a remedial reading clinic to improve their reading speed.   Corresponding speeds in words per minute on standardized material at the start and after a short period of treatment are shown in the following table:

Speed	Student A	B	C	D	E	F	G
Initial	210	235	255	275	300	310	315
After treatment	255	260	310	260	330	355	375

(a) Test for a significant improvement. (b) Why would a two-sided critical region be inappropriate? (c) Does the lack of a control group vitiate the test of significance?

Ans. (a) At the .05 level of significance and a one-sided critical region, the acceptance region for the null hypothesis with 6 degrees of freedom is $t < 1.94$. A significant improvement (on an average) is demonstrated, for the sample yields $t = 3.66$, which falls in the critical region. (b) A two-sided critical region would not be appropriate because it would be incompatible with the working hypothesis that treatment would be beneficial. (c) The test of significance, as such, is independent of a control group, for it depends solely on the probability distribution under the null hypothesis. A control group would, however, strengthen the argument that the established tendency to improve is due to the treatment. But one need not confine his knowledge of a field to the evidence presented by a small sample.

10-25   In the manufacture of files, the teeth are cut on side A first, then on side B. The following data were taken on a random sample of six files and refer to weight in grams of filings removed from standard stock per 1,000 strokes:

		File				
Side	1	2	3	4	5	6
A	8	6	5	9	11	10
B	7	4	3	10	9	12

(a) Do sides A and B differ significantly in the average weights of filings removed? (b) Justify your method of calculation.

Ans. (a) The acceptance region for the null hypothesis at the .05 level of significance is $-2.57 < t < 2.57$, since the critical region is two-sided and there are 5 degrees of freedom. The sample value, $t = 0.94$, falls in the acceptance region, thus sustaining the null hypothesis that the average weights, in the long run, are equal. (b) The method of paired comparisons is used to ensure independence in case the two sides of the same file mutually reflect common causes of variation.

10-26   A farmer grows the same crop on two fields A and B, of equal acreage. On field A he puts an enriched fertilizer, but on field B an equal amount of standard fertilizer. He continues this experiment for 5 years and gets the following yields in bushels per acre:

		Year			
Field	1	2	3	4	5
A	18	16	24	18	23
B	17	14	21	19	20

Does field A produce a significantly higher average yield?

Ans. The critical region being one-sided, in view of the working hypothesis, the acceptance region for t with 4 degrees of freedom is $t < 2.13$ at the .05 level of significance. Hence the sample value, $t = 2.14$, indicates a significantly higher average yield from field A.

10-27 Two random samples of size 10 and size 12 from normal populations gave the results $\bar{x}_1 = 19$, $s_1 = 5$; $\bar{x}_2 = 24$, $s_2 = 4$. Are the means significantly different?

Ans. There are $9 + 11 = 20$ degrees of freedom, and the acceptance region for the .05 level of significance is $-2.09 < t < 2.09$. Thus, the sample value $t = 2.60$ demonstrates a significant difference in the means.

10-28 As a step toward the optimum allocation of sales effort, the customers for a certain line of merchandise were classified according to various factors which seemed likely to affect sales volume by means of an objective system which did not depend on current sales data. The classification scheme was then subjected to various tests; and because of the extreme skewness of the initial sales data, the cube root of amount purchased was used as a normalizing transformation (refer to Prob. 10-18).

Sample results for two of the classes were as follows: $n_1 = 10$, $\bar{x}_1 = 27.3$, $s_1 = 9.1$; $n_2 = 12$, $\bar{x}_2 = 20.1$, $s_2 = 4.6$.

(a) By interpolating on the reciprocals of degrees of freedom, find the .05 critical levels of F with 9 and 11 degrees of freedom and F with 11 and 9 degrees of freedom. (b) Establish the acceptance region for the hypothesis of equal variances at the .10 level of significance, and decide whether the two variances can be considered equal. (c) Test the hypothesis of equal means by the Cochran-Cox method. (d) What simplification could you suggest?

Ans. (a) 2.90, 3.09. (b) $.324 < s_1{}^2/s_2{}^2 < 2.90$, or equally well, $.345 < s_2{}^2/s_1{}^2 < 3.09$. Sample value $s_1{}^2/s_2{}^2 = 3.91$ indicates a significant difference. (c) At the .05 level of significance, the acceptance region for the hypothesis of equal means is $-2.25 < t' < 2.25$. Sample value $t' = 2.27$ indicates a significant difference. (d) Before computing the weighted average a, check significance against the greater of the limits a_1, a_2.

10-29 A random sample of 10 observations from a normal population gave $\bar{x} = 40$, $s = 5$. Determine 95 percent confidence limits for μ. *Ans.* $36.43 \leq \mu \leq 43.57$.

10-30 With reference to the data on weight of filings (Prob. 10-25), (a) test the variation of average weights removed by different files against the variation of weights removed by alternate sides of the same file. (b) What simplification can be employed in computing the sum of squares "within files"?

Ans. (a) Under the null hypothesis that the average weights are equal, the acceptance region at the .05 level of significance for the variance ratio with

Source	Degrees of freedom	Sum of squares	Mean square	Variance ratio
Between files	5	80.67	16.1	10.7
Within files	6	9.00	1.5	
Total	11	89.67		

5 and 6 degrees of freedom is $F < 4.39$. The sample value of 10.7 indicates that the means are significantly different. (b) When there are two observations in a class, the sum of squares within the class is simply half the square of the difference of the two observations.

10-31 Three samples of six each were taken at different times by picking consecutive cigarettes from the production line. The data on individual weights in units of .01 ounce were as follows:

Sample 1   4.08, 3.81, 3.97, 3.78, 3.73, 4.03
Sample 2   4.28, 3.95, 4.15, 4.21, 3.91, 4.04
Sample 3   3.92, 4.05, 3.82, 3.80, 4.00, 3.75

(a) Test the hypothesis of equal variances. Note that natural logarithms can be obtained from an exponential table in case a convenient table of natural logarithms is not at hand. (b) Test the hypothesis of equal means, assuming the samples to be random. (c) Suppose that 30 percent of the total variance of weight could be eliminated by a special attachment which regulates weights automatically. What are the chances that a random sample of 86 independent observations, when referred to a one-sided critical region, would indicate a significant reduction of variance at the .05 level of significance if the initial variance were known? (See note at foot of Table 10-3.)

Ans. (a) At the .10 level of significance with 2 degrees of freedom the acceptance region for the hypothesis of homoscedasticity is $Z < 4.61$. The sample value, $Z = .25$, falls in the acceptance region and sustains the hypothesis.

Sample	$\bar{x}$	$\Sigma(x - \bar{x})^2$	v	$\ln v$
1	3.90	.1056	.0211	−3.859
2	4.09	.1086	.0217	−3.830
3	3.89	.0712	.0142	−4.254
Total	....	.2854	.....	−11.943

$$a = 5, \ A = 15, \ \bar{v} = .0190, \ \ln \bar{v} = -3.963, \ Z = .25$$

(b) Under the null hypothesis that the average weights are equal, the acceptance region at the .05 level of significance for the variance ratio with 2 and 15 degrees of freedom is $F < 3.68$. The sample value of 4.01 indicates that the means are significantly different and, hence, that the samples are not random.

Source	Degrees of freedom	Sum of squares	Mean square	Variance ratio
Between samples	2	.1524	.0762	4.01
Within samples	15	.2854	.0190	
Total	17	.4378		

(c) $\chi_0^2 = 64.5$; $\chi_1^2 = \chi_0^2/.7 = 92$; $P(\chi^2 < 92) \approx .7$.

10-32   A small job shop manufactures custom-built machinery, and a fixed price for each job is agreed upon with the customer before the work is started.   Errors in cost estimation, therefore, cause a variation in profit from job to job, even though a constant-percent profit is intended.   Apparently, the errors are roughly proportional to the contract price, although from one point of view, a downward trend, percentagewise, might be expected.   After the price range was divided into approximately equal intervals, 2 random samples of 10 were selected from each price group and the average price, average profit, and (approximate) average percent profit on the selling price were computed, with the results as shown in the accompanying table.   The percent profit was actually estimated from the ratio of average profit to average price in each sample; but this figure was checked against the average of 10 individual ratios in the lowest price group, and the two estimates were found to agree to the number of figures exhibited.
(a) Spot-check a few values of v and ln v and test the hypothesis of homoscedasticity.   (b) Test the hypothesis of uniform average percent profit.   (c) Assuming that the percent profit on the selling price is normally distributed, what is the probability of suffering a loss on an individual job?   (d) Assume a pricing formula of the type $S = (1 + m)C'$, where C' is the estimated cost, m the markup as a decimal fraction, and S the selling price.   Judging from the data presented, what seems to be the value of m?   (e) How large would m have to be in order to reduce the probability of loss to .01?

| Price Group | Sample A | | Sample B | | Percent profit | | | | Estimates of variance | |
	Mean price	Mean profit	Mean price	Mean profit	A	B	Sum	Difference	v	ln v
1	$ 262	$ 32	$ 297	$ 41	12.2	13.8	26.0	−1.6	1.28	0.247
2	703	83	808	76	11.8	9.4	21.2	2.4	2.88	1.058
3	1,242	166	1,191	167	13.4	14.0	27.4	−0.6	0.18	−1.715
4	1,790	301	1,757	246	16.8	14.0	30.8	2.8	3.92	1.366
5	2,212	230	2,250	351	10.4	15.6	26.0	−5.2	13.52	2.604
6	2,732	418	2,779	381	15.3	13.7	29.0	1.6	1.28	0.247
7	3,270	376	3,220	235	11.5	7.3	18.8	4.2	8.82	2.177
8	3,726	600	3,738	639	16.1	17.1	33.2	−1.0	0.50	−0.693
9	4,250	348	4,321	640	8.2	14.8	23.0	−6.6	21.78	3.081
10	4,776	602	4,681	562	12.6	12.0	24.6	0.6	0.18	−1.175
Totals	·······	·······	·······	·······	128.3	131.7	260.0	−3.4	54.34	6.657

Ans. (a) At the .10 level of significance, the acceptance region for the hypothesis of homoscedasticity with 9 degrees of freedom is $Z < 14.68$.   The sample value of 7.51 falls in the acceptance region and sustains the hypothesis.
(b) At the .05 level of significance, the acceptance region for the variance ratio with 9 and 10 degrees of freedom is $F < 3.02$.   The sample value of 1.79

falls in the acceptance region, and so the hypothesis of uniform average percent profit is tenable.

Source	Degrees of freedom	Sum of squares	Mean square	Variance ratio
Between price groups	9	85.84	9.54	1.79
Within price groups	10	54.34	5.43	
Total	19	140.18		

(c) $P = .04$, (d) $m = .15$, (e) $m = .20$.

10-33 Industrial specifications for a certain manufactured article which must withstand repeated percussion stipulate a core of hard wood, with a protective coating of resilient plastic. Although the coating usually remains intact, it occasionally loses its adhesion to the core; and when the separation exceeds a certain limit, the article is no longer suitable for its purpose. Resistance to separation is tested in the factory by a mechanical device which delivers a given number of blows of fairly uniform intensity. The following data pertain to the unseparated area u per square inch of test surface at the termination of the impact test. For the experimental comparison of three coating processes, random samples of four articles made by the respective processes were subjected to the impact test.

Process A: .55, .44, .51, .42; $\Sigma u = 1.92$, $\Sigma u^2 = 0.9326$
Process B: .47, .58, .66, .53; $\Sigma u = 2.24$, $\Sigma u^2 = 1.2738$
Process C: .54, .59, .69, .74; $\Sigma u = 2.56$, $\Sigma u^2 = 1.6634$

Are the mean unseparated areas significantly different?

   Ans. For the variance ratio with 2 and 9 degrees of freedom, the acceptance region, such that the null hypothesis of equal means will be sustained at the .05 level of significance, is $F < 4.26$. The sample value of 4.13 falls in the acceptance region and thus supports the null hypothesis.

Source	Degrees of freedom	Sum of squares	Mean square	Variance ratio
Between processes	2	.0512	.0256	4.13
Within processes	9	.0554	.0062	
Total	11	.1066		

10-34 A fairly accurate representation of the trend in successive stages of coat separation in the context of Prob. 10-33 may be derived by assuming that the increment of separation per square inch on the nth impact is proportional to the fractional area u_{n-1} still unseparated after the $n - 1$st impact. This means that $u_n - u_{n-1} = -kf_n u_{n-1}$ $(u_0 = 1)$, where k is constant and f_n represents the effective intensity of the nth impact. This factor is subject to chance variations both in the blow delivered and in the material struck. The formal solution of the difference equation is $u_n = (1 - kf_1)(1 - kf_2) \cdots (1 - kf_n)$, provided all $f_i < 1/k$; or $u_n = 0$ if any $f_i \geq 1/k$. We arrive at a simplified probabilistic model by setting $h = 1 - kE(f)$ and defining a random variable y_n such that $u_n = h^n y_n$ $(0 \leq y_n \leq h^{-n})$. Here, the parameter h is a characteristic of the coating process (including the material used), and y_n may be regarded as having practically the same distribution for all coating processes, since the extreme limit h^{-n} would rarely have an effect. Thus it is clear that the variance of u_n will be a function of the coating process. This dependence can be avoided by transforming to logarithms. Define $m = -n \log h$, $z = -\log y_n$, $x = -\log u_n$. Then $x = m + z$ and the variance of x will equal the variance of z, independently of m. By this transformation the following set of data was obtained from that of Prob. 10-33:

Sample A .598, .821, .673, .868; $\Sigma x = 2.960$, $\bar{x} = .740$, $\Sigma x^2 = 2.2380$
Sample B .755, .545, .416, .635; $\Sigma x = 2.351$, $\bar{x} = .588$, $\Sigma x^2 = 1.4433$
Sample C .616, .528, .371, .301; $\Sigma x = 1.816$, $\bar{x} = .454$, $\Sigma x^2 = 0.8865$

Are the mean values of x significantly different?

Ans. The acceptance region at the .05 level of significance is $F < 4.26$. Therefore the sample value of 4.31 indicates significantly different means.

Source	Degrees of freedom	Sum of squares	Mean square	Variance ratio
Between processes	2	.1638	.0819	4.31
Within processes	9	.1712	.0190	
Total	11	.3350		

10-35   When a machine has several cams on one shaft, the resultant eccentricity of the cam system must be kept below a prescribed upper limit for proper mechanical operation. Scaled to appropriate units, the resultant eccentricity x is distributed very nearly as a beta variate with parameters $\alpha = a - 1$, $\lambda = b - 1$, where a and b are usually nonintegral numbers and $a < 1$. In case tables of the incomplete beta function are either inapplicable or not readily available, the probability levels can be estimated with considerable accuracy by using a normalizing transformation due to Edward Paulson and originally derived for F with n_1 and n_2 degrees of freedom.[1] In connection with F, Paulson has shown that the variate u, given by the transformation $u = [(1 - w_2)q - (1 - w_1)]/(w_2 q^2 + w_1)^{1/2}$, where $q = F^{1/3}$, $w_1 = \frac{2}{9}n_1$, and $w_2 = \frac{2}{9}n_2$, is approximately unit normal, provided that $n_2 \geq 3$. If the lower tail of the distribution is also of interest, then a further restriction is that $n_1 \geq 3$.

(a) Show that if x is a beta variate with parameters $\alpha = a - 1$, $\lambda = b - 1$, then the variate $y = bx/a(1 - x)$ is distributed as F, with parameters (equivalent to

[1] Edward Paulson, An Approximate Normalization of the Analysis of Variance Distribution, *Ann. Math. Stat.*, vol. 13, no. 2, p. 233, June, 1942.

degrees of freedom) given by $n_1 = 2a$, $n_2 = 2b$. (b) Hence, show that the same normalizing transformation holds for the beta variate if $q = [bx/a(1 - x)]^{1/3}$, $w_1 = \frac{1}{9}a$, and $w_2 = \frac{1}{9}b$—the restrictions being that $b \geq \frac{3}{2}$ if only the upper tail is required, and $a \geq \frac{3}{2}$ if the lower tail is required as well. (c) Estimate the upper .05 level of x if $a = 0.8$ and $b = 3.2$. That is, find X such that $P(x \geq X) = .05$.

Ans. (a) Univariate change of variable. (b) Direct substitution. (c) $X = .565$.

INDEX